**High-Entropy Materials**

# High-Entropy Materials

From Basics to Applications

*Huimin Xiang, Fu-Zhi Dai, and Yanchun Zhou*

WILEY-VCH

**Authors**

***Dr. Huimin Xiang***
Aerospace Research Inst. of Materials and Processing Technology
No. 1 South Dahongmen Road
100076 Beijing
China

***Dr. Fu-Zhi Dai***
Aerospace Research Inst. of Materials and Processing Technology
No. 1 South Dahongmen Road
100076 Beijing
China

***Dr. Yanchun Zhou***
Aerospace Research Inst. of Materials and Processing Technology
No. 1 South Dahongmen Road
100076 Beijing
China

**Cover Image:** © CasiopeaEstudio/Shutterstock



**Library of Congress Card No.:** applied for

**British Library Cataloguing-in-Publication Data**
A catalogue record for this book is available from the British Library.

**Bibliographic information published by the Deutsche Nationalbibliothek**
The Deutsche Nationalbibliothek lists this publication in the Deutsche Nationalbibliografie; detailed bibliographic data are available on the Internet at http://dnb.d-nb.de.



**Print ISBN:** 978-3-527-35035-3
**ePDF ISBN:** 978-3-527-83718-2
**ePub ISBN:** 978-3-527-83719-9
**oBook ISBN:** 978-3-527-83720-5

**Typesetting** Straive, Chennai, India
**Printing and Binding** CPI Group (UK) Ltd, Croydon, CR0 4YY

C9783527350353_031024

# Contents

# Preface

Years of endeavor on searching for metallic glasses with super-high glass-forming ability led to the invention of single-phase multicomponent alloys by Prof. Cantor and Prof. Yeh in 2004, individually. These unique alloys, having unusually high entropy of mixing, are distinctly different from traditional alloys based on one or two principal elements. That's the reason these alloys were called high-entropy alloys (HEAs) by Prof. Yeh. Soon, there was ever-growing interest from all across the academia in these alloys since the concept of these new materials had broken the inertial design strategy of traditional alloys. It is possible for HEAs to break the properties limits of traditional alloys, and improved hardness, strength, oxidation, and corrosion resistance have been reported in these material systems. Prof. Yeh summarized mainly four core effects for HEAs: (i) thermodynamics: high-entropy effects, (ii) structure: lattice distortion, (iii) kinetics: sluggish diffusion, and (iv) properties: cocktail effects.

Inspired by the concept of HEAs, Rost et al. employed high entropy to stabilize oxides with five different cations in equiatomic fractions, and soon, the concept was employed in the design of high-entropy borides, carbides, and other ceramic systems. Similarly, the four core effects are valid for high-entropy ceramics (HECs), and improved properties are observed compared to traditional ceramic solid solutions. However, the structure of HECs is not as simple as that of HEAs; it covers all types of Bravias lattices except the triclinic system. Moreover, the bonding between atoms in HECs is essentially different from that in HEAs since all types of bonds, including metallic, ionic, and covalent bonds, exist in HECs. Therefore, the chemical space for composition and property is quite abundant in HECs, and enthusiastic research studies have been conducted after 2015.

There have been several excellent books covering a very wide spectra of HEAs, ranging from manufacturing and processing to advanced characterization to mechanical and functional properties and from physical metallurgy to computational modeling. A number of comprehensive review papers regarding these aspects of HEAs and HECs have been published in these years. However, no book or paper has reviewed the connections and differences between these two fascinating high-entropy materials (HEMs), yet. That's the scope of this book: providing a comprehensive review on the crystal, electronic, and phonon structure characteristics of HEAs and HECs that define their appealing properties. More importantly, the differences between these two types of materials in structure characteristics are emphasized to reveal the origin of different processing methods, mechanical and functional properties of HEAs and HECs.

In this book, a brief history about the birth of HEAs and HECs is given in Chapter 1, followed by the four core effects of HEMs summarized by Prof. Yeh. Detailed descriptions on structure features, including crystal structure, electronic structure, and lattice dynamics of HEMs, which play a central role in understanding their processing/structure/properties' relationships, are described in Chapter 2. In Chapter 3, a comprehensive review on the theoretical tools that probe the formation and properties of HEMs is given. Vast computational modeling based on density functional theory (DFT), molecular dynamic (MD) simulations, Monte Carlo (MC) simulations, calculation of phase diagrams (CALPHAD), surrogate models from machine learning (ML) methods, and simplified models based on physical assumptions or other methods are employed in the simulation of HEMs. Then, synthesis and processing routes via liquid, solid, and gas states to obtain powders, dense and porous HEMs are illustrated in Chapter 4. Chapter 5 deals with the characterization of HEMs, including phase identification, microstructure evolution, and some other novel methods. The mechanical and functional properties of HEAs and HECs are given in detail in Chapters 6 and 7. Shaped by their different structure features, HEAs and HECs perform differently in many aspects. The potential applications that are determined by their properties are reviewed in Chapter 8, while conclusions and prospects of these two types of HEMs are given in Chapter 9.

The intended readers of this book are students from colleges and graduate schools and research professionals from academia and industries.

April 2022

Y.C. Zhou
Beijing, China

# 1

# Introduction to High-Entropy Materials

The advancements of human civilization are largely determined by the materials and tools that were manufactured and mastered by humans. During the Stone Age, the ancient human could use only natural materials, including stone, bone, and wood. After that, with the capability of reduction and extraction of copper, tin, lead, and iron from their ores, tools made of copper and iron were widely used in daily life, transportation, and construction during the Bronze Age and the Iron Age. Nowadays, more and more materials with specific properties and performance have been discovered and utilized due to the more efficient metallurgy techniques and material design paradigms, as shown in Figure 1.1 [2]. In the early stage of material design, one principal element is chosen and alloyed with other elements in trace amount to improve the properties of the original element. This design strategy is still dominant in our material design, such as Al alloys, Ni alloys, Fe alloys, and binary solid solution ceramics. Thus, information and understanding are highly developed on materials close to the corners and edges of a multicomponent phase diagram, with little known about those located at the center of the diagram [1, 3]. Recently, crystalline multi-principal element materials located at or near the center of the phase diagram with equal or near-equal atomic fraction of constituting elements have been introduced and have attracted increasing attention due to their unique compositions, microstructure, and unexpected properties. In this chapter, the basic concepts and information about these fascinating materials will be reviewed.

## 1.1 History of High-Entropy Materials

Attempts to synthesize multicomponent alloys were driven by the desire of metallic glasses with super-high glass-forming ability. After examining the works of Inoue et al. and Peker et al. [4, 5], Greer proposed a "confusion principle," which states that the more the number of elements involved, the lower is the chance that the alloy could

*High-Entropy Materials: From Basics to Applications,* First Edition.
Huimin Xiang, Fu-Zhi Dai, and Yanchun Zhou.

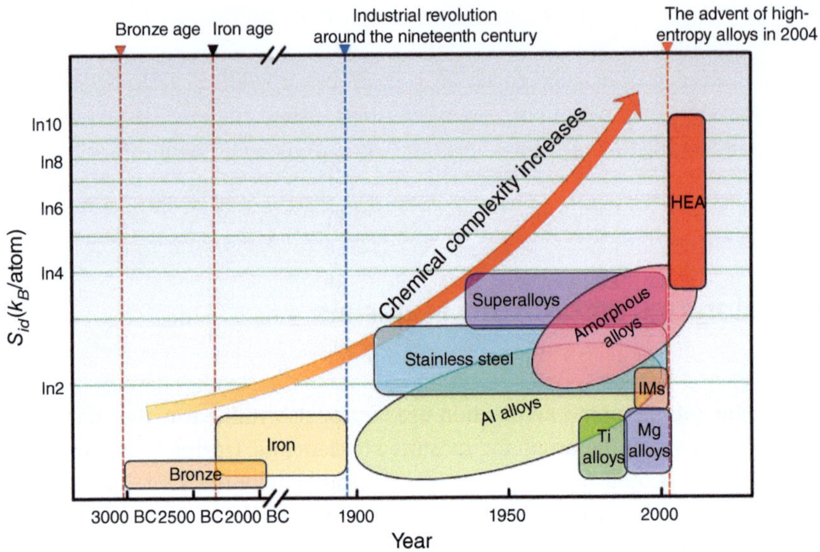


**Figure 1.1** Rising trend of alloy chemical complexity versus time (IMs: intermetallics or metallic compounds, HEA: high-entropy alloy). Source: Reproduced with permission from Cantor [1]/Springer Nature.

select viable crystal structures, and the greater is the chance of glass formation [6]. He also pointed out that the most "confused" elements were those that differed most from each other in size. This principle has boosted the subsequent search of metallic glasses in multicomponent alloys with large atomic size difference [7–9].

In 2004, two independent papers regarding multicomponent alloys have been published by Yeh et al. and Cantor et al. [10, 11]. Their results seemed to falsely verify the principle of confusion. In Cantor's paper, entitled "Microstructural development in equiatomic multicomponent alloys," the authors performed induction melting and melt quenching rapid solidification experiments on an equiatomic mixture of 20 elements (Mn, Cr, Fe, Co, Ni, Cu, Ag, W, Mo, Nb, Al, Cd, Sn, Pb, Bi, Zn, Ge, Si, Sb, and Mg) and 16 elements (Mn, Cr, Fe, Co, Ni, Cu, Ag, W, Mo, Nb, Al, Cd, Sn, Pb, Zn, and Mg) [11]. Surprisingly, instead of glassy phases, these alloys were multiphase, crystalline, and brittle, even after melt spinning. Moreover, they found that these alloys consisted predominantly of a single face-centered cubic (FCC) primary phase containing many elements but particularly rich in transition metals, notably Cr, Mn, Fe, Co, and Ni. Based on these results, they designed and prepared an equimolar $Cr_{20}Mn_{20}Fe_{20}Co_{20}Ni_{20}$ alloy with a single FCC structure and a typical dendritic microstructure in the as-cast condition. They also found that the total number of phases was always well below the maximum equilibrium number allowed by the Gibbs phase rule and even further below the maximum number allowed under non-equilibrium solidification conditions. They argued that the confusion principle might not apply and other factors were more important in promoting glass formation rather than chemical complexity.

Simultaneously, Yeh et al. published the paper entitled "Nanostructured high-entropy alloys with multiple principal elements: novel alloy design concepts and outcomes," and clearly defined the term of high-entropy alloys (HEAs) [10]. By rough estimation of mixing entropies of multiple-principal-element alloys, they found that the mixing entropy of alloys with five or more constituent elements was larger than that of metal fusion and formation enthalpies of strong intermetallic (IM) compounds, indicating the preferred tendency of formation of random solid solutions during solidification of these alloys. Due to the central role of mixing entropy in stabilizing the solid solution, they defined these alloys as HEAs. They took the as-cast $CuCoNiCrAl_xFe$ alloy system as an example to illustrate that how simple the structure of these chemically complex alloys could be. In the same paper, they also concluded some conspicuous properties of HEAs, such as nanostructures resulting from difficulty in substitutional diffusion of elements, high hardness induced by multi-strengthening mechanisms, excellent resistance to annealing softening, wear, and oxidation. These promising properties made HEAs potentially suitable for many applications such as tools, molds, dies, mechanical parts as well as anticorrosive high-strength parts in chemical plants, integrated circuit (IC) foundries, and even marine applications. Moreover, unlike the commonly used traditional alloy systems, the number of HEAs was countless, even after excluding chemically incompatible elements that produced liquid immiscibility, and the concept of HEAs led to a whole new and uncharted territory, in which many possible new materials, phenomena, theories, and applications were awaited [9].

These two innovative works have ignited enthusiastic research on HEAs. It is worth mentioning that, before they published their groundbreaking works, years of related and ground works have been conducted. Cantor started the study of multicomponent alloys consisting of a large number of constituents in equal or near-equal proportions with his undergraduate students since the beginning of 1980s [12], while the exploration on multicomponent alloys by Yeh and his students could be traced back to 1995 [13].

The attempts to search high-entropy ceramics (HECs) were started by Chen et al. in their pursuit of promoting the hardness of HEA films by introducing non-metallic elements, such as nitrogen and oxygen [14–16]. They found that introducing nitrogen into $Al_{0.5}CoCrCuFeNi$, $Al_2CoCrCuFeNi$, and AlCrNiSiTi resulted in the amorphization and hardening of the nanostructured metallic solid solution [14, 15]. The oxidation of $Al_xCoCrCuFeNi$ ($x = 0.5, 1, 2$) HEAs led to the formation of a single cubic-spinel crystalline phase [16]. Although they synthesized the ceramics with multi-principal elements, the concept of high-entropy-stabilized ceramics was not used to describe these materials until Rost et al. published their work on high-entropy oxides (HEOs) with a single rock-salt phase and five different cations in equiatomic fractions [17]. This work has led to a significant increase in the number of publications about HEOs, and up to now, a large part of the publications is related to HEOs due to their fascinating mechanical and functional properties. Inspired by this work, high-entropy borides, high-entropy carbides, and other non-oxide HECs have been discovered and reported subsequently [18–21].

## 1.2 Definition of High-Entropy Materials

There are two generally accepted and widely used definitions for HEAs. One is based on composition and the other is based on configurational entropy. The first one was proposed in 2004 by Yeh et al. based on the compositional requirements [10]. They defined the HEAs as those containing at least five major principal elements with the concentration of each element being between 35 and 5 at.%. In their definition, the ratio of the major elements was not restricted to be equimolar, taking $CuCo_{0.5}Ni_{1.2}CrAlFe_{1.5}Ag_{0.02}B_{0.1}C_{0.15}$ as an example, leading to countless high entropy (HE) alloys with multiple principal elements. Moreover, HEAs were not required to be a single-phase solid solution according to this definition, and $CuCoNiCrAl_xFe$ alloys with $0.8 < x < 2.8$ was an illustration.

The second definition is based on the magnitude of mixing entropy. According to the Boltzmann's thermodynamic statistics principle, the quantitative relationship between the entropy and complexion of the system is given by:

$$\Delta S_{conf} = k \ln \omega \tag{1.1}$$

where $k$ is the Boltzmann's constant, and $\omega$ is the thermodynamic probability, which represents the total number of microscopic states contained in the macroscopic state. With the increase in the number of microscopic states, the configuration entropy of the system increases. For a solid solution consisting of $n$ kinds of atoms, the configurational entropy can be estimated by:

$$\Delta S_{conf} = -R\left[c_1 \ln c_1 + c_2 \ln c_2 + \cdots + c_n \ln c_n\right] = -R\sum_{i=1}^{n} c_i \ln c_i \tag{1.2}$$

where $R$ is the gas constant, $c_i$ is the mole fraction of the $i$th element, and $n$ is the total number of components. According the extreme value theorem, the configurational entropy reaches its maximum value when the mole fraction of all elements is equal, which is $R\ln n$. Yeh calculated the configuration entropies of equimolar alloys with constituent elements up to 13 [22]. Comparing the configurational entropy of a three-element equimolar alloy and a five-element alloy, they believed that $1.5R$ was large enough to compete with mixing enthalpy as well as to be used as a border line between HEAs and medium-entropy alloys (MEAs). In addition, $1R$ was used to separate the MEAs and low-entropy alloys (LEAs). Accordingly, the alloy world could be categorized, as shown in Figure 1.2. They also suggested a practical upper bound of 13 principal elements for HEAs, since the benefit from the high-entropy effect of more principal elements was quite limited with the cost of complexity in handling raw materials or recycling the alloys.

Generally, the HEAs defined by these two definitions overlap with each other in most cases. Still, some exceptions exist. For $FeCoNiAl_{0.2}Si_{0.2}$ alloy [23], it can be recognized as HEA by the composition definition since it contains five principal elements with the atomic percentage larger than 5 at.%. However, the configurational entropy of this alloy is approximately $1.41R$, which is located in the MEA

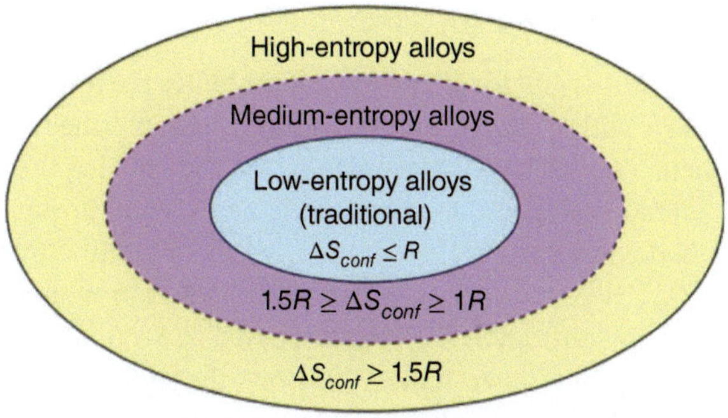


**Figure 1.2** Alloy world based on configurational entropy. Source: Reproduced with permission from Yeh [22]/Springer Nature.

territory according to the entropy definition. On the other hand, with the development of novel synthesis method, the number of constituent elements that could mixed in one alloy progressively increases. Extreme mixing of 15 and 21 metals in one alloy has been recently reported [24, 25]. For the alloy containing 21 elements, the concentration of each element is smaller than 5 at.%, which cannot fit the composition definition but still can be regarded as HEA due to its large mixing entropy. As stated by Yeh, it is not easy to give a clear-cut definition for HEAs, and these two definitions are just guidelines for the design of HEAs [22].

The concept and definition of HECs are inherited from those of HEAs. Due to the existence of more than one sublattice in the structure, the molar configurational entropy of HECs should be calculated by [26]:

$$\Delta S_{conf} = -R\sum_{j=1}^{n}\sum_{i=1}^{m} c_i^{\,j} \ln c_i^{\,j} \tag{1.3}$$

where $c_i^j$ represents the mole fraction of elements in the $j$th sublattice. Similarly, with the mixing entropy larger than $1.5R$, the ceramic solid solution can be classified as HECs. From Eq. (1.3), it is quite clear that for ceramics with chemical disorder in more than one sublattice, the mixing entropy is significantly larger than that of HEAs. We should emphasize that, for both HEAs and HECs, the composition and entropy definitions are just guidelines without the guarantee of formation of a single-phase solid solution.

## 1.3 Core Effects of HEMs

Based on their ground research studies, Yeh summarized mainly four core effects for HEAs, which were as follows: (i) thermodynamics: high-entropy effects, (ii) structure: lattice distortion, (iii) kinetics: sluggish diffusion, and (iv) properties: cocktail effects [22, 27].

### 1.3.1 High-Entropy Effect

The high-entropy effect is the most important for HEAs since it promotes the formation of random solid solution alloys with a simple structure rather than intermetallic compounds in multicomponent alloys. Thermodynamically, the state having the lowest mixing free energy $\Delta G_{mix}$ among all possible states would be the equilibrium state. The mixing free energy is determined by the competitive mixing enthalpy $\Delta H_{mix}$ and entropy $\Delta S_{mix}$ by: $\Delta G_{mix} = \Delta H_{mix} - T\Delta S_{mix}$. For a solid solution with a large number of elements, the mixing entropy is high, as discussed in Section 1.2, and it becomes more dominant at higher temperatures. Therefore, the significantly higher mixing entropy of HEAs for the random solution state is expected to significantly extend the solubility range for terminal solutions and form simple multielement solution phases, especially at higher temperatures. The formation of single-phase CrMnFeCoNi alloy is a vivid illustration of competition of enthalpy and entropy. For this alloy system, the as-cast ingot exhibited a typical dendrite structure from its melts with a single FCC structure [11], indicating the random occupation of all elements in a solid solution. However, minor precipitates on the grain boundaries of samples annealed at temperatures lower than 800 °C were found and determined to be intermetallics [28, 29]. The metastable nature of CrMnFeCoNi HEA demonstrated the significant role of mixing entropy in the formation of this single phase at high temperature. This effect has also been used to understand the formation of single phase HEOs, (CoCuMgNiZn)O [17]. Rost et al. found an endothermic reaction upon cyclic heat treatment of rocksalt (CoCuMgNiZn)O HEO. They attributed this to the enthalpic penalties of Zn and Cu for crystallization of a rocksalt structure. To lower the barrier of formation of a single solid solution, these two elements were excluded from HEO. However, this was not the case, exclusion of Zn and/or Cu led to the formation of multiphases. This was a strong indication that the role of mixing entropy was prominent, since it decreased when one cation was removed.

Although high-entropy effect is important for the formation of HEMs, it is not the only factor that determines the phase formation for multicomponent alloys and ceramics, as will be discussed in Chapter 2.4. Nevertheless, it is still the core effect for HEM formation and reduces the number of phases as predicted by Gibbs phase rule that permits the number of phases in equilibrium to increase with the number of components.

### 1.3.2 Lattice Distortion

This structure feature is quite obvious due to the random distribution of different atoms with different atomic size in one structure, as schematically illustrated in Figure 1.3 [30]. Also, due to the different electronegativity of atoms, the bonding strength and bond length between atoms differ from each other, which is also responsible for the lattice distortion. This crystal feature was proposed by Yeh et al. in the investigation of CuCoNiCrAlFeTiV alloys [31], and the effects on the X-ray diffraction (XRD) intensities of CuNiAlCoCrFeSi alloy were mathematically

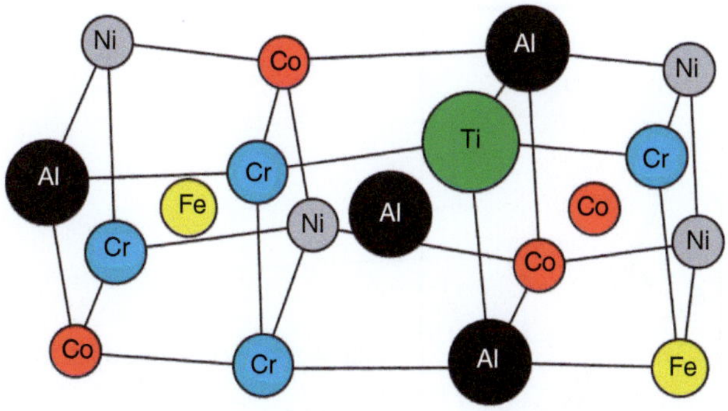


**Figure 1.3** A schematic illustration of serious distorted $AlCoCrFeNiTi_{0.5}$ lattice. Source: Reproduced with permission from Zhang et al. [30]/John Wiley & Sons.

determined [32]. With the help of various characterization techniques, lattice distortion was observed in multiple HEAs and ceramics [33–37].

In addition to the reduction in the intensity of XRD pattern, lattice distortion is also used to explain the variation in other properties. Lattice distortion can hinder dislocation motion and result in significant solid solution strengthening effects [10]. It is also responsible for the severe scattering of electrons and phonons, leading to the significant decrease in electric and thermal conductivity and weak temperature dependency of the transport properties [38]. It might also lead to the amorphization when sufficiently large atomic size differences collapse the distorted lattice. The influence of lattice distortion on the properties of HEMs will be discussed in Chapter 7.

### 1.3.3 Sluggish Diffusion

Atom diffusion in the matrix is an important process in the formation of new phases and grain growth, and it is closely related to the high-temperature strength and structure stability. This effect has been used to explain the formation of nanoprecipitates in the as-cast HEA matrix [10], and easier formation of amorphous structure for a higher number of elements in HEAs [27].

To verify this effect, Tsai et al. examined the diffusion phenomenon of constituent elements in a quinary HEA, $CoCrFeMn_{0.5}Ni$ [39]. Through theoretical analysis, they were able to determine the intrinsic diffusion coefficients from quasi-binary diffusion couples and validate the comparison of their data with other tracer-based diffusion results. The measured diffusion rate of elements in $CoCrFeMn_{0.5}Ni$ decreased in the sequence of Mn, Cr, Fe, Co, and Ni. The extrapolation values of the diffusion coefficients for each element at the melting point, $D_{T_m}$, were the smallest in the Co–Cr–Fe–Mn–Ni alloy system, comparing with other FCC matrices, including Fe–Cr–Ni–(Si) alloys, Fe, Co, and Ni. In addition, the activation energies normalized by melting points, $Q/T_m$, of all elements in HEA were the largest. They concluded that these results were direct evidence for the sluggish diffusion effect in HEA. The origin of the sluggish diffusion effect was also identified to be the fluctuation in lattice

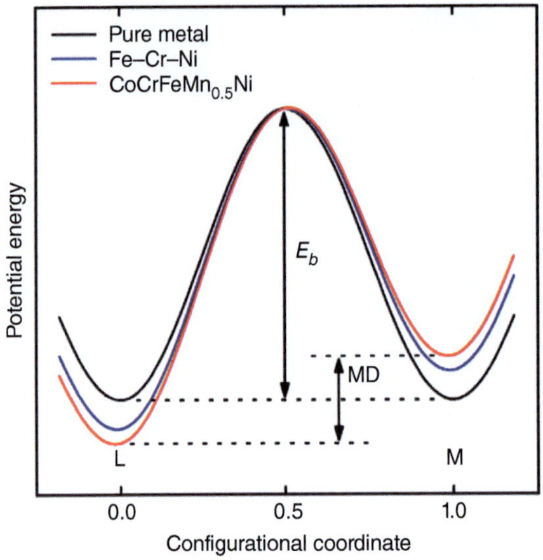


**Figure 1.4** Schematic diagram of the variation in lattice potential energy (LPE) and mean difference (MD) during the migration of a Ni atom in different matrices. The MD for pure metals is zero, whereas that for HEA is the largest. Source: Reproduced from Tsai et al. [39]/ with permission of Elsevier.

potential energy (LPE) by calculating the so-called seven-bond interaction energy (SBIE, which represents the energy difference before and after the atomic migration) using a pair potential or quasi-chemical model. They found that the mean difference (MD) values of SBIE in different matrices were quite different. The calculated MD for Ni in HEA was nearly 1.5 times of those in Fe–Cr–Ni ternary alloys, indicating that a diffusing Ni atom experienced significantly greater LPE fluctuation in HEA than in traditional alloys. As illustrated in Figure 1.4, due to the existence of MD in LPE of HEAs and traditional alloys, the transition frequencies between adjacent lattice sites were asymmetric, and the atoms preferred to stay at lattice sites with lower LPE, which served as atomic traps that hinder diffusion. The larger the LPE fluctuation, the stronger is the trapping effect, and the lower the diffusion rate would be.

On the other hand, Miracle et al. argued that the measured diffusion coefficients in $CoCrFeMn_{0.5}Ni$ were not essentially different from diffusion in elements and conventional alloys [40]. The data they extracted from the experiments demonstrated that diffusion coefficients in $CoCrFeMn_{0.5}Ni$ were generally higher than in conventional materials when compared at the same temperature, as shown in Figure 1.5. For instance, at 1173 K, the diffusion coefficients of Ni in $CoCrFeMn_{0.5}Ni$, Fe–15Cr–20Ni, and pure Fe were $14.3\times10^{-18}$ m$^2$/s, $6.56\times10^{-18}$ m$^2$/s, and $3.12\times10^{-18}$ m$^2$/s, respectively. Moreover, the higher activation energies normalized by melting points was based on the empirical observation that diffusion coefficients at $T_m$, $D_{T_m}$, were roughly equal for alloys with similar crystal structure and bonding type, and the value of $D_{T_m}$ for all elements in HEA fell in the range of $D_{T_m}$ values in pure metals and binary alloys, thus demonstrating no significant difference between HEA and conventional alloys in diffusion when the data were normalized at $T_m$.

Controversy exists on this structure characteristic, and additional works on the intrinsic diffusion coefficients in HEAs may be needed, which is challenging

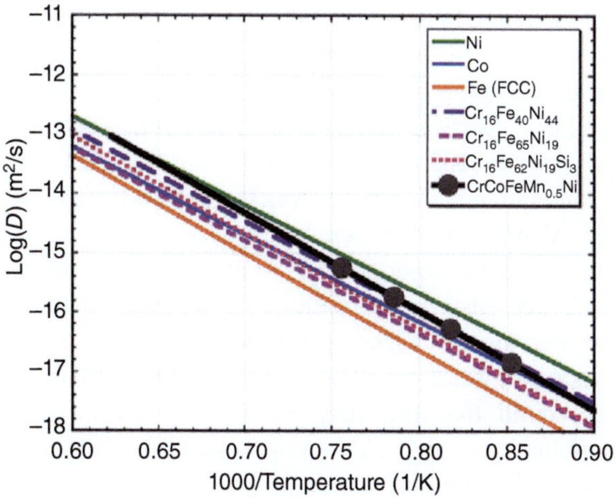


**Figure 1.5** Diffusion coefficients of Ni in FCC elements, stainless steel alloys (compositions in the legend are shown in at.%) and $CoCrFeMn_{0.5}Ni$ as a function of inverse absolute temperature. Source: Reproduced from Miracle et al. [40]/with permission of Elsevier.

in experiment. Nevertheless, more atomic diffusion relation phenomena have been reported, such as good irradiation resistance and slow grain growth [41, 42].

### 1.3.4 Cocktail Effect

The initial intent of Ranganathan using the phrase of "multimetallic cocktails" was to emphasize alloy pleasures in alloy design and development [43]. In HEMs, cocktail effect is used to emphasize a synergistic effect of the whole properties from the overall contribution of the constituent elements. Considering the expanded solubility of different elements in HEMs, the design freedom of HEMs is huge, which gives great opportunity for property adjustment in HEMs based on the mixture rule since different elements with different chemical states and properties can be selected. Moreover, because of the mutual interactions among the elements, unexpected results could be obtained through synergistic effect.

The central idea of cocktail effect is that the properties of HEMs are designable by adding proper ingredient according to the mixture rule. For example, the hardness of body-centered cubic (BCC) structured material is normally higher than that of FCC structured material due to the limited slip systems. Thus, by adjusting the content of BCC phases, the hardness of HEAs changes accordingly. Al element is a strong BCC stabilizer, and adjusting the content of Al can control the content of BCC phases. As demonstrated by Yeh et al., with the Al content increasing in $CuCoNiCrAl_xFe$ HEA, the phases changed from FCC to BCC + FCC and finally to BCC structure [10]. The lattice constant for both the FCC and BCC structures increased as well as the hardness of the HEA. This situation was also observed in Cu-free $Al_xCoCrFeNi$ HEA [44]. Another example on the application of cocktail effect is the property optimization of a soft magnetic HEA, $FeCoNi(AlSi)_x$ alloys [45].

For an applicable soft magnetic material, high saturation magnetization, electrical resistivity, and malleability are required. Due to their ferromagnetic nature, Ni, Co, and Fe elements are positive to the improvement of saturation magnetization. However, experimentally proved low electrical resistivity of NiCoFe MEAs brought large eddy current loss. To optimize the combinational properties of NiCoFe MEA, Al and Si were chosen. First, the nonmetal element Si was known to increase the electrical resistivity, while large Al atoms led to severe lattice distortion that could scatter the electrons efficiently. Secondly, co-doping of smaller Si and larger Al atoms enhanced the crystal lattice distortion, resulting in a solid solution-strengthening effect that could promote the mechanical property. Thus, mixing of Ni, Co, Fe, Al, and Si could lead to a satisfactory result. By careful adjustment of Al and Si content, Zhang et al. found that the alloy with the composition of $FeCoNi(AlSi)_{0.2}$ had the best combination of $M_s$(1.5 T), $H_c$(1400 A/m), $\rho$(69.5 μΩ·cm), and $\varepsilon_p$ (>50%) for application as a soft magnetic material [45].

Some deviations from the mixture rule are often observed in HEMs due to the synergistic effect. In the hardness measurements of high-entropy borides, Gild et al. found that the measured hardness of six single-phase high-entropy metal diborides were generally greater than the averages of the hardness values measured from individual metal diborides [18], as shown in Figure 1.6. In addition, thermal expansions that depart from their average values have also been reported in HEAs and HECs, which will be discussed in Chapter 7 [46–48].

Due to the existence of cocktail effect, designing HEMs with desired properties is feasible. It is important for one to comprehensively consider various factors involved before selecting appropriate composition and processes. Moreover, one has to keep in mind that unexpected result that can come from the unusual synergistic interactions between the selected elements.

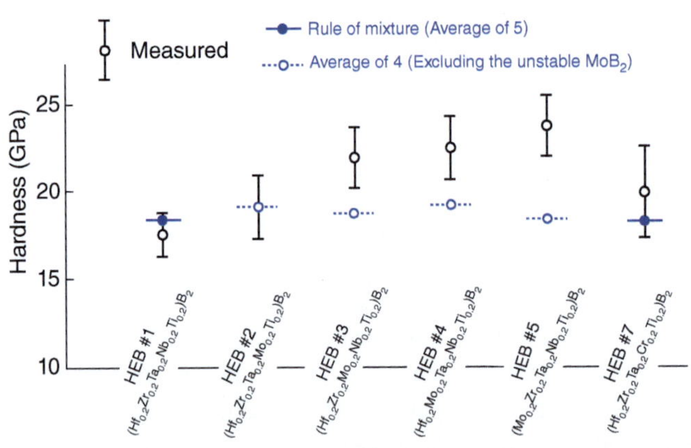


**Figure 1.6** Measured hardness of high-entropy borides and the average values from the rule of mixture of individual metal diborides. Source: Gild et al. [18]/Springer Nature/ CC BY-4.0.

## 1.4 Development of the HEMs

In the early stage of HEAs investigations, most of the works aimed to achieve a single phase with superior properties from equal or near-equal atomic fraction of constituting elements. With the development of HEAs, it is found that HEAs with non-equimolar compositions or multiphase structure prevail in mechanical properties. For instance, Li et al. developed a new class of transformation-induced plasticity-assisted dual-phase (TRIP-DP) $Fe_{50}Mn_{30}Co_{10}Cr_{10}$ HEA, which overcame the strength–ductility trade-off, as shown in Figure 1.7 [49]. By departing from equiatomic composition, the stability of this HEA decreased with two key benefits: interface hardening due to a dual-phase microstructure and transformation-induced hardening, leading to the simultaneous increase in strength and ductility. Lu et al. proposed a novel strategy to design HEAs using the eutectic alloy concept to solve the strength–ductility paradox, inferior castability, and segregation issues of HEAs [50]. The eutectic-like microstructure that consisted of an alternate layer of soft FCC and brittle BCC phases was developed in $AlCrFeCoNi_{2.1}$ HEA. This alloy exhibited unusual work hardening behavior, attributing to the fine lamellar structure. They also proposed a strategy, specifically, through calculation of mixing enthalpy, to locate eutectic compositions in HEAs [51], and designed a series of eutectic HEAs containing Zr, Hf, Nb, and Ta. The complete eutectic

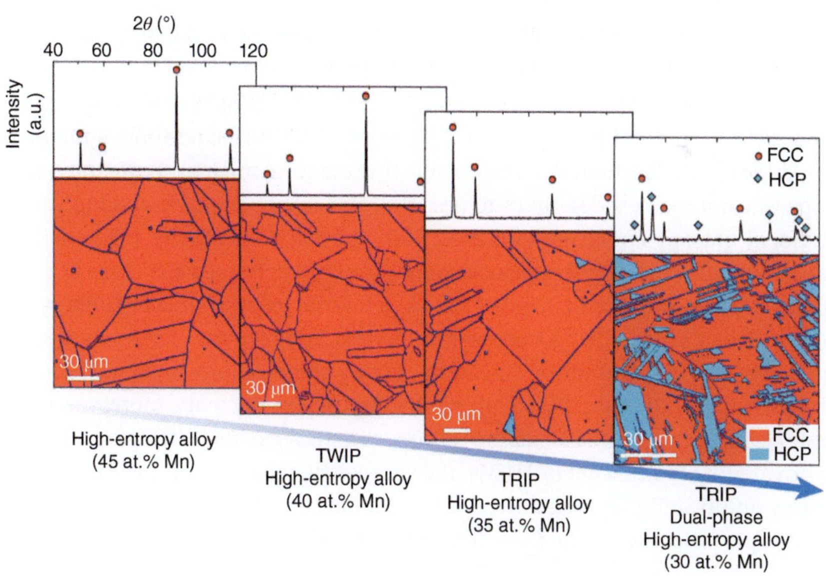


**Figure 1.7** XRD patterns and electron backscatter diffraction (EBSD) phase maps of $Fe_{80-x}Mn_xCo_{10}Cr_{10}$ ($x$ = 45 at.%, 40 at.%, 35 at.%, and 30 at.%) HEAs. The Mn content plays an important part in phase constitution and tuning phase stability for the activation of specific displacing transformation mechanisms. Source: Reproduced with permission from Li et al. [49]/Springer Nature.

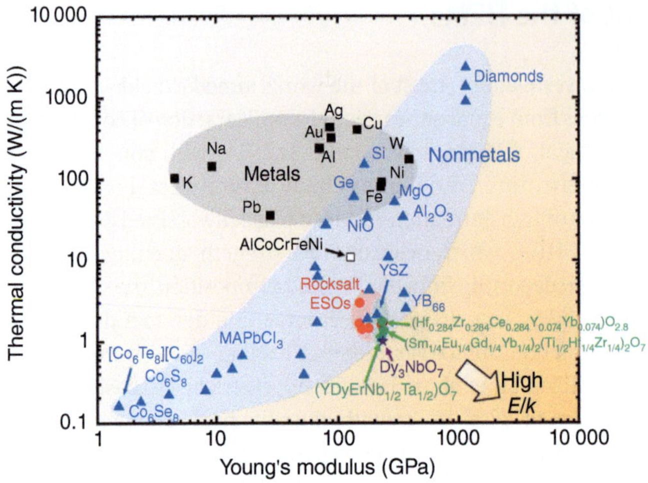


**Figure 1.8** Thermal conductivity ($\kappa$) versus Young's modulus ($E$) of several types of materials to feature several HECs/CCCs with a high $E/\kappa$ ratio. ESO = entropy-stabilized oxide. Source: Reproduced with permission from Wright et al. [55]/Springer Nature.

structures were observed in $Zr_{0.6}CoCrFeNi_{2.0}$, $Nb_{0.74}CoCrFeNi_{2.0}$, $Hf_{0.55}CoCrFeNi_{2.0}$, and $Ta_{0.65}CoCrFeNi_{2.0}$ HEAs. These results demonstrate impressive comprehensive performance of HEAs with non-equimolar compositions or multiphase structure, which is considered as the next-generation HEAs [52].

Inspired by the exciting discovery in the field of HEAs, Wright et al. proposed to expand HECs to compositionally complex ceramics (CCCs) to include non-equimolar compositions [53]. They found in their works that the ideal mixing (configurational) entropy was not the best descriptor to describe thermal conductivity and high configurational entropy could not guarantee low thermal conductivity [53, 54]. For instance, the nominal oxygen vacancy concentration determined the thermal conductivity of Yttria-stabilized zirconia (YSZ)-like fluorite oxides, while in cubic pyrochlore oxides, size disorder governed the thermal conductivity. These combined studies demonstrated that various medium-entropy compositions could outperform their high-entropy counterparts in tuning the thermal conductivity. Moreover, high $E/\kappa$ ratios could be achieved in medium-entropy ceramics, as shown in Figure 1.8 [55].

By broadening the concept of HEMs, the compositional space for these materials increases significantly, and the degree of freedom in tuning the properties, particularly multiple properties at the same time, is rendered.

## References


1 Cantor, B. (2007). Stable and metastable multicomponent alloys. *Ann. Chim. Sci. Mat.* 32: 245–256.

2 He, Q., Ding, Z., Ye, Y. et al. (2017). Design of high-entropy alloy: a perspective from nonideal mixing. *JOM* 69: 2092–2098.

3 Zhang, Y., Zuo, T., Tang, Z. et al. (2014). Microstructures and properties of high-entropy alloys. *Prog. Mater. Sci.* 61: 1–93.

4 Inoue, A., Zhang, T., and Masumoto, T. (1990). Production of amorphous cylinder and sheet of $La_{55}Al_{25}Ni_{20}$ alloy by a metallic mold casting method. *Mater. Trans., JIM* 31: 425–428.

5 Peker, A. and Johnson, W.L. (1993). A highly processable metallic glass: $Zr_{41.2}Ti_{13.8}Cu_{12.5}Ni_{10.0}Be_{22.5}$. *Appl. Phys. Lett.* 63: 2342–2344.

6 Greer, A.L. (1993). Confusion by design. *Nature* 366: 303–304.

7 Lin, X. and Johnson, W.L. (1995). Formation of Ti–Zr–Cu–Ni bulk metallic glasses. *Appl. Phys. Lett.* 78: 6514–6519.

8 Lin, X., Johnson, W.L., and Rhim, W.K. (1997). Effect of oxygen impurity on crystallization of an undercooled bulk glass forming Zr–Ti–Cu–Ni–Al alloy. *Mater. Trans., JIM* 38: 473–477.

9 Cantor, B., Kim, K.B., and Warren, P.J. (2002). Novel multicomponent amorphous alloys. *Mater. Sci. Forum* 386–388: 27–32.

10 Yeh, J., Chen, S., Lin, S. et al. (2004). Nanostructured high-entropy alloys with multiple principal elements: novel alloy design concepts and outcomes. *Adv. Eng. Mater.* 6: 299–303.

11 Cantor, B., Chang, I.T.H., Knight, P. et al. (2004). Microstructural development in equiatomic multicomponent alloys. *Mater. Sci. Eng., A* 375–377: 213–218.

12 Cantor, B. (2014). Multicomponent and high entropy alloys. *Entropy* 16: 4749–4768.

13 Gao, M.C., Yeh, J., Liaw, P.K. et al. (2016). *High-Entropy Alloys: Fundamentals and Applications*. Cham, Switzerland: Springer International Publishing.

14 Chen, T., Shun, T., Yeh, J. et al. (2004). Nanostructured nitride films of multi-element high-entropy alloys by reactive DC sputtering. *Surf. Coat. Technol.* 188, 189: 193–200.

15 Chen, T., Wong, M., Shun, T. et al. (2005). Nanostructured nitride films of multi-element high-entropy alloys by reactive DC sputtering. *Surf. Coat. Technol.* 200: 1361–1365.

16 Chen, T. and Wong, M. (2007). Structure and properties of reactively-sputtered $Al_x$CoCrCuFeNi oxide films. *Thin Solid Films* 516: 141–146.

17 Rost, C.M., Sachet, E., Borman, T. et al. (2015). Entropy-stabilized oxides. *Nat. Commun.* 6: 8485.

18 Gild, J., Zhang, Y., Harrington, T. et al. (2016). High-entropy metal diborides: a new class of high-entropy materials and a new type of ultrahigh temperature ceramics. *Sci. Rep.* 6: 37946.

19 Yan, X.L., Constantin, L., Lu, Y.F. et al. (2018). $(Hf_{0.2}Zr_{0.2}Ta_{0.2}Nb_{0.2}Ti_{0.2})C$ high-entropy ceramics with low thermal conductivity. *J. Am. Ceram. Soc.* 101: 4486–4491.

20 Qin, Y., Liu, J., Li, F. et al. (2019). A high entropy silicide by reactive spark plasma sintering. *J. Adv. Ceram.* 8: 148–152.

21 Guan, J., Li, D., Yang, Z. et al. (2020). Synthesis and thermal stability of novel high-entropy metal boron carbonitride ceramic powders. *Ceram. Int.* 46: 26581–26589.

22 Yeh, J. (2013). Alloy design strategies and future trends in high-entropy alloys. *JOM* 65: 1759–1771.

**23** Zuo, T., Ren, S., Liaw, P.K. et al. (2013). Processing effects on the magnetic and mechanical properties of $FeCoNiAl_{0.2}Si_{0.2}$ high entropy alloy. *Int. J. Miner. Metall. Mater.* 20: 549–555.

**24** Yao, Y., Huang, Z., Hughes, L.A. et al. (2021). Extreme mixing in nanoscale transition metal alloys. *Matter* 4: 2340–2353.

**25** Liao, Y., Li, Y., Zhao, R. et al. (2022). High-entropy-alloy nanoparticles with 21 ultra-mixed elements for efficient photothermal conversion. *Nat. Sci. Rev.* nwac041. https://doi.org/10.1093/nsr/nwac041.

**26** Sarkar, A., Wang, Q., Schiele, A. et al. (2019). High-entropy oxides: fundamental aspects and electrochemical properties. *Adv. Mater.* 31: 1806236.

**27** Yeh, J. (2006). Recent progress in high-entropy alloys. *Ann. Chim. Sci. Mat.* 31: 633–648.

**28** Pickering, E.J., Muñoz-Moreno, R., Stone, H.J. et al. (2016). Precipitation in the equiatomic high-entropy alloy CrMnFeCoNi. *Scr. Mater.* 113: 106–109.

**29** Otto, F., Dlouhý, A., Pradeep, K.G. et al. (2016). Decomposition of the single-phase high-entropy alloy CrMnFeCoNi after prolonged anneals at intermediate temperatures. *Acta Mater.* 112: 40–52.

**30** Zhang, Y., Zhou, Y., Lin, J. et al. (2008). Solid-solution phase formation rules for multi-component alloys. *Adv. Eng. Mater.* 10: 534–538.

**31** Yeh, J., Chen, S., Gan, J. et al. (2004). Formation of simple crystal structures in Cu–Co–Ni–Cr–Al–Fe–Ti–V alloys with multiprincipal metallic elements. *Metall. Mater. Trans. A* 35A: 2533–2536.

**32** Yeh, J., Chang, S., Hong, Y. et al. (2007). Anomalous decrease in X-ray diffraction intensities of Cu–Ni–Al–Co–Cr–Fe–Si alloy systems with multi-principal elements. *Mater. Chem. Phys.* 103: 41–46.

**33** Owen, L.R., Pickering, E.J., Playford, H.Y. et al. (2017). An assessment of the lattice strain in the CrMnFeCoNi high-entropy alloy. *Acta Mater.* 122: 11–18.

**34** Jiang, B., Bridges, C.A., Unocic, R.R. et al. (2021). Probing the local site disorder and distortion in pyrochlore high-entropy oxides. *J. Am. Chem. Soc.* 143: 4193–4204.

**35** Oh, H., Duancheng, M., Leyson, G. et al. (2016). Lattice distortions in the FeCoNiCrMn high entropy alloy studied by theory and experiment. *Entropy* 18: 321.

**36** Rost, C.M., Rák, Z., Brenner, D.W. et al. (2017). Local structure of the $Mg_xNi_xCo_xCu_xZn_xO$ ($x = 0.2$) entropy-stabilized oxide: an EXAFS study. *J. Am. Ceram. Soc.* 100: 2732–2738.

**37** Zou, Y., Maiti, S., Steurer, W. et al. (2014). Size-dependent plasticity in an $Nb_{25}Mo_{25}Ta_{25}W_{25}$ refractory high-entropy alloy. *Acta Mater.* 65: 85–97.

**38** Jin, K., Sales, B.C., Stocks, G.M. et al. (2016). Tailoring the physical properties of Ni-based single-phase equiatomic alloys by modifying the chemical complexity. *Sci. Rep.* 6: 20159.

**39** Tsai, K., Tsai, M., and Yeh, J. (2013). Sluggish diffusion in Co–Cr–Fe–Mn–Ni high-entropy alloys. *Acta Mater.* 61: 4887–4897.

**40** Miracle, D.B. and Senkov, O.N. (2017). A critical review of high entropy alloys and related concepts. *Acta Mater.* 122: 448–511.

41 Jin, K., Lu, C., Wang, L.M. et al. (2016). Effects of compositional complexity on the ion-irradiation induced swelling and hardening in Ni-containing equiatomic alloys. *Scr. Mater.* 119: 65–70.

42 Zhao, Z., Xiang, H., Dai, F. et al. (2019). $(La_{0.2}Ce_{0.2}Nd_{0.2}Sm_{0.2}Eu_{0.2})_2Zr_2O_7$: a novel high-entropy ceramic with low thermal conductivity and sluggish grain growth rate. *J. Mater. Sci. Technol.* 35: 2647–2651.

43 Ranganathan, S. (2003). Alloyed pleasures: multimetallic cocktails. *Curr. Sci.* 85: 1404–1406.

44 Kao, Y., Chen, T., Chen, S. et al. (2009). Microstructure and mechanical property of as-cast, -homogenized, and -deformed $Al_xCoCrFeNi$ ($0 \leq x \leq 2$) high-entropy alloys. *J. Alloys Compd.* 488: 57–64.

45 Zhang, Y., Zuo, T.T., Cheng, Y.Q. et al. (2013). High-entropy alloys with high saturation magnetization, electrical resistivity, and malleability. *Sci. Rep.* 3: 1455.

46 Laplanche, G., Gadaud, P., Bärsch, C. et al. (2018). Elastic moduli and thermal expansion coefficients of medium-entropy subsystems of the CrMnFeCoNi high-entropy alloy. *J. Alloys Compd.* 746: 244–255.

47 Zhao, Z., Chen, H., Xiang, H. et al. (2020). High-entropy $(Nd_{0.2}Sm_{0.2}Eu_{0.2}Y_{0.2}Yb_{0.2})_4Al_2O_9$ with good high temperature stability, low thermal conductivity and anisotropic thermal expansivity. *J. Adv. Ceram.* 9: 595–605.

48 Ridley, M., Gaskins, J., Hopkins, P. et al. (2020). Tailoring thermal properties of multi-component rare earth monosilicates. *Acta Mater.* 195: 698–707.

49 Li, Z., Pradeep, K.G., Deng, Y. et al. (2016). Metastable high-entropy dual-phase alloys overcome the strength-ductility trade-off. *Nature* 534: 227–230.

50 Lu, Y., Dong, Y., Guo, S. et al. (2014). A promising new class of high-temperature alloys: eutectic high-entropy alloys. *Sci. Rep.* 4: 6200.

51 Lu, Y., Jiang, H., Guo, S. et al. (2017). A new strategy to design eutectic high-entropy alloys using mixing enthalpy. *Intermetallics* 91: 124–128.

52 Zhang, Y. (2019). *High-Entropy Materials: A Brief Introduction*. Singapore: Springer Nature Singapore Pvt Ltd.

53 Wright, A.J., Wang, Q., Huang, C. et al. (2020). From high-entropy ceramics to compositionally-complex ceramics: a case study of fluorite oxides. *J. Eur. Ceram. Soc.* 40: 2120–2129.

54 Wright, A.J., Wang, Q., Ko, S.-T. et al. (2020). Size disorder as a descriptor for predicting reduced thermal conductivity in medium- and high-entropy pyrochlores. *Scr. Mater.* 181: 76–81.

55 Wright, A.J. and Luo, J. (2020). A step forward from high-entropy ceramics to compositionally complex ceramics: a new perspective. *J. Mater. Sci.* 55: 9812–9827.

# 2

# Structural Features and Thermodynamics of High-Entropy Materials

## 2.1 Structural Features of High-Entropy Materials

### 2.1.1 Crystal Structure of High-Entropy Alloys

Generally, the crystal structure of a solid solution is largely determined by the structure of the matrix. In high-entropy materials (HEMs), where multi-principal elements with equal or near-equal atomic ratios share the same Wyckoff site, it is arduous to identify the matrix. All individual constituents contribute to the phase formation of the HEMs. In high-entropy alloys (HEAs), in spite of diversity in crystal structure of multiple components, the resultant phases are rather simple; most reported HEAs are in simple face-centered cubic (FCC) or body-centered cubic (BCC) structure. Initially, Cantor et al. reported a five-component $Fe_{20}Cr_{20}Mn_{20}Ni_{20}Co_{20}$ alloy forming a single-phase FCC dendritic structure in their first published paper entitled"Microstructural development in equiatomic multicomponent alloys" [1]. This FCC alloy is capable of dissolving substantial amounts of other transition metals such as Nb, Ti, and V, which are not FCC. Besides, more electronegative elements such as Cu and Ge are also dissolvable in the FCC dendrites with considerably less solubility. Contemporaneously, in their classic paper, Yeh et al. proposed the concept of HEA and reported a phase-transformable Al-containing HEA system, $CuCoNiCrAl_xFe$ [2]. Those alloys with aluminum contents from $x = 0$ to $x = 0.5$ showed a simple FCC structure. As $x$ exceeded 0.8, a BCC structure appeared in addition to the FCC. A single BCC structure was subsequently obtained for $x > 2.8$, as shown in Figure 2.1. Since then, the field of HEAs has been boosted. Since most of the material systems in the succeeding research studies are variants of the initial ones, the crystal structures of those HEAs are cubic.

Cubic HEAs are the majority; however, some other types of solid solutions are also found. Lilensten et al. reported an orthorhombic structure in the $Ti_{35}Zr_{27.5}Hf_{27.5}Ta_5Nb_5$ alloy [3]. Although whether this alloy can be categorized as HEM is still questionable (the entropy of this alloy is only 1.35$R$, which cannot satisfy the

*High-Entropy Materials: From Basics to Applications,* First Edition.
Huimin Xiang, Fu-Zhi Dai, and Yanchun Zhou.

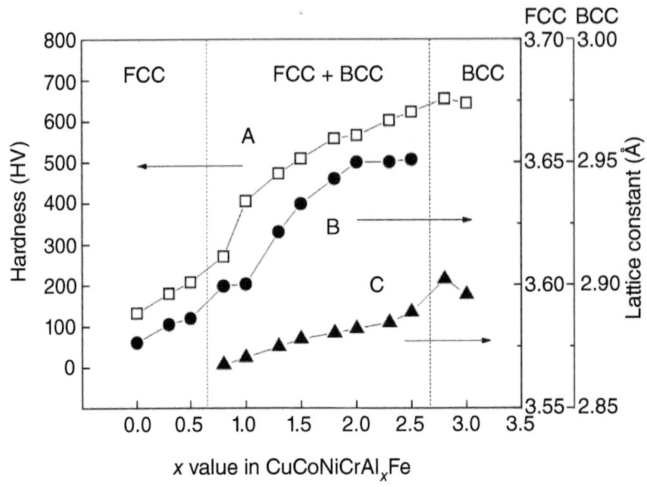


**Figure 2.1** Hardness and lattice constants of the CuCoNiCrAl$_x$Fe alloy system with different $x$ values, demonstrating the phase transition from FCC to BCC. (a) Hardness of CuCoNiCrAl$_x$Fe alloy. (b) Lattice constant of FCC phase. (c) Lattice constant of BCC phase. Source: Reproduced with permission from Yeh et al. [2]/John Wiley & Sons.

criterion of HEA [2]), the possibility of HEAs with structures beyond FCC and BCC is exhibited. Another possible choice for HEAs is hexagonal close-packed (HCP) structure. Zhang et al. predicted that an HEA with an HCP structure can be formed from rare-earth elements with an HCP or double HCP structure due to their similar atomic sizes and isomorphous binary solid solution formability [4]. Based on this conjecture, Takeuchi et al. selected Be, Mg, Sc, Y, Ti, Zr, Hf, Tc, Re, Ru, Os, Co, Zn, Cd, Tl, and heavy lanthanides excluding Yb as candidates for the formation of HCP HEA [5]. Using the proposed alloy design strategy, they finally narrowed the composition down to GdTbDyTmLu and YGdTbDyLu, and successfully fabricated two HEAs with HCP structure by arc-melting in an Ar atmosphere. Single HCP HEAs with the composition of HoDyYGdTb, CoFeReRu, and GdHoLaTbY were also found, sequentially [6]. Although alloys with structures other than cubic have been reported, it is still an open question whether more HEA systems with different structures can be found.

### 2.1.2 Crystal Structure of High-Entropy Ceramics

Unlike HEAs, the crystal structures of high-entropy ceramics (HECs) are diverse and complex, since the constituents of HECs are not constrained in cubic or hexagonal structures and two or more sublattices often coexist in one lattice. Moreover, the database of inorganic ceramics is vast, the combination of compositions and structures for possible HECs would be much more abundant than HEAs, as illustrated in Figure 2.2 [7]. So far, the most investigated HECs are high-entropy oxides (HEOs) with cubic structures due to their appealing properties and promising applications.

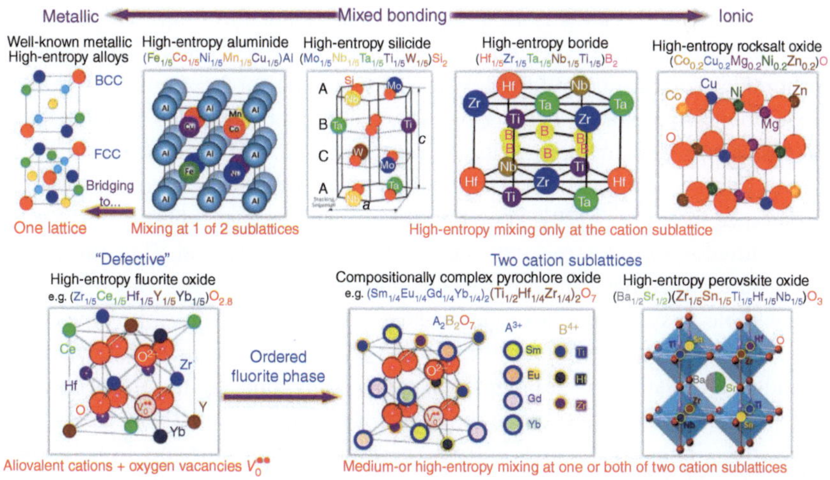


**Figure 2.2** Schematic illustration of selected high-entropy ceramics (HECs) fabricated and reported recently. Source: Reproduced with permission from Wright and Luo [7]/ Springer Nature.

Rost et al. reported the first bulk HEO (MgCoNiCuZn)O with rock salt structure in 2015 [8]. They also proposed a rationale of selection on individual constituents for the search of new HECs as follows: no uniform crystal structure exhibited in the ensemble of binary compound, electronegativity or cation coordination, and existence of pairs that do not exhibit extensive solubility. Using the X-ray absorption fine structure method, they proved that the crystal structure model of this HEO fit with the experimental scattering data for the rock salt structure and the cations were randomly distributed on an *Fm-3m* sublattice with minor local disorder [9]. Further investigations demonstrated that the structure of this HEO was stable when different synthesis methods, stoichiometry, raw materials, and oxygen pressure were employed [10–13].

The rock salt (MgCoNiCuZn)O reported by Rost et al. is a binary oxide with one metal sublattice where the metal atoms were distributed randomly. Inspired by these pioneer works, other binary HEOs with one metal sublattice, such as fluorite and bixbyite structure, were discovered continually [14–17]. Different from binary oxides, ternary oxides, such as those with perovskite structure, spinel structure, pyrochlore structure, etc. [18–27], are structurally more complex. For example, in perovskite structure with a general formula of $ABO_3$, there are two sublattices (A-site and B-site), and they can crystallize in cubic, orthorhombic, or tetragonal crystal structure according to the Goldschmidt's tolerance factor ($t$), which can be calculated from the ionic radii [28]:

$$t = \frac{r_A + r_O}{\sqrt{2}\left(r_B + r_O\right)} \tag{2.1}$$

where $r_A$ and $r_B$ are the ionic radii of the cation at A-site and B-site, respectively, and $r_O$ is the radius of the oxygen ion. A cubic phase is formed when $t = 1$. For $t < 1$, orthorhombic structure is preferred, whereas tetragonal or hexagonal structures are preferably formed for $t > 1$. Thus, it is feasible to manipulate the structure and properties of high-entropy perovskites by proper selection of cations. Sarkar et al. explored 11 different multicomponent equiatomic perovskite-based oxide systems with tolerance factor $t < 1$ [29]. They found that regardless of whether the A-site was substituted with five rare-earth cations (Gd, La, Nd, Sm, and Y) or the B-site was substituted with five transition metal elements (Co, Cr, Fe, Mn, and Ni), it was always possible to synthesize a single-phase, orthorhombic, perovskite-type compound. For the system where both the A- and B-sites were replaced by five rare-earth and five transition metal elements, respectively, phase-pure orthorhombic perovskite was also guaranteed due to its higher entropy. Jiang et al. successfully synthesized 6 single-phase, high-entropy, perovskite oxides among 13 compositions that they had examined [23]. Since the tolerance factor $t$ of these six compounds was close to unity, cubic structure was obtained. More complex ternary oxides with monoclinic structures are also reported, e.g. monosilicates [30, 31], disilicates [32–34], aluminates [35–37], and phosphates [38, 39]. These HEMs show excellent thermal stability and enhanced properties compared to their constituents.

Other than oxides, HECs in other systems with different structures began to emerge since the discovery of high-entropy metal diborides in 2016 [40]. For all reported high-entropy carbides and nitrides [41, 42], rock salt structure is formed, whereas for high-entropy borides (HEBs), the structure is dependent on the boron content or B—B bonding, i.e. monoborides are orthorhombic [43, 44], diborides are hexagonal [40], tetraborides are tetragonal [45], while hexaborides are cubic [46, 47]. Till today, no matured conclusions have been drawn on the relationship of structure between HECs and single-component systems. However, the diversity in crystal structure of HECs has provided an orthogonal strategy to discover new crystalline matters and untapped opportunities for property engineering [8].

## 2.1.3 Atomic Distribution

Ideally, in a solid solution, all elements distribute homogeneously on their own sites. However, in real HEMs, the absolutely random distribution of atoms cannot be achieved due to the diffusion kinetics [48]. Down to the atomic level, there is a major difference between HEAs and HECs, that is, the existence of an ordered anion/cation sublattice in HECs, as demonstrated by Rost et al. [8] in Figure 2.3. This characteristic makes a difference in the atomic arrangement in HEAs and HECs.

### 2.1.3.1 Atomic Distribution in HEAs

In real HEAs, inhomogeneous distribution of atoms is manifested as local fluctuation in composition or chemical short-range order (CSRO), which was initially discovered in the melts of multicomponent alloys. Gao and Alman simulated the partial correlation functions of $Al_{1.3}CoCrCuFeNi$ at 1873 K using *ab initio* molecular

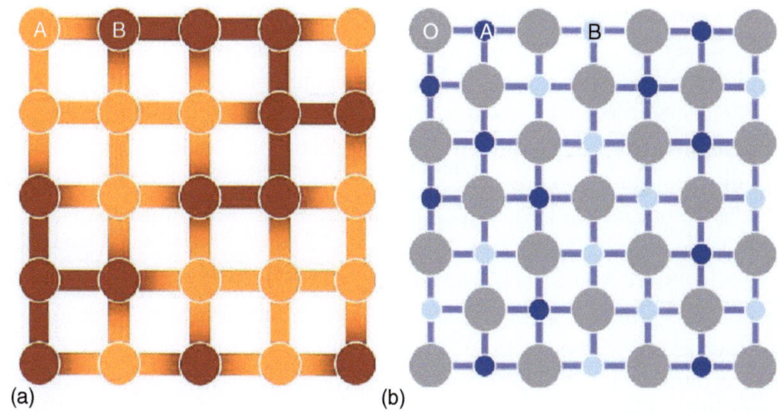


**Figure 2.3** A schematic representation of two lattices illustrating how the first-near-neighbor environments between species having different electronegativity (the darker, the more negative charge localized) for (a) a random binary metal alloy and (b) a random pseudo-binary mixed oxide. Source: Rost et al. [8]/Springer Nature/CC BY-4.0.

dynamics (AIMD) [49] and noted that the preferred correlation for both Al and Ni was the Al–Ni pair, the Cr–Fe pair for both Cr and Fe, the Co–Cr pair for Co, and the Cu–Cu pair for Cu. Those results suggest the existence of preferred short-range ordering of Al–Ni, Co–Cr, and Cr–Fe pairs in the liquid and are coincident with the observation on the cast AlCoCrCuFeNi HEA by three-dimensional atom probe that Cu-rich precipitates, Al–Ni-rich plates, and Cr–Fe-rich interplates exist [50]. These predictions were verified by Santodonato et al. [51]. They observed a complex microstructure of $Al_{1.3}CoCrCuFeNi$ alloy at room temperature (RT): the matrix consisted of BCC/B2 phase mixture and two types of submicron Cu-rich precipitates. The BCC phase was Cr–Fe-enriched alloy and the B2 phase was Al–Ni-enriched one, as shown in Figure 2.4. The matrix was formed by spinodal decomposition. With the help of high-temperature neutron scattering and AIMD, they identified the preferred nearest-neighbor pair correlations in the liquid of multicomponent alloy and the contribution of these correlations to the formation of complex RT microstructure. Shortly, the CSRO in Ni-based HEAs was simulated using different theoretical methods, separately [52, 53]. The CSRO is predicted to have effects on properties such as alloy formation energy, atomic mean square displacements, electronic density of states (DOSs), and configurational entropy. Direct experimental evidence of CSRO in HEAs was provided by Zhang et al. [54]. They examined the local structure of NiCoCr alloy with neutron and X-ray pair distribution function analysis and extended X-ray absorption fine structure (EXAFS) measurements. The EXAFS analysis revealed that Cr in the solid solution alloys has a tendency to bond with Ni and Co and form short-range ordering.

As pointed out by Varvenne et al. [55], the strengthening of HEAs is mainly due to the dislocation interactions with the random local concentration fluctuations around the average composition. The existence of CSRO in HEAs changes the local concentration and may shape the dislocation behaviors and related properties.

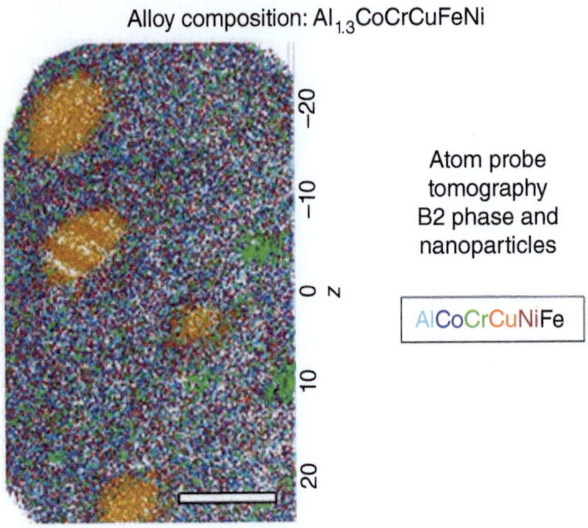


**Figure 2.4** An atom map shows the atomic distribution within the B2 phase of $Al_{1.3}CoCrCuFeNi$ alloy, which contains Cu-rich and Cr-rich nanoprecipitates. The chemical short-range order is clearly demonstrated. Scale bar, 0.01 μm. Source: Santodonato et al. [51], Reproduced with permission from Springer Nature.

Ding et al. [56] found that CoCrFeNiPd had much higher tensile strength than CoCrFeNiMn and attributed it to the increased concentration wave (or CSRO) in the solid solution. However, Yin and Curtin [57] argued that the strengthening of CoCrFeNiPd should be attributed to the large misfit volume of Pd in CoCrFeNi. Thus, it is still debatable whether CSRO could strengthen HEAs, and further investigations are still awaited.

Another phenomenon regarding the elements distribution in HEAs is the newly discovered possible segregation of elements at the grain boundary (GB), which is brought out in the investigation of phase stability of HEAs after prolonged annealing. Otto et al. found precipitates at the GB of CrMnFeCoNi HEA after annealing for 500 days at 500 and 700 °C [58]. Pickering et al. discovered a similar phenomenon when they annealed CrMnFeCoNi HEA at 700 °C for 1000 hours [59]. Two precipitates had been identified as an $M_{23}C_6$ carbide and the topologically close-packed $\sigma$ phase. Their results have demonstrated that CrMnFeCoNi is thermodynamically stable as a single solid solution at all temperatures below its solidus can no longer be considered valid, and the element distribution at the GB is dissimilar to that within the grain. Ming et al. carefully examined the microstructure of CrMnFeCoNi at RT using atom probe tomography (APT) and found the fast nanosegregation of principal elements Ni, Cr, and Mn separately into the GBs [60]. The segregation was responsible for the ductility loss of CrMnFeCoNi HEA at intermediate temperatures. Li et al. employed hybrid Monte Carlo (MC)/molecular dynamics (MD) simulations to simulate the element segregation in NbMoTaW system, and clear segregation of Nb to the GBs could be observed [61]. All these results have demonstrated that the element segregation in HEAs might be a common feature.

#### 2.1.3.2 Atomic Distribution in HECs

The situation in HECs is different. As presented by Rost et al. [8], with the existence of an ordered sublattice, the differentiation between every cation lattice site is small since each has the same immediate surroundings. A large number of quasi-equivalent sites in a crystal structure with an intermediate sublattice will increase the distribution homogeneousness and configurational entropy of cations and favor the formation of single-phase solid solution. The dispersion of atoms in HECs has been experimentally observed by electron microscopy coupled with energy dispersive X-ray spectroscopic system, and this technique is now considered as a standard tool to probe the distribution homogeneousness [39, 41, 46, 47]. Delicate techniques to determine element distribution, such as scanning transmission electron microscopy (STEM) and APT, have also been employed, and the atomic scale homogeneous distribution is reported and no significant aggregation in the grains is found (Figure 2.5) [62–64]. Nevertheless, the segregation of elements at the GB in ceramic solid solutions also exists. Wang et al. [65] explored the atomic structure and chemistry of W- and W-Mo-containing HEBs by atomic-resolution transmission electron microscope (TEM) imaging, electron diffraction, and chemical analysis. They found co-segregation and precipitation at the GB in both HEBs (Figure 2.6), and extra hardening was achieved. A similar phenomenon is also found in high-entropy carbonitrides and perovskite [66, 67]. For solid solutions, the effects of GB segregation on the mechanical and functional properties have been explored extensively; however, such efforts have barely been devoted in HECs. More delicate works are needed to probe the element distribution around the imperfections and their effects on the properties of HECs.

### 2.1.4 Lattice Distortion

As stated in Section 1.3, the lattice distortion is one of the main core effects of HEMs. It is from the difference of constituent elements in atomic size, electronegativity, and bonding energy with the neighboring atoms. This structure feature was proposed by Yeh et al. in their early work on CuCoNiCrAlFeTiV alloys [68], where they found that not only were all X-ray diffraction (XRD) peak intensities of alloys with multi-principal elements markedly lower than the corresponding ones of conventional alloys under the same XRD measurement conditions, but also the intensities at high angles decreased much more rapidly than those at low angles. Further mathematical treatment was introduced to understand the effect of lattice distortion on the XRD intensities in CuNiAlCoCrFeSi alloy systems [69]. The XRD intensities of these alloys decreased with the increasing number of incorporated elements, much beyond the thermal effect, and an intrinsic lattice distortion effect caused by the addition of multi-principal elements with different atomic radii is expected for the anomalous decrease in the XRD intensities. Guo et al. also found that, in ZrNbHf alloy, the thermal factors determined by the Rietveld analysis were larger than what was expected for phonon thermal factors due to the existence of lattice distortion [70]. Using high-resolution transmission electron microscopy (HRTEM), Zou et al. displayed

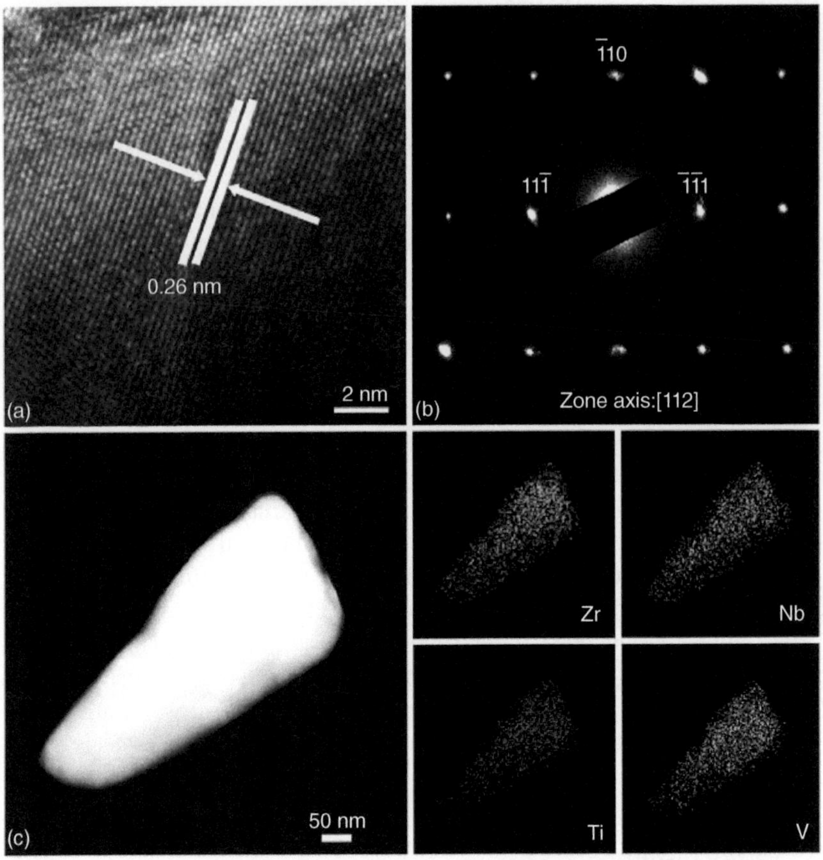


**Figure 2.5** STEM image and the corresponding energy dispersive spectroscopy (EDS) compositional maps of $(Zr_{0.25}Nb_{0.25}Ti_{0.25}V_{0.25})C$, demonstrating the homogeneous distribution of elements. (a) High-resolution transmission electron microscope image. (b) Selected area electron diffraction pattern. (c) Scanning transmission electron microscope image Source: Ye et al. [62], Reproduced with permission from Elsevier.

mismatches and local distortions in the Fourier-filtered images of NbMoTaW HEA (shown in Figure 2.7) [71]. However, as pointed out by He and Yang [72], one should be cautious about the interpretation of an HRTEM image since a curved atomic plane could also be caused by a strain field around defects, such as dislocations. Moreover, the scale of lattice distortion is closely related to the deviation of atomic radii. Tong et al. quantitatively analyzed the local lattice distortion in CoCrFeNi, CoCrFeMnNi, and CoCrFeNiPd HEAs by X-ray total scattering and extended X-ray absorption fine structure (EXFAS) methods [73]. They found that the lattice distortion in CoCrFeNi and CoCrFeMnNi was very small while that in CoCrFeNiPd was larger owing to the relatively large size of Pd. Besides the above experimental efforts, theoretical models were also developed. Song et al. proposed a supercell model with similar local atomic environment (SLAE) to simulate a random solid solution with

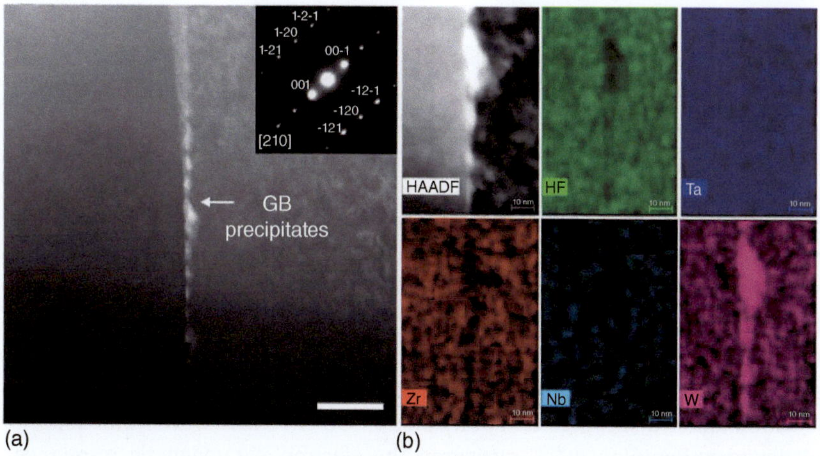


**Figure 2.6** Grain boundary (GB) segregation and precipitation in $(Zr_{0.2}Hf_{0.2}Nb_{0.2}Ta_{0.2}W_{0.2})B_2$. (a) A high angle annular dark field- scanning transmission electron microscope (HAADF-STEM) image of a GB and (b) EDS maps of Hf, Ta, Zr, Nb, and W across the GB. Source: Wang et al. [65], Reproduced with permission from Elsevier.

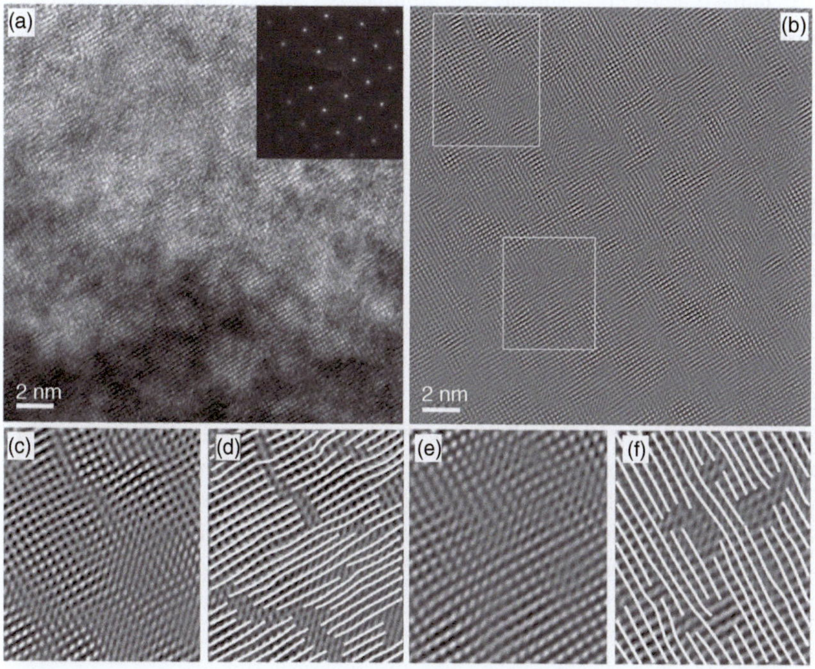


**Figure 2.7** (a) A bright-field TEM image oriented in [100] zone axis and the corresponding electron diffraction pattern; (b) an inverse fast Fourier transform image of the area a; (c) and (e) are enlarged images of the indicated boxes in (b). High-resolution TEM images of the Nb–Mo–Ta–W HEA. Lattice fringes in (d) and (f) are traced to indicate the regions with lattice distortions. Source: Zou et al. [71], Reproduced with permission from Elsevier.

multiple elements [74]. Their results demonstrated that the local lattice distortion of the refractory HEAs was much more significant than that of the 3*d* HEAs, and the atomic size mismatch evaluated with the atomic radius was not accurate enough to predict the local lattice distortion effects. Ye et al. carried out extensive density functional theory (DFT) calculations on a series of equal-molar binary to quinary chemical complex alloys [75]. Notable local atomic-scale distortions could be seen in the distorted lattice, and the diffraction peak profile of the distorted lattice exhibited peak broadening and even peak splitting, particularly so for the high angle peaks.

Regarding HECs, lattice distortion also exists. Using fitted EXAFS data, Rost et al. quantified the local atomic structure of (MgNiCoCuZn)O on an element-by-element basis [8, 9]. Model-fitted experimental data and DFT calculations demonstrated that the anion sublattice displaced from ideal locations to accommodate the distortions, while cation atoms distributed randomly on an FCC sublattice with minimal positional disorder. Berardan et al. measured the electron paramagnetic resonance (EPR) signal of copper ions in (MgCoNiZn)CuO lattices. They found that the signal was distorted, thus indicating deviation of lattice from the ideal rock salt structure [10]. DFT and MD calculation on the structure of the (MgCoCuNiZn)O system also verified that the lattice distortion was derived from the effect of a combination of both Jahn–Teller and/or simple size arguments [76]. Theoretically, Dai et al. employed a deep learning potential (DLP) to investigate the structure and properties of high-entropy (ZrHfTiNbTa)C ceramic [77]. They found that the carbon atoms deviated more significantly from the ideal positions in comparison with metal atoms, which was also proved by Ye et al. using the DFT method [78]. Using the same DLP method, Dai et al. also explored the structure of high-entropy diboride $(TiZrHfNbTa)B_2$, as illustrated in Figure 2.8 [79]. Their results revealed that lattice distortions in high entropy (HE) diboride were one order of magnitude smaller than those in HE carbide. Nevertheless, the boron atoms deviated more significantly from the ideal positions in high-entropy diboride.

Lattice distortion is one of the main structure features in HEMs, and it shapes the unique combination of mechanical and thermal properties of HEMs. For example, lattice distortion hinders the movement of dislocation and results in significant solid solution strengthening effects in BCC HEAs [4]. It is also the main reason for the phase transformation in HEAs [4]. It has a significant impact on the transport of electrons and phonons, leading to the decrease in electrical conductivity and thermal conductivity [4]. The details of effects of lattice distortion on the properties of HEMs will be discussed in Chapter 7.

## 2.2 Electronic Structure and Band Gap Engineering

### 2.2.1 Electronic Structure of HEAs

The electronic structure of a material is closely related to the stability, transport, optical, and magnetic properties. For HEMs, more than five principal elements provide huge chemical space for tuning the electronic structure and tailoring the related

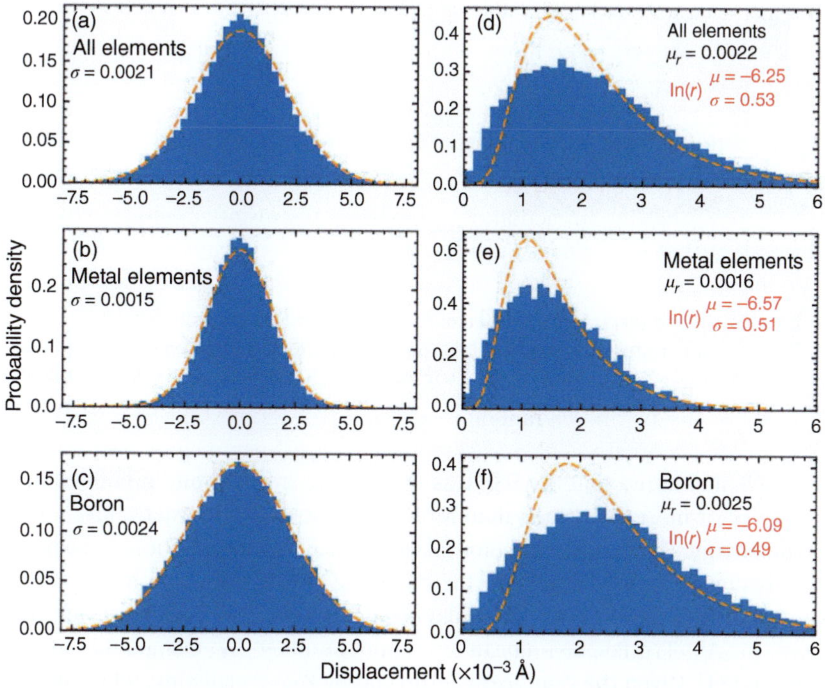


**Figure 2.8** Distributions of displacements of atoms from their ideal positions in high-entropy borides. Displacements along *c* direction for (a) all of the elements, (b) all of the metal elements, and (c) boron. Displacements in the basal plane for (d) all of the elements, (e) all of the metal elements, and (f) boron. Source: Reproduced with permission from Dai et al. [79]/ Elsevier.

properties. The largest difference between HEAs and HECs in electronic structure is the electric conductivity of HEAs, which means no band gap between conduction and valence band, and it is impossible to tune the band gap in HEAs by composition control. However, by introducing different constituents, it is feasible to regulate the electron transfer and band structure around the Fermi surface to realize specific properties.

In the pursuit of electronic structure of materials, theoretical modeling is a natural choice for its convenience and visualization. The available state-of-the-art theoretical methods in the framework of *ab initio* calculations based on DFT for simulating disorder can be roughly categorized into two classes, those based on supercell models and those based on the coherent-potential approximation (CPA) [80–82]. Both classes of methods have their advantages and disadvantages, and it depends on the specific type of application to which they are better suited. Troparevsky et al. demonstrated the power of Korringa–Kohn–Rostoker coherent-potential approximation (KKR-CPA) in the simulation of effects of disorder on the electronic properties, such as electronic, magnetic transport, and formation energies [83]. They utilized KKR-CPA to calculate the configurationally averaged DOSs and Bloch spectral function (BSF) of two five-component HEAs, CoCrFeMnNi and

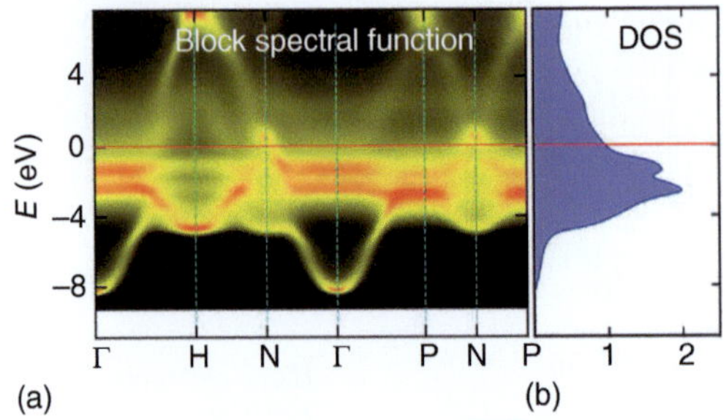


**Figure 2.9** Calculated electronic structure of the $Al_{0.5}TiZrPdCuNi$ crystalline random alloys. The smearing in the DOS is quite obvious. (a) Block spectral function of $Al_{0.5}TiZrPdCuNi$. (b) Density of state of $Al_{0.5}TiZrPdCuNi$. Source: Reproduced with permission from Odbadrakh et al. [84]/AIP Publishing LLC.

MoNbTaVW, and found that the BSF was much more smeared out in both energy and wave-vector in CoCrFeMnNi than MoNbTaVW for occupied states just below the Fermi energy. The major reason for this difference was the additional magnetic disorder scattering from elements in the middle of the 3*d*-transition metal series, which was absent in MoNbTaVW. Similar smearing near the Fermi level was also obtained in $Al_{0.5}TiZrPdCuNi$ crystalline random alloy by Odbadrakh et al. (Figure 2.9) [84]. Given the significant Fermi energy BSF broadening, which corresponded to a short mean free path, they speculated that these alloys should have high residual resistivity.

As an electronic-originated property, magnetic moment is also closely related to the electronic structure of HEAs. Using the exact muffin-tin orbitals (EMTO) method in combination with CPA, Tian et al. calculated the DOS of the hypothetical nonmagnetic and paramagnetic FCC solid solutions of NiCoFeCr, CuNiCoFeCr, and NiCoFeCrTi [85]. They found that paramagnetism played an important role in the understanding of the magnetic state of these 3*d*-HEAs and accurate estimation of the equation of state. A similar procedure was also employed by Wang et al. in the investigation of ferrimagnetic (FM) origination of CrMnFeCoNi alloy [86]. By examining the DOS of this HEA, they concluded that the primary contribution to the total magnetic moment ($2.14\mu_B$) had arisen from the spin polarization of Fe *d* states and Mn *d* states. In Wang's work, the special quasi-random structures (SQSs) were introduced to construct the disordered structure of HEA to reduce the scale and difficulty of calculations using a simple supercell. The concept of SQS [87, 88] is to build a special periodic structure with a small number of atoms per unit cell whose correlation functions for the first few nearest-neighbor shells are as close to those of a target random alloy as possible such that the periodicity errors only exist between more distant neighbors.

Besides, the electronic structure can also be used to evaluate the stability of the investigated phases. Leong et al. examined the electronic structure of a number of five-component HEA alloys based on the equiatomic CoCrFeNi composition using a combined modeling and experimental approach to improve the accuracy of the predictions of the majority phase [89]. They constructed a SQS supercell of these alloys with different structures and calculated the total energy difference between these structures to identify the stable phase. By integrating the spin-polarized total DOSs, the energy difference was related to the exact valence electron concentration (VEC) of the system, which was tunable by element adjustment. Using this model, they predicted a new magnetic HEA system (CoFeNiV), which was verified experimentally.

### 2.2.2 Electronic Structure of HECs

Unlike HEAs, HECs could be a conductor, a semiconductor, or an insulator. The interaction between ceramics and incident energy is variable, which ensures their applications in multiple energy-related fields. Their variable applications originate from the abundant and tunable electronic structures of HECs. Up to now, few theoretical works have been conducted on the electronic structure of HECs due to the exhausting calculations. Rák et al. constructed a 480 atom rock salt supercell with a 3×4×5 size to model the random structure in the extensive DFT calculations on three entropic oxides, (MgCoCuNiZn)O, (MgCoCuNiZnLi)O, and (MgCoCuNiZnSc)O [90]. Analysis on the Bader charges and DOSs indicated that addition of Sc to (MgCoCuNiZn)O resulted in the reduction and movement from their ideal lattice sites of a majority of Cu cations, while the addition of Li to (MgCoCuNiZn)O resulted in oxidation of Co, Ni, and Cu atoms with no substantial displacement from their ideal lattice sites. Using the same model, they calculated the density of (MgCoCuNiZn)O and found that the $CuO_6$ octahedra were in both elongated and compressed geometry [76]. The elongated and compressed Cu-centers adopted different electronic ground states with half-filled $d_{x2-y2}$ or $d_{z2}$ orbitals, respectively, as illustrated in the projected DOS of Cu *d*-states near Fermi level. The displacement of O atoms was responsible for the experimentally observed Jahn–Teller distortion in (MgCoCuNiZn)O.

The SQS is also approved to be efficient in avoiding the time-consuming calculations for HECs. Using this technique, Ye et al. and Jiang et al. [62, 91] constructed the 2×2×2 SQS supercell of high-entropy carbide (TiZrHfNbTa)C with 64 atoms. The predicted DOS indicated that the metallic conductivity of high-entropy carbides was preserved, and the presence of a pseudogap at the Fermi level suggested strong covalent bonding between metal atoms and C. Similar results were also reported by Wang et al. in the calculation of electronic structure of $(TiZrHfNbTa)B_2$ using SQS supercell [92]. Liu et al. [93] constructed a SQS supercell containing 100 atoms for $Zn_{0.1}Ca_{0.1}Sr_{0.4}Ba_{0.4}ZrO_3$. They found that the lattice distortion resulted from high-entropy effect induced chemical bonds inhomogeneity, i.e. Zn–O and Ca–O bonds were weaker than Sr–O and Ba–O bonds, which was responsible for the lower stiffness and mechanical properties of HEC than their single-component counterparts.

Another excitement about the electronic structure of HECs is the tunable band gap by composition control. Sarkar et al. [94] prepared high-entropy rare-earth oxides (HE-REOs) containing three to seven rare-earth cations (Ce, Gd, La, Nd, Pr, Sm, and Y) in equiatomic amounts by nebulized spray pyrolysis. Band gap

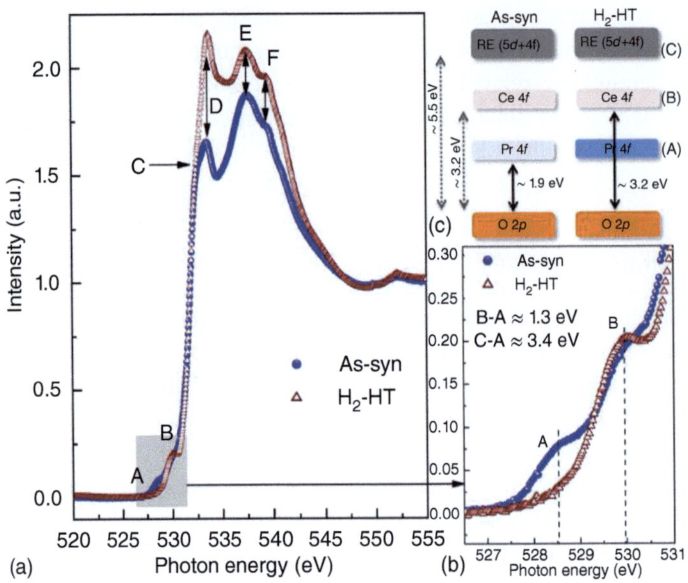


**Figure 2.10** (a) O K edge X-ray absorption spectra of as-synthesized and reduced by $H_2$-heat treatment (HT) high-entropy oxide $(Ce_{0.2}La_{0.2}Pr_{0.2}Sm_{0.2}Y_{0.2})O_{2-\delta}$ and (b) magnification of the pre-edge region. (c) Schematic of the electronic band diagram to illustrate electron transition from an occupied state to an unoccupied state before and after the reduction. Source: Sarkar et al. [95]/AIP Publishing/CC BY-4.0.

measurement with ultraviolet–visible (UV–Vis) absorption spectra revealed narrowing of the gap compared to that for $CeO_2$. They speculated that the multivalent element Pr was responsible for the band gap narrowing, and thus, enabled HE-REOs to absorb light over the entire visible spectral range. To verify their own hypothesis, they further obtained element-resolved charge state and electronic structure of $(CeLaPrSmY)O_{2-\delta}$ by x-ray absorption spectroscopy (XAS) and electron energy loss spectroscopy (EELS) [95], and the ability of Pr in tuning the band gap energy of HE-REO by redox reaction was confirmed, as demonstrated in Figure 2.10. Bérardan et al. synthesized HEO (MgNiCoCuZn)O with colossal dielectric permittivity [96]. The lithium content had a large impact on the observed dielectric constants and the band gap. The lesser the content of lithium in the sample, the higher was its intrinsic resistance and also the higher was its maximum permittivity. Although the electronic structure feature of HECs has not been extensively explored, these pioneering works have demonstrated their potential applications as catalyst, energy storage, electric, and optical materials by tuning the band structure of HECs, which is achievable simply by choosing suitable additional elements.

What we have to mention is, although SQS is efficient in constructing the quasi-random structure and easing the calculation burden, one should pay attention to the usage of this method in the simulation of HEAs and HECs. This is because, in SQS, the interactions between distant neighbors are considered to contribute much less to the system, which may not be the case for ionic crystals where long-range interactions may still be important under a given unit cell size constraint. Moreover, the energy of the SQS of some systems might be sensitive to the atomic configurations, and the uncertainty in the energy of those systems is too high, especially for anisotropic lattices [97].

## 2.3 Lattice Dynamics and Phonon Dispersion

The phase stability, thermodynamic properties, and thermal conductivity of a material are linked intrinsically to lattice vibrations, which is also an essential structure feature. Accurate modeling of phonon excitations and their interactions plays a decisive role in materials design, exploration, and optimization. For HEMs, the lattice vibrations are quite unique for their chemical disorder and mass fluctuation. Theoretical quantization of their unique lattice vibration feature is quite a challenge since the estimation of interatomic force constants requires the perturbation of all atoms in the supercell, which is quite time-consuming. For efficient calculations, different strategies have been utilized.

Körmann et al. constructed the SQS cells for the equimolar 2-, 3-, 4-, and 5-component alloys containing 128, 54, 128, and 125 atoms with the elements of Hf, Mo, Nb, Ta, Ti, V, and W [98]. For all considered alloys, very strong broadening, reaching widths of up to a few THz, in the mid- and high-frequency region (above ~4 THz) was found, while the long wavelength limit (region near Γ) was unaffected, as shown in Figure 2.11. The partial phonon spectral functions of individual contributions from each element revealed that the highest frequencies for each alloy were predominantly determined by the lightest elements, whereas the heavier ones determine the low-frequency parts. Their results suggested that the lighter element had smaller force constants and the heavier one had larger force constants, which

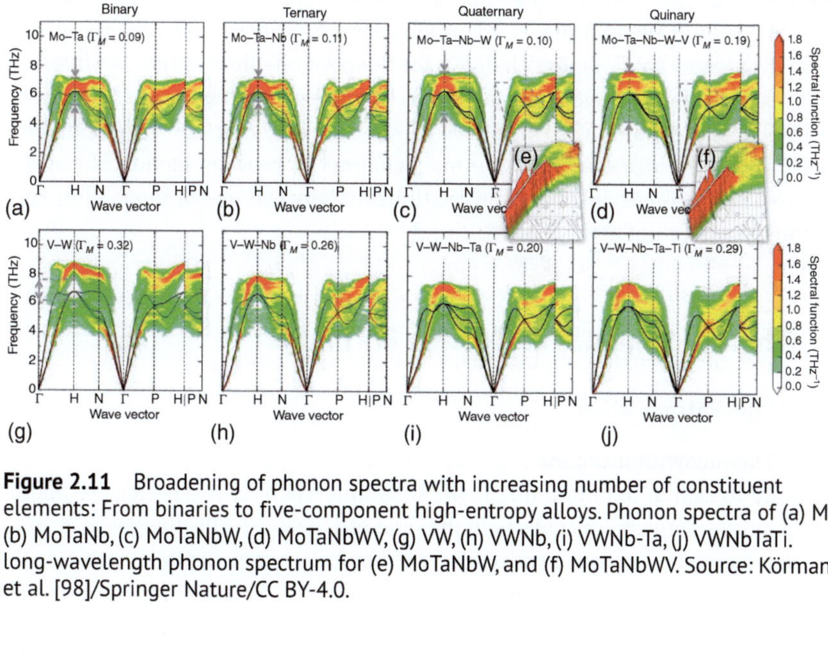


**Figure 2.11** Broadening of phonon spectra with increasing number of constituent elements: From binaries to five-component high-entropy alloys. Phonon spectra of (a) MoTa, (b) MoTaNb, (c) MoTaNbW, (d) MoTaNbWV, (g) VW, (h) VWNb, (i) VWNb-Ta, (j) VWNbTaTi. long-wavelength phonon spectrum for (e) MoTaNbW, and (f) MoTaNbWV. Source: Körmann et al. [98]/Springer Nature/CC BY-4.0.

implied that the force constant fluctuations counteracted the mass disorder induced fluctuations and contracted the band gap and the overall broadening induced by mass disorder. Thus, for quantitative predictions, force constant fluctuations cannot be neglected or artificial phonon band gaps may appear. Their work also proved that the number of constituents was less important in determining the phonon spectra than general impression.

In HECs, since there are more than one sublattices, the situation is more complicated. Esters et al. calculated the lattice dynamics of (HfNbTaTiR)C (R = V, W, Zr) and binary carbides using the automatic flow (AFLOW) framework [100]. In their work, the commonly used SQS was abandoned to avoid an extra layer of uncertainty on the driving forces stabilizing the disorder. They adopted a partial occupation (POCC) algorithm where statistically weighted structures were used to produce ensembles determining physical properties [99]. The phonon density of states (pDOSs) was then ensemble-averaged to calculate vibration properties through the following expression [99]:

$$g^{avg}(\omega) = \sum_i P_i(T_{POCC})\, g_i(\omega) \tag{2.2}$$

where $g^{avg}(\omega)$ is the ensemble-averaged pDOS, $P_i(T_{POCC})$ is the probabilities of the relaxed ordered representatives, and $g_i(\omega)$ is the individual pDOS for every derivative structure. The total and projected pDOS of high-entropy carbides using this technique were broad and largely featureless due to structural and chemical disorder, while those of the binary carbides showed sharp features. The phonon gap between metal and carbon contributions was preserved in the pDOS of HECs. The phonon broadening in the carbon was the consequence of force constant disorder, whereas, for metals, it was governed by both force constant and mass fluctuations, which was consistent with that in HEAs.

In HEAs and HECs, broadening of phonon dispersion induced by mass disorder and force constant fluctuations is a common structural feature. Smearing in the phonon dispersion implies more channels for the phonon–phonon (*ph–ph*) interaction and short phonon mean free path, which will reshape the thermal transport properties of these materials. This effect will be enormous for HECs without free carriers, since the lattice vibration is the main channel for heat conduction. However, up to now, efforts on establishing the relationship between *ph–ph* interaction and heat transport have not been devoted.

## 2.4 Thermodynamics and Phase Formation

### 2.4.1 High-Entropy Alloys

#### 2.4.1.1 Thermodynamic Criteria

The ability to search for more unknown stable members or to predict the phase selection is one of the main subjects in the field of HEMs. Ideally, the phase formation is thermodynamically controlled by the Gibbs free energy, *G*, which is related to the enthalpy, $H$, and the entropy, $S$, via the following equation for alloys from mixing elemental components:

$$\Delta G_{mix} = \Delta H_{mix} - T\Delta S_{mix} \tag{2.3}$$

where $\Delta G_{mix}$ is the Gibbs free energy of mixing, $\Delta H_{mix}$ is the enthalpy of mixing, $\Delta S_{mix}$ is the entropy of mixing, and $T$ is the temperature. Thus, the competition between $\Delta H_{mix}$ and $T\Delta S_{mix}$ determines the phase selection. In their preliminary work, Yeh et al. had emphasized the role of configuration entropy in stabling the solid solutions [2]. Following Boltzmann's hypothesis, the entropy of a system is linearly related to the number of possible micro-states corresponding to the macroscopic states of a system:

$$S = k \ln \omega \tag{2.4}$$

where $k$ is the Boltzmann's constant. During the formation of a random solid solution from $n$ elements with equimolar fractions, the configurational entropy $\Delta S_{conf}$ of mixing is [2]:

$$\Delta S_{conf} = R \ln n \tag{2.5}$$

where $R$ is the gas constant. The configurational entropy $\Delta S_{conf}$ of equimolar alloys with three elements is $1.1R$, which is larger than that of metal fusion [2]. Consequently, alloys with a higher number of principal elements will more easily yield the formation of random solid solutions during solidification. Despite the importance of mixing entropy in stabilizing the HEAs, they also point out that for those with very large heats of formation, the phase selection would be different.

To include the contribution of mixing enthalpy, Zhang et al. extended the Hume-Rothery rule and introduced two factors that would affect the formation of the solid solution in alloys [101]. The first is the size effect of component atoms, $\delta$, as follows:

$$\delta = \sqrt{\sum_{i=1}^{N} c_i \left(1 - r_i / \bar{r}\right)^2} \tag{2.6}$$

where $N$ is the number of the components, $c_i$ is the atomic fraction of the $i$th component, and $r_i$ is the atomic radius. The second is the chemical compatibility between the components, i.e. the electronegativity difference or the enthalpy of mixing, $\Delta H_{mix}$:

$$\Delta H_{mix} = \sum_{i=1, i\neq j}^{n} \Omega_{ij} c_i c_j = \sum_{i=1, i\neq j}^{n} 4\Delta H_{ij}^{mix} c_i c_j \tag{2.7}$$

where $\Delta H_{ij}^{mix}$ is the enthalpy of mixing for the binary liquid between the $i$th and $j$th elements at an equiatomic composition. After summarizing the reported data, they found that the solid solution only formed in the zone where the component atomic-size difference was relatively small and $\Delta H_{mix}$ was not negative enough for alloys to form a compound, as shown in Figure 2.12a. To be quantitative, Zhang et al. explored the phase composition in various HEAs [102]. The values of $\Delta S_{mix}$, $\Delta H_{mix}$, and $\delta$ were calculated and displayed in Figure 2.12b. They found

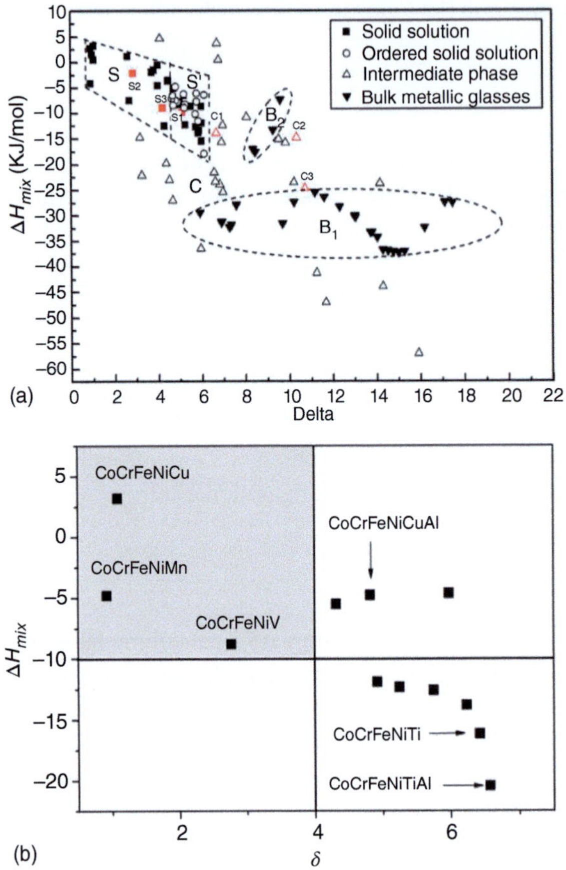


**Figure 2.12** (a) Relationship between Delta and $\Delta H_{mix}$ for high-entropy alloys and typical multicomponent bulk metallic glasses, and (b) relationship between $\Delta S_{mix}$ and $\delta$ values of some high-entropy alloy systems. Source: (a) Reproduced with permission from Zhang et al. [101]/John Wiley & Sons and (b) Reproduced with permission from Zhang and Fu [102]/Elsevier.

that the $\Delta H_{mix}$ and $\delta$ of simple solid solutions were located at the upper-left corner. Therefore, the formation criterion for HEAs should be the $\Delta H_{mix}$ and $\delta$ values falling into the upper-left district in Figure 2.12b under the fulfillment of high mixing entropy ($\Delta S_{mix} \geq 1.61R$), which could be expressed numerically: $\Delta S_{mix} \geq 13.38\,\text{J/(mol·K)}$, $-10\,\text{kJ/mol} \leq \Delta H_{mix} \leq 5\,\text{kJ/mol}$, and $\delta \leq 4\%$. Similar conclusion has also been drawn with the formation zone of HEAs being slightly different [103].

As stated above, the competition between $\Delta H_{mix}$ and $T\Delta S_{mix}$ determines the phase selection. When the effect of high entropy of mixing overcomes the effect of the enthalpy of mixing at high temperatures, the driving force will prevail the resistance, and the solid solution could be prior to the ordered phase to form. Thus, Yang et al. defined a new parameter $\Omega$ to guide the design of HEAs [104]:

$$\Omega = \frac{T_m \Delta S_{mix}}{|\Delta H_{mix}|} \tag{2.8}$$

where $T_m$ is the average melting point, which is defined as: $T_m = \sum c_i(T_m)_i$. Here, $(T_m)_i$ is the melting point of the $i$th component of alloy. $\Omega > 1$ indicates that the effect of the mixing entropy is greater than that of the enthalpy at the melting temperature, and the high-entropy solid solution tends to form. The criteria they proposed for forming solid solution phase were $\Omega \geq 1.1$ and $\delta \leq 6.6\%$, as shown in Figure 2.13. It is worth mentioning that, in Figure 2.13, there were overlaps between the value of $\Omega$ for single phases and multiphases, which suggested that this descriptor should not be used solely to predict the phase selection in HEAs [105].

Except configurational entropy, the $\Delta S_{mix}$ in Eq. (2.3) should include all entropy sources such as vibrational, electronic, and magnetic contributions. Compared to the configurational one $\Delta S_{conf}$, the contributions from electronic and magnetic might be small. Whereas vibrational contribution is comparable to $\Delta S_{conf}$, even higher at certain circumstances for HEMs. Thus, a proper consideration of finite-temperature excitations is more important for the alloys' phase stability. As stated in Section 2.3, it is a challenge to accurately compute the contributions from those other than configurational entropy, especially for the vibrational one. Corresponding studies for multicomponent alloys are therefore very rare. Ma et al. used a finite-temperature *ab initio* method to compute the electronic, vibrational, and magnetic free energy contributions for a five-component prototype HEA CoCrFeMnNi [106]. Their analysis had revealed that entropy contributions beyond the configurational one were important for determining phase stability and other properties of HEAs.

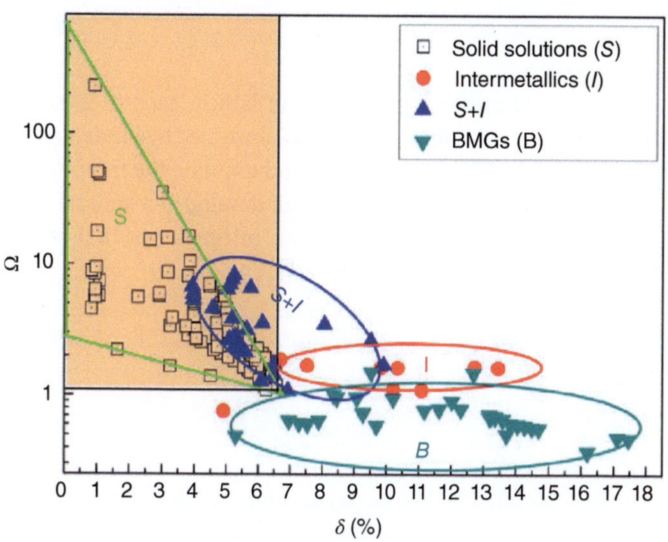


**Figure 2.13** Phase-formation rules based on $\Omega$ and $\delta$ for multicomponent alloys. *S* stands for single solid solution, *I* represents intermetallic compound, and *B* indicates amorphous phase. Source: Reproduced with permission from Yang and Zhang [104]/Elsevier.

Electronic and magnetic entropies contributed up to 50% of the configurational entropy value. The vibrational entropy rose quickly in absolute value with temperature up to four times of the configurational one. The importance of entropy contributions beyond the configurational one was likewise pointed out by Li et al. in their investigation of the FCC-HCP phase stability of $CoCrFe_{1-x}MnNi_x$ HEAs based on a similar approach [107].

#### 2.4.1.2 Valence Electron Concentration Criteria

The well-established Hume-Rothery rule has pointed out that electron concentration is one of the main reasons that affect the stability of solid solutions. Guo et al. extended this criterion to the field of HEAs by introducing the parameter of *VEC* [108], which is defined as the weighted average from VEC of the constituent components $(VEC)_i$:

$$VEC = \sum_{i=1}^{n} c_i (VEC)_i \tag{2.9}$$

They studied the relationship between the VEC and the stability of FCC and BCC solid solution in HEAs, and they found that the BCC phase was stable at lower VEC (<6.87) while the FCC phase was stable at higher VEC (≥8). The same conclusion was also obtained by Poletti et al. [109]. Ye et al. analyzed the VEC and phase selections of ~90 cast HEAs holistically and found that the VEC of the alloys played a crucial role in determining the crystallinity of the solid solution phases [103]. Within the region of the single-phase solid solution, FCC prevails around a VEC of $8.5 \pm 1.0$, BCC around a VEC of $5 \pm 0.7$, and HCP around a VEC of $2.8 \pm 0.2$.

#### 2.4.1.3 Residual Strain Criteria

In general, mixing different-sized atoms into a common lattice leads to intrinsic residual strains, which could result in lattice instability, as illustrated by Lindemann's criterion [48]. To estimate the residual strain in HEAs and evaluate the importance of residual strain in the phase stability of HEAs, Ye et al. developed a general self-contained geometric model [110]. In their model, the intrinsic residual strain around a central atom in the HEAs is defined as follows:

$$\varepsilon_i = \frac{\sum_{j=1}^{n} \omega_{ij} c_j}{\sum_{k=1}^{n} A_{ik} c_k} - \frac{4\pi\bar{\eta}}{N_i \sum_{k=1}^{n} A_{ik} c_k} \tag{2.10}$$

where $\omega_{ij}$ is the solid angle subtended by the atom $j$ around the central atom $i$, $c_i$ is the atomic fraction of element $i$, $N_i$ is the coordinate number of atom $i$, $\bar{\eta}$ is the equilibrium value of atomic packing efficiency, and $A_{ij} = \frac{2\pi x_{ij}}{(x_{ij}+1) 2\sqrt{x_{ij}(x_{ij}+2)}}$, where $x_{ij} = r_i/r_j$ is the ratio of atomic size. Inspired by the Lindemann's criterion,

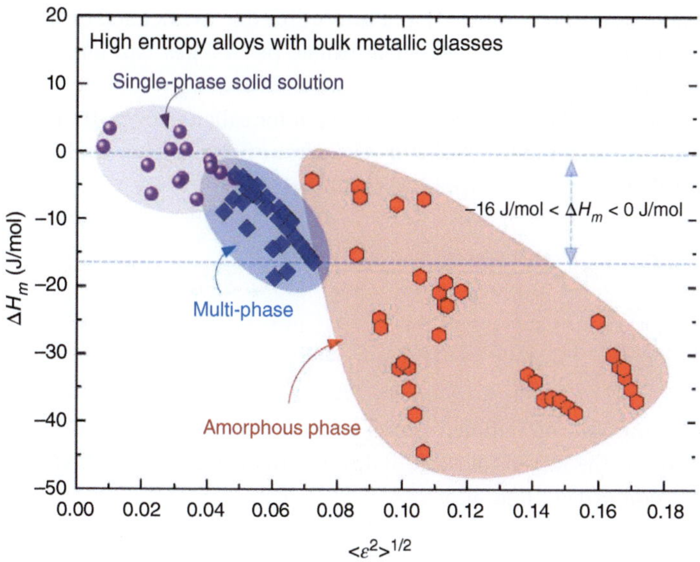


**Figure 2.14** Phase selection with the root mean square residual strain being the descriptor. Source: Reproduced with permission from Ye et al. [110]/Elsevier.

they correlated the phase stability in a HEA with the root mean square (RMS) residual strain derived from the residual strains of individual elements by the following expression:

$$\varepsilon_{RMS}^2 = \sum_{j=1}^{n} c_j \varepsilon_j^2 \tag{2.11}$$

After examining various HEA systems, they proved that the RMS strain $\varepsilon_{RMS}$ could be taken as an indicator for phase stability. Regardless of the chemical complexity of multicomponent alloys, upon solidification, a single-phase solid solution prevailed in as-cast HEAs when their RMS residual strain was less than 5% or an amorphous structure prevailed when their RMS residual strain was above 10%. In between, the as-cast HEAs tended to form multiphased structures, as shown in Figure 2.14. Nevertheless, as pointed out by the authors, the criteria based on residual strains are necessary conditions as the chemical effects are neglected.

### 2.4.2 High-Entropy Ceramics

#### 2.4.2.1 Thermodynamic Criteria

Similar to HEAs, thermodynamically, the phase stability of HECs is also determined by the Gibbs free energy. As shown in Eq. (2.3), larger entropy favors lower Gibbs free energy and hence phase stability. Rost et al. initially pointed out the critical role of configurational entropy in stabilizing (MgCoCuNiZn)O [8], in spite of an enthalpic penalty associated with the structure transition of CuO and ZnO from tenorite and wurtzite to rock salt. This effect is confirmed by the endothermic reaction of

(MgCoCuNiZn)O and MD simulation [111] and thus utilized to predict and explain the stability of other high-entropy systems such as oxides, perovskites, carbides, and borides [15, 43, 46, 94].

Sarker et al. [112] developed an entropy-related descriptor called entropy forming ability (EFA) to address the synthesizability of HECs from first-principles calculation. The descriptor they proposed is expressed as follows:

$$EFA = \left( \sqrt{\frac{\sum_{i=1}^{n} g_i \left(H_i - H_{mix}\right)^2}{\left(\sum_{i=1}^{n} g_i\right) - 1}} \right)^{-1} \tag{2.12}$$

where $n$ is the total number of sampled geometrical configurations, $g_i$ are their degeneracies, and $H_{mix}$ is the mixed-phase enthalpy approximated by averaging the enthalpies $H_i$ of the sampled configurations:

$$H_{mix} = \frac{\sum_{i=1}^{n} g_i H_i}{\sum_{i=1}^{n} g_i} \tag{2.13}$$

For a given composition, large EFA corresponds to large entropy. They generated 56 five-metal carbides in the simulations and synthesized nine out of 56, and six composites were single-phase high-entropy carbides. Based on the simulation and experimental results, they proposed a potential threshold for five-metal carbides that could form a single phase: EFA larger than 50 $(\text{eV/atom})^{-1}$.

Pitike et al. [113] developed other entropy-related descriptors from the statistical mean and standard deviations of local mixing enthalpies of HEOs. In their simulations, a nearest neighbor model (NNM) was adopted to reduce the computational costs associated with directly studying a five-component oxide. The strategy of this model, simple yet powerful, is to discard the interaction between cations beyond the first nearest neighbor cation coordination shell in the estimation of local energy of a cation. The local mixing enthalpy in the local phase $P_i$ is estimated through:

$$\Delta H_i\left[FCO, P_i\right] = \frac{1}{n}\sum_{j=1}^{n} \Delta H_{mix}\left[\left(A^i A^j\right)O, P_i\right] \tag{2.14}$$

where $\Delta H_{mix}[(A^i A^j)O,P_i]$ is the average mixing enthalpy of $A^i$ with an $A^j$ cation species, which can be calculated by:

$$\begin{aligned}\Delta H_{mix}\left[\left(A^i A^j\right)O, P\right] &= E_{DFT}\left[\left(A^i A^j\right)O, P\right] - \frac{1}{2m} E_{DFT}\left[A^i{}_m O_x, G\right] \\ &- \frac{1}{2n} E_{DFT}\left[A^j{}_n O_x, G\right] - \frac{1}{2}\left(1 - \frac{x}{2m} - \frac{y}{2n}\right) E_{DFT}\left[O_2, G\right]\end{aligned} \tag{2.15}$$

where $E_{DFT}[A^i{}_mO_x,G]$ and $E_{DFT}[A^j{}_nO_x,G]$ are the DFT total energies of the $A^i{}_mO_x$ and $A^j{}_nO_x$ oxides in their respective ground-state phases. $E_{DFT}[(A^iA^j)O,P]$ is the DFT total energy of the $(A^iA^j)O$ two-component oxide in phase $P$. They then introduced a set of descriptors that quantified the enthalpy and configurational entropy for a HEC; that was the mean, $\mu_{local}$, and standard deviation, $\sigma_{local}$, of the local mixing enthalpy distributions and defined as:

$$\mu_{local} = \langle \Delta H_{local} \rangle \tag{2.16}$$

$$\sigma_{local} = \sqrt{\langle \Delta H_{local}{}^2 \rangle - \langle \Delta H_{local} \rangle^2} \tag{2.17}$$

where $\langle \Delta H_{local} \rangle = \frac{1}{N}\sum_{i=1}^{N}(\Delta H_i[HEO,P_i])$ and $\langle \Delta H_{local}{}^2 \rangle = \frac{1}{N}\sum_{i=1}^{N}(\Delta H_i[HEO,P_i])^2$ are the mean values of the local mixing enthalpies and squared local mixing enthalpies, respectively. A multicomponent system with low positive mean (low mixing enthalpy) and low standard deviation (high configurational entropy) should readily form a single-phase high-entropy compound. Using these proposed descriptors correctly, they identified the experimentally realized HEC, (CoCuMgNiZn)O, as the composition most likely to form a high-entropy rock salt phase.

Similar to HEAs, the lattice vibration has its own contribution to the stability of HECs, whereas, unlike HEAs, this contribution shows less importance. Esters et al. investigated the effect of lattice vibration in three high-entropy carbides to their phase stability [100]. They found that, in spite of their rather small values compared to configurational ones, the vibrational entropy and vibrational free energy did contribute to the phase stability of HECs. The small values could be explained by the atomic environments in these materials. Despite the structural distortions, the first coordination shells were very similar between the binary and the high-entropy carbides, resulting in only small changes in the vibrational entropy. They also found that the effect of stabilization from vibrational contribution was more significant when the symmetry of the precursor was lower than the symmetry of the HEMs, whereas the price was the additional enthalpy penalties because the nearest atomic environments were different than in their ground state.

Other than DFT, MD and calculation of phase diagrams (CALPHAD) can also be used to explore the stability. The advantage of MD calculations is the ability to deal with large number of atoms. However, the accuracy of this method depends on the accuracy of the interatomic potentials. Anand et al. [114] explored the stability of (MgCoNiCuZn)O and its derivatives and the competition between enthalpy and entropy of these materials using MD, where the Buckingham interatomic potential was used to depict ionic interaction. Nevertheless, the usage of MD is still limited due to the lack of highly accurate interatomic potentials to capture the chemical complexity for different material systems, especially for HECs, since the chemical bonding and interatomic potentials are more complex. With the development of machine learning and artificial intelligence, a new paradigm emerged recently. In this new paradigm, machine learning methods are adopted to fit interatomic potentials from experimental or theoretical data sets. The fitted potentials are called machine learning potentials [78, 80]. These potentials show high accuracy comparable to DFT calculations and high efficiency in dealing with large and complex chemical systems. To the author's knowledge, the application of machine learning potentials to the stability of HECs has not yet been conducted. CALPHAD describes the Gibbs free energy of a system using a large amount of thermodynamic data [115]. In the perovskite $LaMnO_{3\pm}\delta$ system [116], combined with defects chemistry, CALPHAD was used to investigate the effects of oxygen partial pressure on the thermochemical properties and the correlation between thermochemical properties and defects mixing behavior.

#### 2.4.2.2 Other Criteria

Besides configurational entropy, the phase stability of HECs could be affected by the size and charge of the constituent elements. Djenadic et al. [117] found the important role of Ce in the formation of fluorite-structured HE-REOs. The systems without Ce could not form single-phase fluorite. They suggested that the oxidation state (+4) and size ratio of Ce to oxygen atom were the main factors in the stabilization of fluorite structure. Dąbrowa et al. found that, with the addition of Mo, the structure of Ce–Gd–Nd–Pr–Sm–O and Ce–Gd–La–Nd–Pr–O systems transformed from bixbyite ($Ia\overline{3}$) to fluorite structure [118]. High oxidation state of Mo (+6) seemed to be responsible for the structure change.

As described in Section 2.4.1, the atomic size difference ($\delta$) is one of the main factors that influence the formation of single phases in HEAs, where the threshold is 6.6% [104]. Nevertheless, in HECs, single phases are still observed in the systems with $\delta$ larger than 6.6%. Jiang et al. [23] found that the Goldschmidt tolerance factor ($t$) was a more useful parameter than atomic size difference in predicting the stability of high-entropy perovskites. They found that the systems with unity tolerance factor tended to form single phase in spite of large size difference. With regard to other HEC systems, no mature descriptors have been proposed yet.

Compared to HEAs, HECs is a newly emerged field. The investigations on the thermodynamics and phase stability are insufficient. More delicate efforts on the theoretical and empirical descriptors are still awaited.

## References


1 Cantor, B., Chang, I.T.H., Knight, P. et al. (2004). Microstructural development in equiatomic multicomponent alloys. *Mater. Sci. Eng., A* 375–377: 213–218.

2 Yeh, J., Chen, S.K., Lin, S.J. et al. (2004). Nanostructured high-entropy alloys with multiple principal elements: novel alloy design concepts and outcomes. *Adv. Eng. Mater.* 6: 299–303.

3 Lilensten, L., Couzinié, J.P., Perrière, L. et al. (2014). New structure in refractory high-entropy alloys. *Mater. Lett.* 132: 123–125.

4 Zhang, Y., Zuo, T.T., Tang, Z. et al. (2014). Microstructures and properties of high-entropy alloys. *Prog. Mater. Sci.* 61: 1–93.

5 Takeuchi, A., Amiya, K., Wada, T. et al. (2014). High-entropy alloys with a hexagonal close-packed structure designed by equi-atomic alloy strategy and binary phase diagrams. *JOM* 10 (66): 1984–1992.

6 Feuerbacher, M., Heidelmann, M., and Thomas, C. (2014). Hexagonal high-entropy alloys. *Mater. Res. Lett.* 3: 1–6.

7 Wright, A.J. and Luo, J. (2020). A step forward from high-entropy ceramics to compositionally complex ceramics: a new perspective. *J. Mater. Sci.* 55: 9812–9827.

8 Rost, C.M., Sachet, E., Borman, T. et al. (2015). Entropy-stabilized oxides. *Nat. Commun.* 6: 8485.

9 Rost, C.M., Rak, Z., Brenner, D.W. et al. (2017). Local structure of the $Mg_xNi_xCo_xCu_xZn_xO$ ($x = 0.2$) entropy-stabilized oxide: an EXAFS study. *J. Am. Ceram. Soc.* 100: 2732–2738.

10 Berardan, D., Meena, A., Franger, S. et al. (2017). Controlled Jahn–Teller distortion in (MgCoNiCuZn)O-based high entropy oxides. *J. Alloys Compd.* 704: 693–700.

11 Chen, J., Liu, W., Liu, J. et al. (2019). Stability and compressibility of cation-doped high-entropy oxide $MgCoNiCuZnO_5$. *J. Phys. Chem. C* 123: 17735–17744.

12 Dupuy, A.D., Wang, X., and Schoenung, J.M. (2019). Entropic phase transformation in nanocrystalline high entropy oxides. *Mater. Res. Lett.* 7: 60–67.

13 Sarkar, A., Djenadic, R., Usharani, N.J. et al. (2017). Nanocrystalline multicomponent entropy stabilised transition metal oxides. *J. Eur. Ceram. Soc.* 37: 747–754.

14 Chellali, M.R., Sarkar, A., Nandam, S.H. et al. (2019). On the homogeneity of high entropy oxides: an investigation at the atomic scale. *Scr. Mater.* 166: 58–63.

15 Gild, J., Samiee, M., Braun, J.L. et al. (2018). High-entropy fluorite oxides. *J. Eur. Ceram. Soc.* 38: 3578–3584.

16 Tseng, K.P., Yang, Q., McCormack, S.J. et al. (2020). High-entropy, phase-constrained, lanthanide sesquioxide. *J. Am. Ceram. Soc.* 103: 569–576.

17 Sun, Y., Xiang, H., Dai, F. et al. (2021). Preparation and properties of CMAS resistant bixbyite structured high-entropy oxides $RE_2O_3$ (RE = Sm, Eu, Er, Lu, Y, and Yb): promising environmental barrier coating materials for $Al_2O_{3f}/Al_2O_3$ composites. *J. Adv. Ceram.* 10: 596–613.

18 Xiang, H., Xing, Y., Dai, F. et al. (2021). High-entropy ceramics: present status, challenges, and a look forward. *J. Adv. Ceram.* 10: 385–441.

19 Teng, Z., Zhu, L., Tan, Y. et al. (2020). Synthesis and structures of high-entropy pyrochlore oxides. *J. Eur. Ceram. Soc.* 40: 1639–1643.

20 Wright, A.J., Wang, Q., Ko, S.T. et al. (2020). Size disorder as a descriptor for predicting reduced thermal conductivity in medium- and high-entropy pyrochlore oxides. *Scr. Mater.* 181: 76–81.

21 Li, F., Zhou, L., Liu, J. et al. (2019). High-entropy pyrochlores with low thermal conductivity for thermal barrier coating materials. *J. Adv. Ceram.* 8: 576–582.

22 Zhang, K., Li, W., Zeng, J. et al. (2020, 2020). Preparation of $(La_{0.2}Nd_{0.2}Sm_{0.2}Gd_{0.2}Yb_{0.2})_2Zr_2O_7$ high-entropy transparent ceramic using combustion synthesized nanopowder. *J. Alloys Compd.* 817: 153328.

**23** Jiang, S., Hu, T., Gild, J. et al. (2018). A new class of high-entropy perovskite oxides. *Scr. Mater.* 142: 116–120.

**24** Witte, R., Sarkar, A., Kruk, R. et al. (2019). High-entropy oxides: an emerging prospect for magnetic rare-earth transition metal perovskites. *Phys. Rev. Mater.* 3: 034406.

**25** Grzesik, Z., Smoła, G., Miszczak, M. et al. (2020). Defect structure and transport properties of (Co, Cr, Fe, Mn, $Ni)_3O_4$ spinel-structured high entropy oxide. *J. Eur. Ceram. Soc.* 40: 835–839.

**26** Mao, A., Xie, H., Xiang, H. et al. (2020). A new class of spinel high-entropy oxides with controllable magnetic properties. *J. Magn. Magn. Mater.* 496: 165884.

**27** Parida, T., Karati, A., Guruvidyathri, K. et al. (2020). Novel rare-earth and transition metal-based entropy stabilized oxides with spinel structure. *Scr. Mater.* 178: 513–517.

**28** Goldschmidt, V.M. and Krystallochemie, D.G. (1926). Die Gesetze der Krystallochemie. *Naturwissenschaften* 14: 477–485.

**29** Sarkar, A., Djenadic, R., Wang, D. et al. (2018). Rare earth and transition metal based entropy stabilised perovskite type oxides. *J. Eur. Ceram. Soc.* 38: 2318–2327.

**30** Chen, H., Xiang, H., Dai, F. et al. (2020). High entropy $(Yb_{0.25}Y_{0.25}Lu_{0.25}Er_{0.25})_2SiO_5$ with strong anisotropy in thermal expansion. *J. Mater. Sci. Technol.* 36: 134–139.

**31** Ridley, M., Gaskins, J., Hopkins, P. et al. (2020). Tailoring thermal properties of multi-component rare earth monosilicates. *Acta Mater.* 195: 698–707.

**32** Dong, Y., Ren, K., Lu, Y. et al. (2019). High-entropy environmental barrier coating for the ceramic matrix composites. *J. Eur. Ceram. Soc.* 39: 2547–2579.

**33** Sun, L., Luo, Y., Ren, X. et al. (2020). A multicomponent γ-type $(Gd_{1/6}Tb_{1/6}Dy_{1/6}Tm_{1/6}Yb_{1/6}Lu_{1/6})_2Si_2O_7$ disilicate with outstanding thermal stability. *Mater. Res. Lett.* 8: 424–430.

**34** Sun, L., Luo, Y., Tian, Z. et al. (2020). High temperature corrosion of $(Er_{0.25}Tm_{0.25}Yb_{0.25}Lu_{0.25})_2Si_2O_7$ environmental barrier coating material subjected to water vapor and molten calcium-magnesium-aluminosilicate (CMAS). *Corros. Sci.* 175: 108881.

**35** Zhao, Z., Chen, H., Xiang, H. et al. (2020). High-entropy $(Y_{0.2}Nd_{0.2}Sm_{0.2}Eu_{0.2}Er_{0.2})AlO_3$: a promising thermal/environmental barrier material for oxide/oxide composites. *J. Mater. Sci. Technol.* 47: 45–51.

**36** Chen, H., Zhao, Z., Xiang, H. et al. (2020). High entropy $(Y_{0.2}Yb_{0.2}Lu_{0.2}Eu_{0.2}Er_{0.2})_3Al_5O_{12}$: a novel high temperature stable thermal barrier material. *J. Mater. Sci. Technol.* 48: 57–62.

**37** Zhao, Z., Chen, H., Xiang, H. et al. (2020). High-entropy $(Nd_{0.2}Sm_{0.2}Eu_{0.2}Y_{0.2}Yb_{0.2})_4Al_2O_9$ with good high temperature stability, low thermal conductivity and anisotropic thermal expansivity. *J. Adv. Ceram.* 9: 595–605.

**38** Zhao, Z., Xiang, H., Dai, F. et al. (2019). $(TiZrHf)P_2O_7$: an equimolar multicomponent or high entropy ceramic with good thermal stability and low thermal conductivity. *J. Mater. Sci. Technol.* 35: 2227–2231.

**39** Zhao, Z., Chen, H., Xiang, H. et al. (2019). $(La_{0.2}Ce_{0.2}Nd_{0.2}Sm_{0.2}Eu_{0.2})PO_4$: a high-entropy rare-earth phosphate monazite ceramic with low thermal conductivity and good compatibility with $Al_2O_3$. *J. Mater. Sci. Technol.* 35: 2892–2896.

**40** Gild, J., Zhang, Y., Harrington, T. et al. (2016). High-entropy metal diborides: a new class of high-entropy materials and a new type of ultrahigh temperature ceramics. *Sci. Rep.* 6: 37946.

**41** Yan, X., Constantin, L., Lu, Y. et al. (2018). $(Hf_{0.2}Zr_{0.2}Ta_{0.2}Nb_{0.2}Ti_{0.2})C$ high-entropy ceramics with low thermal conductivity. *J. Am. Ceram. Soc.* 101: 4486–4491.

**42** Lewin, E. (2020). Multi-component and high-entropy nitride coatings – a promising field in need of a novel approach. *J. Appl. Phys.* 127: 160901.

**43** Zhang, H., Zhao, B., Dai, F. et al. (2021). $(Cr_{0.2}Mn_{0.2}Fe_{0.2}Co_{0.2}Mo_{0.2})B$: a novel high-entropy monoboride with good electromagnetic interference shielding performance in K-band. *J. Mater. Sci. Technol.* 77: 58–65.

**44** Qin, M., Yan, Q., Wang, H. et al. (2020). High-entropy monoborides: towards superhard materials. *Scr. Mater.* 189: 101–105.

**45** Qin, M., Yan, Q., Wang, H. et al. (2021). High-entropy rare earth tetraborides. *J. Eur. Ceram. Soc.* 41: 2968–2973.

**46** Chen, H., Zhao, Z., Xiang, H. et al. (2020). Effect of reaction routes on the porosity and permeability of porous high entropy $(Y_{0.2}Yb_{0.2}Sm_{0.2}Nd_{0.2}Eu_{0.2})B_6$ for transpiration cooling. *J. Mater. Sci. Technol.* 38: 80–85.

**47** Zhang, W., Zhao, B., Xiang, H. et al. (2021). One-step synthesis and electromagnetic absorption properties of high entropy rare earth hexaborides (HE $REB_6$) and high entropy rare earth hexaborides/borates (HE $REB_6$/HE $REBO_3$) composite powders. *J. Adv. Ceram.* 10: 62–77.

**48** Zhang, Y. (2019). *High-Entropy Materials: A Brief Introduction*. Singapore: Springer Nature Singapore Pvt Ltd.

**49** Gao, M.C. and Alman, D.E. (2013). Searching for next single-phase high-entropy alloy compositions. *Entropy* 15: 4504–4519.

**50** Singh, S., Wanderka, N., Murty, B.S. et al. (2011). Decomposition in multi-component AlCoCrCuFeNi high-entropy alloy. *Acta Mater.* 59: 182–190.

**51** Santodonato, L.J., Zhang, Y., Feygenson, M. et al. (2015). Deviation from high-entropy configurations in the atomic distributions of a multi-principal-element alloy. *Nat. Commun.* 6: 5964.

**52** Singh, P., Smirnov, A.V., and Johnson, D.D. (2015). Atomic short-range order and incipient long-range order in high-entropy alloys. *Phys. Rev. B* 91: 224204.

**53** Tamm, A., Aabloo, A., Klintenberg, M. et al. (2015). Atomic-scale properties of Ni-based FCC ternary, and quaternary alloys. *Acta Mater.* 99: 307–312.

**54** Zhang, F.X., Zhao, S., Jin, K. et al. (2017). Local structure and short-range order in a NiCoCr solid solution alloy. *Phys. Rev. Lett.* 118: 205501.

**55** Varvenne, C., Luque, A., and Curtin, W.A. (2016). Theory of strengthening in fcc high entropy alloys. *Acta Mater.* 118: 164–176.

**56** Ding, Q., Zhang, Y., Chen, X. et al. (2019). Tuning element distribution, structure and properties by composition in high-entropy alloys. *Nature* 574: 223–229.

**57** Yin, B. and Curtin, W.A. (2020). Origin of high strength in the CoCrFeNiPd high-entropy alloy. *Mater. Res. Lett.* 8: 206–215.

**58** Otto, F., Dlouhý, A., Pradeep, K.G. et al. (2016). Decomposition of the single-phase high-entropy alloy CrMnFeCoNi after prolonged anneals at intermediate temperatures. *Acta Mater.* 112: 40–52.

**59** Pickering, E.J., Muñoz-Moreno, R., Stone, H.J. et al. (2016). Precipitation in the equiatomic high-entropy alloy CrMnFeCoNi. *Scr. Mater.* 113: 106–109.

**60** Ming, K., Li, L., Li, Z. et al. (2019). Grain boundary decohesion by nanoclustering Ni and Cr separately in CrMnFeCoNi high-entropy alloys. *Sci. Adv.* 5: 0639.

**61** Li, X., Chen, C., Zheng, H. et al. (2020). Complex strengthening mechanisms in the NbMoTaW multi-principal element alloy. *npj Comput. Mater.* 6: 70.

**62** Ye, B.L., Wen, T.Q., Nguyen, M.C. et al. (2019). First-principles study, fabrication and characterization of $(Zr_{0.25}Nb_{0.25}Ti_{0.25}V_{0.25})C$ high-entropy ceramics. *Acta Mater.* 170: 15–23.

**63** Harrington, T.J., Gild, J., Sarker, P. et al. (2019). Phase stability and mechanical properties of novel high entropy transition metal carbides. *Acta Mater.* 166: 271–280.

**64** Liu, R., Chen, H., Zhao, K. et al. (2017). Entropy as a gene-like performance indicator promoting thermoelectric materials. *Adv. Mater.* 29: 1702712.

**65** Wang, C., Qin, M., Lei, T. et al. (2021). Synergic grain boundary segregation and precipitation in W- and W-Mo-containing high-entropy borides. *J. Eur. Ceram. Soc.* 41: 5380–5387.

**66** Qiu, S., Li, M., Shao, G. et al. (2021). (Ca, Sr, Ba)$ZrO_3$: a promising entropy-stabilized ceramic for titanium alloys smelting. *J. Mater. Sci. Technol.* 65: 82–88.

**67** Peng, Z., Sun, W., Xiong, X. et al. (2021). Novel refractory high-entropy ceramics: transition metal carbonitrides with superior ablation resistance. *Corros. Sci.* 184: 109359.

**68** Yeh, J., Chen, S., Gan, J. et al. (2004). Formation of simple crystal structures in Cu–Co–Ni–Cr–Al–Fe–Ti–V alloys with multiprincipal metallic elements. *Metall. Mater. Trans. A* 35A: 2533–2536.

**69** Yeh, J., Chang, S., Hong, Y. et al. (2007). Anomalous decrease in X-ray diffraction intensities of Cu–Ni–Al–Co–Cr–Fe–Si alloy systems with multi-principal elements. *Mater. Chem. Phys.* 103: 41–46.

**70** Guo, W., Dmowski, W., Noh, J. et al. (2013). Local atomic structure of a high-entropy alloy: an X-ray and neutron scattering study. *Metall. Mater. Trans. A* 44: 1994–1997.

**71** Zou, Y., Maiti, S., Steurer, W. et al. (2014). Size-dependent plasticity in an $Nb_{25}Mo_{25}Ta_{25}W_{25}$ refractory high-entropy alloy. *Acta Mater.* 65: 85–97.

**72** He, Q. and Yang, Y. (2018). On lattice distortion in high entropy alloys. *Front. Mater.* 5: 42.

**73** Tong, Y., Velisa, G., Yang, T. et al. (2017). Probing local lattice distortion in medium-and high-entropy alloys. *arXiv* [preprint] arXiv:1707.07745, https://doi.org/10.48550/arXiv.1707.07745.

**74** Song, H., Tian, F., Hu, Q. et al. (2017). Local lattice distortion in high-entropy alloys. *Phys. Rev. Mater.* 1: 023404.

**75** Ye, Y., Zhang, Y., He, Q. et al. (2018). Atomic-scale distorted lattice in chemically disordered equimolar complex alloys. *Acta Mater.* 150: 182–194.

**76** Rák, Z., Maria, J.P., and Brenner, D.W. (2018). Evidence for Jahn–Teller compression in the (Mg, Co, Ni, Cu, Zn)O entropy-stabilized oxide: a DFT study. *Mater. Lett.* 217: 300–303.

**77** Dai, F., Wen, B., Sun, Y. et al. (2020). Theoretical prediction on thermal and mechanical properties of high entropy $(Zr_{0.2}Hf_{0.2}Ti_{0.2}Nb_{0.2}Ta_{0.2})C$ by deep learning potential. *J. Mater. Sci. Technol.* 43: 168–174.

**78** Ye, B., Wen, T., Huang, K. et al. (2019). First-principles study, fabrication, and characterization of $(Hf_{0.2}Zr_{0.2}Ta_{0.2}Nb_{0.2}Ti_{0.2})C$ high-entropy ceramic. *J. Am. Ceram. Soc.* 102: 4344–4352.

**79** Dai, F., Sun, Y., Wen, B. et al. (2021). Temperature dependent thermal and elastic properties of high entropy $(Ti_{0.2}Zr_{0.2}Hf_{0.2}Nb_{0.2}Ta_{0.2})B_2$: molecular dynamics simulation by deep learning potential. *J. Mater. Sci. Technol.* 72: 8–15.

**80** Ikeda, Y., Grabowski, B., and Körmann, F. (2019). *Ab initio* phase stabilities and mechanical properties of multicomponent alloys: a comprehensive review for high entropy alloys and compositionally complex alloys. *Mater. Charact.* 147: 464–511.

**81** Vitos, L. (2001). Total-energy method based on the exact muffin-tin orbitals theory. *Phys. Rev. B* 64: 014107.

**82** Vitos, L., Abrikosov, I.A., and Johansson, B. (2001). Anisotropic lattice distortions in random alloys from first-principles theory. *Phys. Rev. Lett.* 87: 156401.

**83** Troparevsky, M.C., Morris, J.R., Daene, M. et al. (2015). Beyond atomic sizes and Hume-Rothery rules: understanding and predicting high-entropy alloys. *JOM* 67: 2350–2363.

**84** Odbadrakh, K., Enkhtor, L., Amartaivan, T. et al. (2019). Electronic structure and atomic level complexity in $Al_{0.5}TiZrPdCuNi$ high-entropy alloy in glass phase. *J. Appl. Phys.* 126: 095104.

**85** Tian, F., Varga, L.K., Chen, N. et al. (2013). *Ab initio* investigation of high-entropy alloys of 3d elements. *Phys. Rev. B* 87: 075144.

**86** Wang, S., Zhang, T., Hou, H. et al. (2018). The magnetic, electronic, and thermodynamic properties of high entropy alloy CrMnFeCoNi: a first-principles study. *Phys. Status Solidi B* 255: 1800306.

**87** Zunger, A., Wei, S., Ferreira, L.G. et al. (1990). Special quasirandom structures. *Phys. Rev. Lett.* 65: 353–356.

**88** Wei, S.H., Ferreira, L.G., Bernard, J.E. et al. (1990). Electronic properties of random alloys: special quasirandom structures. *Phys. Rev. B* 42: 9622–9649.

**89** Leong, Z., Wróbel, J.S., Dudarev, S.L. et al. (2017). The effect of electronic structure on the phases present in high entropy alloys. *Sci. Rep.* 7: 39803.

**90** Rák, Z., Rost, C.M., Lim, M. et al. (2016). Charge compensation and electrostatic transferability in three entropy-stabilized oxides: results from density functional theory calculations. *J. Appl. Phys.* 120: 095105.

**91** Jiang, S., Shao, L., Fan, T. et al. (2020). Elastic and thermodynamic properties of high entropy carbide (HfTaZrTi)C and (HfTaZrNb)C from *ab initio* investigation. *Ceram. Int.* 46: 15104–15112.

**92** Wang, Y., Gan, G., Wang, W. et al. (2018). *Ab initio* prediction of mechanical and electronic properties of ultrahigh temperature high-entropy ceramics ($Hf_{0.2}Zr_{0.2}Ta_{0.2}M_{0.2}Ti_{0.2})B_2$ (M = Nb, Mo, Cr). *Phys. Status Solidi B* 255: 1800011.

**93** Liu, Y., Jia, D., Zhou, Y. et al. (2020). $Zn_{0.1}Ca_{0.1}Sr_{0.4}Ba_{0.4}ZrO_3$: a non-equimolar multicomponent perovskite ceramic with low thermal conductivity. *J. Eur. Ceram. Soc.* 40: 6272–6277.

**94** Sarkar, A., Loho, C., Velasco, L. et al. (2017). Multicomponent equiatomic rare earth oxides with a narrow band gap and associated praseodymium multivalency. *Dalton Trans.* 46: 12167–12176.

**95** Sarkar, A., Eggert, B., Velasco, L. et al. (2020). Role of intermediate 4f states in tuning the band structure of high entropy oxides. *APL Mater.* 8: 051111.

**96** Bérardan, D., Franger, S., Dragoe, D. et al. (2016). Colossal dielectric constant in high entropy oxides. *Phys. Status Solidi RRL* 10: 328–333.

**97** Gao, M.C., Yeh, J., Liaw, P.K. et al. (2016). *High-Entropy Alloys*. Cham: Springer International Publishing.

**98** Körmann, F., Ikeda, Y., Grabowski, B. et al. (2017). Phonon broadening in high entropy alloys. *npj Comput. Mater.* 3: 36.

**99** Yang, K., Oses, C., and Curtarolo, S. (2016). Modeling off-stoichiometry materials with a high-throughput *ab-initio* approach. *Chem. Mater.* 28: 6484–6492.

**100** Esters, M., Oses, C., Hicks, D. et al. (2020). Settling the matter of the role of vibrations in the stability of high-entropy carbides. *Nat. Commun.* 12: 5747.

**101** Zhang, Y., Zhou, Y.J., Lin, J.P. et al. (2008). Solid-solution phase formation rules for multi-component alloys. *Adv. Eng. Mater.* 10: 534.

**102** Zhang, K. and Fu, Z. (2012). Effects of annealing treatment on phase composition and microstructure of $CoCrFeNiTiAl_x$ high-entropy alloys. *Intermetallics* 22: 24–32.

**103** Ye, Y., Wang, Q., Lu, J. et al. (2017). High-entropy alloy: challenges and prospects. *Mater. Today* 19: 349–362.

**104** Yang, X. and Zhang, Y. (2012). Prediction of high-entropy stabilized solid-solution in multi-component alloys. *Mater. Chem. Phys.* 132: 233–238.

**105** Zhang, Y., Lu, Z., Ma, S. et al. (2014). Guidelines in predicting phase formation of high-entropy alloys. *MRS Commun.* 4: 57–62.

**106** Ma, D., Grabowski, B., Körmann, F. et al. (2015). *Ab initio* thermodynamics of the CoCrFeMnNi high entropy alloy: importance of entropy contributions beyond the configurational one. *Acta Mater.* 100: 90–97.

**107** Li, Z., Körmann, F., Grabowski, B. et al. (2017). *Ab initio* assisted design of quinary dual-phase high-entropy alloys with transformation-induced plasticity. *Acta Mater.* 136: 262–270.

**108** Guo, S., Ng, C., Lu, J. et al. (2011). Effect of valence electron concentration on stability of fcc or bcc phase in high entropy alloys. *J. Appl. Phys.* 109: 103505.

**109** Poletti, M.G. and Battezzati, L. (2014). Electronic and thermodynamic criteria for the occurrence of high entropy alloys in metallic systems. *Acta Mater.* 75: 297–306.

**110** Ye, Y., Liu, C., and Yang, Y. (2015). A geometric model for intrinsic residual strain and phase stability in high entropy alloys. *Acta Mater.* 94: 152–161.

**111** Biesuz, M., Spiridigliozzi, L., Dell'Agli, G. et al. (2018). Synthesis and sintering of (Mg, Co, Ni, Cu, Zn)O entropy-stabilized oxides obtained by wet chemical methods. *J. Mater. Sci.* 53: 8074–8085.

**112** Sarker, P., Harrington, T., Toher, C. et al. (2018). High-entropy high-hardness metal carbides discovered by entropy descriptors. *Nat. Commun.* 9: 4980.

**113** Pitike, K.C., Santosh, K.C., Eisenbach, M. et al. (2020). Predicting the phase stability of multicomponent high-entropy compounds. *Chem. Mater.* 32: 7507–7515.

**114** Anand, G., Wynn, A.P., Handley, C.M. et al. (2018). Phase stability and distortion in high-entropy oxides. *Acta Mater.* 146: 119–125.

**115** Saunders, N. and Miodownik, A.P. (1998). *CALPHAD: Calculation of Phase Diagrams – A Comprehensive Guide*. Amsterdam: Elsevier.

**116** Zhong, Y., Sabarou, H., Yan, X. et al. (2019). Exploration of high entropy ceramics (HECs) with computational thermodynamics – a case study with $LaMnO_{3\pm\delta}$. *Mater. Des.* 182: 108060.

**117** Djenadic, R., Sarkar, A., Clemens, O. et al. (2017). Multicomponent equiatomic rare earth oxides. *Mater Res. Lett.* 5: 102–109.

**118** Dąbrowa, J., Szymczak, M., Zajusz, M. et al. (2020). Stabilizing fluorite structure in ceria-based high-entropy oxides: influence of Mo addition on crystal structure and transport properties. *J. Eur. Ceram. Soc.* 40: 5870–5881.

# 3

# Theoretical Design Aspects in High-Entropy Materials

## 3.1 Introduction

High-entropy materials (HEMs), including high-entropy alloys (HEAs), high-entropy ceramics (HECs), and other materials with high chemical configurational entropy, are a new set of materials with their design space concentrated in the center of multicomponent phase diagrams. It means that there are multicomponents in equal or near-equal contents in HEMs, which is different from conventional materials with one or two elements as the principal element(s), whereas minor other elements are added to modify materials' properties. The distinct multiprincipal element nature of HEMs makes people believe that materials with new properties and physics may be discovered, e.g. overcoming the strength-ductility trade-off [1], promising retention of strength at elevated temperatures [2], excellent resistance to irradiation [3], colossal dielectric constant [4], lithium superionic conductivity [5], and amorphous-like thermal conductivity [6]. Great efforts have been made to design new HEMs to meet the needs in different domains, e.g. energy, transportation, and national defense. At the early stage of HEMs' studies, attention was focused on developing single-phase solid solutions (SSs). However, it is not only the high-entropy single-phase state but also the medium-entropy state or multi-phase state may provide intriguing properties. Therefore, the development of HEMs is on its way to compositionally complex materials [7–10].

Inverse design of materials based on computation has long been the objective in materials research. The proposition of Materials Genome Initiative (MGI) has greatly promoted the development of this goal, which challenges the traditional "try and error" research schema toward a new paradigm that accelerates the discovery, design, development, and deployment of materials by synergistically combining experiment, theory, and computation in a tightly integrated, high-throughput manner. For the demand-driven design of HEMs, two fundamental aspects need to be addressed, as shown in Figure 3.1. One is predicting the formation ability of HEMs with respect to composition and processing, while the other is exploring the "composition–structure–property" relations and developing property prediction models. Computational modeling can employ density functional theory (DFT)-based

*High-Entropy Materials: From Basics to Applications,* First Edition.
Huimin Xiang, Fu-Zhi Dai, and Yanchun Zhou.

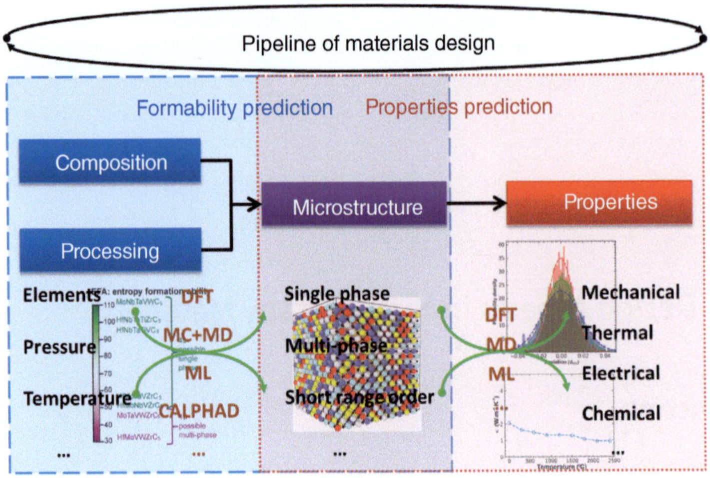


**Figure 3.1** Pipeline of materials design. Source: Xiang et al. [11]/Springer Nature/ CC BY-4.0.

methods, molecular dynamic (MD) simulations, Monte Carlo (MC) simulations, calculation of phase diagrams [CALPHAD], surrogate models from machine learning (ML) methods, simplified models based on physical assumptions, or other methods. The vast combination of elements and the variation in their concentration make the exploration of new HEMs particular hard, which demands computational tools for fast screening the vast composition space. In this chapter, we will introduce how computational tools are applied in HEMs.

## 3.2 Formability Prediction

Since the discovery of HEMs, tremendous efforts have been made to develop theories or models to predict the formation ability of HEMs, i.e. how to select the combination of elements and their concentrations to give rise to uniform SSs. The corresponding methods can be divided into two categories:

1) The first category is empirical models based on descriptors (also called fingerprints or features). In this category, the stability of a single-phase SS or its preferred phase structure is identified based on a set of descriptors. The separation criteria on the descriptors are usually determined either manually or by ML methods based on experimental data. With the increase in experimental data, ML methods are becoming more popular nowadays, which provides better classification accuracy.
2) The second category is thermodynamic computations. In this category, phase equilibria for a given system are evaluated based on the CALPHAD approach. In this approach, the thermodynamic properties are either calculated from CALPHAD parametrizations (database) or from DFT-based methods. MC

simulations are also usually adopted to simulate second-order phase transformation, e.g. order–disorder transformation and ferromagnetic–paramagnetic transformation, in SSs. It should be clear that thermodynamic equilibrium may not always be attained in experiments due to kinetic limitations, especially at low temperatures. Nevertheless, thermodynamic computation may still be the best way to study the stability of HEMs.

Due to the low computational cost of empirical models and the CALPHAD approach based on parametrizations, these methods can screen the vast design space of HEMs based on element combination and concentration variation. Methods based on DFT calculations are computationally demanding, which are usually adopted to study a particular HEM in detail. With the increase in computational capability, the high-throughput and ML-assisted approaches are playing increasing roles nowadays. Due to the fact that most of the models were derived from HEAs, the following discussions are focused on HEAs.

### 3.2.1 Empirical Models

The most well-known empirical model to predict whether a SS can form in physical metallurgy is the Hume-Rothery rules [12], which was proposed in 1920s. According to the Hume-Rothery rules, the formation of a SS is controlled by atomic size mismatch, electronegativity mismatch, and electron concentration among the constituent elements. For example, it is usually speculated that the atom radius mismatch should be less than 15% for a binary SS. Similar to this criterion for binary SSs, many descriptors were proposed to predict the stability of SSs in multicomponent materials [13–22]. Boundaries of these descriptors that separate SSs from other states, e.g. bulk metallic glasses (BMGs) and intermetallic compounds (IMCs), were determined based on experimental data [13–15, 18–20]. Then, the boundaries were taken as criteria to predict new stable SSs.

The descriptors are derived from fundamental properties of elements, e.g. atom radius ($r$), electronegativity ($\chi$), and valence electron concentration ($VEC$), or their thermodynamic properties, e.g. mixing enthalpy ($\Delta H_{mix}$), mixing entropy ($\Delta S_{mix}$), or their combinations. Due to the multicomponent nature of HEMs, the descriptors are usually taken as the mean value or the standard deviation of the properties of the constituted elements to indicate their overall intensity and similarity. Some typical descriptors are listed as follows.

1) Atom radius mismatch $\delta$, which is the scaled standard deviation of the atom radius of the constitute elements: $\delta = \sqrt{\sum_{i=1}^{n} c_i \left(1 - \frac{r_i}{\bar{r}}\right)^2}$, where $\bar{r} = \sum_{i=1}^{n} c_i r_i$ is the mean atom radius, $c_i$ and $r_i$ are the concentration and radius of element $i$, respectively;
2) Mixing enthalpy $\Delta H_{mix}$, which is the mean value of the binary mixing enthalpy: $\Delta H_{mix} = \sum_{i,j=1}^{n} 4 c_i c_j \Delta H_{ij}$, where $\Delta H_{ij}$ is the binary mixing enthalpy between element $i$ and $j$ at $c_i = c_j = 0.5$;

3) A dimensionless ratio value $\Omega = \frac{\overline{T_m}\Delta S_{mix}}{|\Delta H_{mix}|}$, where $\overline{T_m} = \sum_{i=1}^{n} c_i (T_m)_i$ is the mean value of the melting point of constitute elements $(T_m)_i$, $\Delta S_{mix} = -R\sum_{i=1}^{n} c_i \ln c_i$ is the ideal mixing entropy and $R$ is the gas constant;
4) Mean valence electron concentration $\overline{VEC} = \sum_{i=1}^{n} c_i VEC_i$, where $VEC_i$ is the valence electron concentration of element $i$.

Atom radius mismatch $\delta$, mixing enthalpy $\Delta H_{mix}$, and $\Omega$ play a decisive role in determining the stability of SSs. For example, Yang and Zhang [13] used $\Omega$ and $\delta$ to distinguish the stability of SSs and suggested that high-entropy SSs stabilizes when $\Omega > 1.1$ and $\delta \leq 6.6\%$ (see, Figure 3.2a). Guo et al. [14] analyzed experimental data and suggested (see, Figure 3.2b): SSs can form when $\delta$ is small ($\delta < {\sim}6.6\%$), and $\Delta H_{mix}$ is either slightly positive or insignificantly negative ($-11.6$ kJ/mol $< \Delta H_{mix} < 3.2$ kJ/mol); in contrast, BMGs form when $\delta$ is large ($\delta > 6.4\%$) and

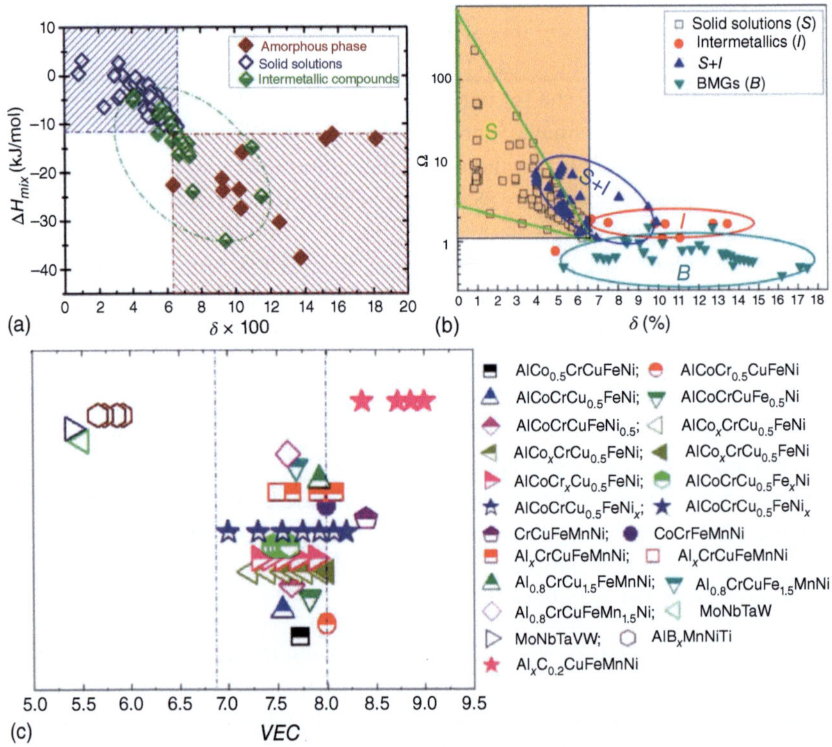


**Figure 3.2** (a) Mixing enthalpy $\Delta H_{mix}$ versus atom radius mismatch $\delta$ plot for different alloys. Source: Reproduced with permission from Guo et al. [14]/Elsevier. (b) Dimensionless ratio value $\Omega$ versus atom radius mismatch $\delta$ plot for different alloys. Source: Reproduced with permission from Yang and Zhang [13]/Elsevier. (c) Plot of $\overline{VEC}$ for different HEAs. In (c), solid points are alloys with FCC phase, hollow points are alloys with BCC phase, and half-filled points are alloys with FCC and BCC duplex phases. Source: Reproduced with permission from Guo et al. [15]/AIP Publishing.

$\Delta H_{mix}$ is significantly negative $\Delta H_{mix} < -12.2$ kJ/mol. In addition, the crystal structure of a high-entropy SS is also an important concern, e.g. a lattice with face-centered cubic (FCC) usually has better plasticity. Guo et al. [15] pointed out that the mean valence electron concentration $\overline{VEC}$ is a good indicator of crystal structure, which suggested that FCC phases are found to be stable at higher $\overline{VEC}$ (>8), whereas body-centered cubic (BCC) phases are stable at lower $\overline{VEC}$ (<6.87) (see, Figure 3.2c). This conclusion is consistent with the general trend of *VEC* versus crystal structure for pure metals in the periodic table.

In HEAs, the competition states to SSs are BMGs and IMCs. Mixing enthalpy $\Delta H_{mix}$ is a measure of the attraction or separation tendency between elements in an alloy. There is no doubt that $\Delta H_{mix}$ is important for SSs to form. If it is a big positive, it means that elements tend to separate from each other; the alloy will then split into different SSs with enrichment of different elements. If it is a substantial negative, it means that strong chemical bonds form between some of the elements; the alloy will then form IMCs or BMGs. It should be clarified that a BMG is not an equilibrium state. The formation of a BMG is due to the rapid quenching of its liquid phase, where the relaxation of local atomic structure is suppressed. In principle, if the quenching speed is big enough, any alloy may form a BMG or a multi-phase structure. However, in reality, only when an alloy meets some criteria, it can form BMG due to the limited quenching speed. To slow down the relaxation process of local atomic structures, local strong chemical bonds are necessary, e.g. covalent-like bonds, which is corresponding to a substantial negative $\Delta H_{ij}$. It is one of the reasons why carbon and ceramics can form amorphous easier than alloys. Different from $\Delta H_{mix}$, where only the enthalpy effect is considered, $\Omega$ takes the competition of entropy and enthalpy into consideration. It indicates whether the entropy effect can overcome the attraction or separation tendency caused by enthalpy. Except for $\Delta H_{mix}$ and $\Omega$, some other descriptors related to thermodynamic properties were also proposed, e.g. comparing the formation tendency between SSs and IMCs directly by considering the approximated Gibbs free energy $\Delta G$ by neglecting either entropy or enthalpy contribution [16]. The entropy descriptor in HECs indicates the distribution of configurational-dependent enthalpy [21–24]. When considering the mixing enthalpy of a multicomponent system, there is an important simplification, that is, the configurational energy can be approximated by only considering pairwise interactions, e.g. $\Delta H_{ij}$. Though the simplification loses some accuracy, it reduces the computational consumption greatly, which enables us fast scanning on a vast combination space of elements and a large compositional space. In alloys, there is a simple Miedema's model [25] to estimate $\Delta H_{ij}$, which is proven to show similar tendency in comparison to DFT results reported by King et al. [17]. Alternatively, $\Delta H_{ij}$ can also be more accurately evaluated by DFT-based calculations [16]. However, in ceramics, $\Delta H_{ij}$ usually needs to be evaluated by DFT-based calculations [22]. In ceramics, calculating thermodynamic descriptors by DFT-based methods to assist the design of new HECs is common nowadays, and this approach has been successfully adopted to design high-entropy super-hard carbides and rock salt-type high-entropy cathodes for Li-ion batteries [23, 24]. Figure 3.3 shows how Wang et al. [23] designed super-hard carbides from DFT-calculated enthalpy formation and entropy

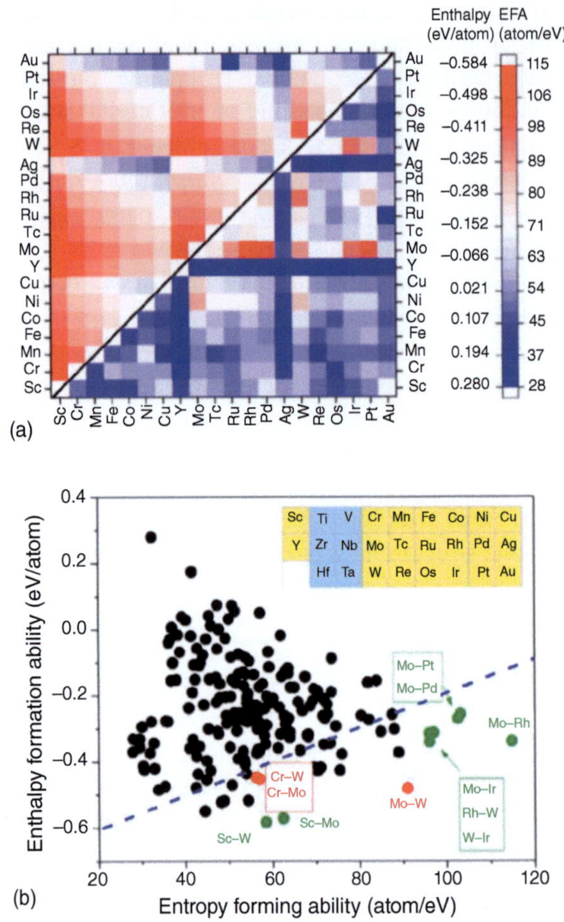


**Figure 3.3** (a) DFT-calculated formation enthalpy and EFA of eight-cation high-entropy carbides with 190 possible compositions. Six of the elements are fixed (the blue elements in (b) inset: Ti, Zr, Hf, V, Nb, Ta), while the left two are elements selected from the yellow elements in (b) inset. (b) Selection of compositions based on calculated enthalpy and EFA. The dashed line is the boundary separating single- and multi-phase, which is estimated from previous experiments. The red dots are the composition being experimentally fabricated. Source: Wang et al. [23]/John Wiley & Sons/CC BY-4.0.

formation ability (EFA). During calculation, a set of random configurations were first generated, and their energy was calculated. Formation enthalpy is the mean value of these configurational-dependent energy, while EFA is the distribution of these configurational-dependent energy. The narrower the distribution (the smaller the EFA) is, the easier it is for a material to accept all the configurations, which then tends to form a material with high configurational entropy.

Atom radius mismatch $\delta$ may be related to the elastic energy stored in an alloy when multicomponents are mixed together to form one phase. The larger the $\delta$ is, the higher is the elastic energy stored in the alloy. The elastic energy stored in an alloy serves as a barrier for SS formation. Therefore, SSs prefer small $\delta$. Some efforts have been made to search for more accurate descriptors that can characterize the stored elastic energy. For example, Ye et al. [18] derived a root-mean-square residual strain $\langle \varepsilon^2 \rangle^{1/2}$ based on a geometrical packing model and proven that $\langle \varepsilon^2 \rangle$ is proportional to the dimensionless elastic energy storage $u_e$. In addition, they suggested that the transition from a single- to multi-phase SS takes place in crystalline HEAs when $\langle \varepsilon^2 \rangle^{1/2}$ approaches ~5%, and amorphization occurs when $\langle \varepsilon^2 \rangle^{1/2}$ rises above ~10%, which is in line with the Lindemann's criterion for lattice instability [26]. Wang et al. [19] proposed an effective atomic size parameter $\gamma$ as the solid angle ratio between the smallest and the largest element in multicomponent systems and pointed out that the boundary between single- and multi-phase SS is $\gamma = \sim 1.175$. This criterion reduces to atomic mismatch role in the Hume–Rothery rules for a binary alloy (atom radius mismatch being 15% is equivalent to $\gamma = 1.167$). Toda-Caraballo and Rivera-Díaz-del-Castillo [20] estimated lattice distortions of HEAs based on elastic theory and proved that $\delta$ approximately was a representation of interatomic spacing mismatch in an alloy.

In the above empirical models, the criteria that separate different regions were estimated manually from experimental data. Although these models are more suitable for people to understand, their prediction accuracy is relatively low, which cannot effectively screen the vast space of element combination and the adjustable concentration. ML methods have been adopted to predict the formability based on descriptors in recent years. Different from traditional empirical models, where boundaries were determined manually, ML methods determine the separation boundaries between different phases by computer algorithms, e.g. k-nearest neighbors (KNNs), support vector machine (SVM), Gaussian processes (GP), artificial neural network (ANN), and random forest (RF).

Some efforts have been made to apply ML algorithms to obtain classification models for HEAs [27–31], where evident improvements in prediction accuracy were achieved. For example, Zhang et al. [29] presented a novel example on systematic predicting phases in HEAs by ML methods. They prepared a descriptor space (including 70 descriptors) and a ML algorithm pool (nine ML algorithms) and determined the best combination of descriptors and ML algorithm as a predictive model with a genetic algorithm wrapper. The process is illustrated in Figure 3.4. In their work, they treated two classification problems: (i) solid-solution and non-solid-solution (SS/NSS) classification and (ii) FCC, duplex phase (DP), and BCC (FCC/DP/BCC) classification for a given alloy. In the SS/NSS classification task, an SVM model with radial basis function (SVM.rbf) allows the best accuracy of up to 88.7%, which is much higher than traditional empirical models, e.g. $\delta - \Omega$ and $\delta - \Delta H_{mix}$ models mentioned above, which only exhibit classification accuracy of ~75% (see, Figure 3.5). In the FCC/DP/BCC classification task, they reported that a N net (neural network) model reaches the best accuracy of 91.3%, which is much better than the predictivity of $\overline{VEC}$ (see, Figure 3.5). Furthermore, they showed how the ML model could be iteratively improved by employing an active learning approach, which could guide the augmentation of the dataset by informing which experiment to be selected next.

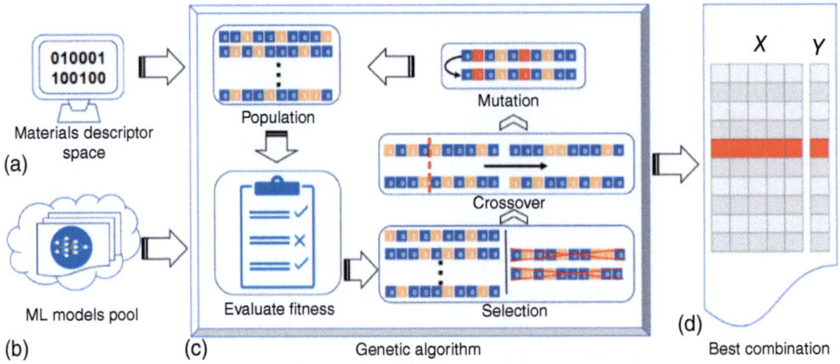


**Figure 3.4** Flowchart of the strategy to search for the best combination of descriptors and ML model with a genetic algorithm wrapper. (a) a descriptor space; (b) a ML model pool; (c) a genetic algorithm iteratively searching for the optimal combination of descriptors and ML algorithm; (d) the best model. (a) and (b) are inputs of (c), and (d) is the output of (c). Source: Reproduced with permission from Zhang et al. [29]/Elsevier.

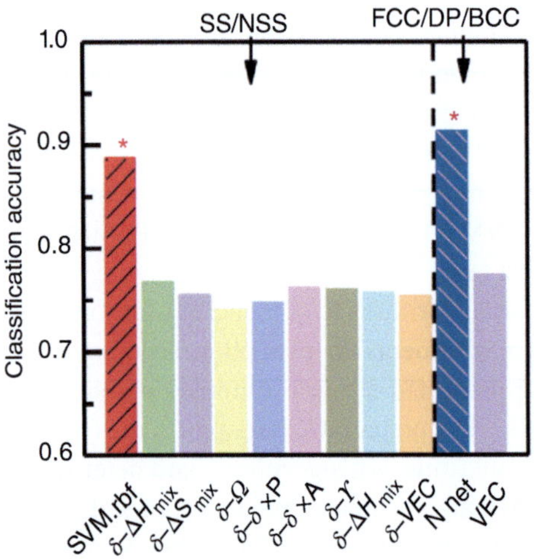


**Figure 3.5** Prediction accuracy of ML models versus empirical models. Source: Reproduced with permission from Zhang et al. [29]/Elsevier.

Identifying the correlation between the inputs and outputs is an important concern in ML methods, which provides explainable information instead of just providing a black box model. Usually, qualitative judgment on the descriptors in determining the phases is analyzed. For example, Zhou et al. [30] trained an ANN model to predict phases in HEAs and analyzed the correlation between descriptors and appearance of phases by a linearizing approximation of the ANN model. It was found that certain descriptors played decisive roles in determining the phases. As shown in Figure 3.6, the decisive descriptors for BMGs and IMCs are similar, but in

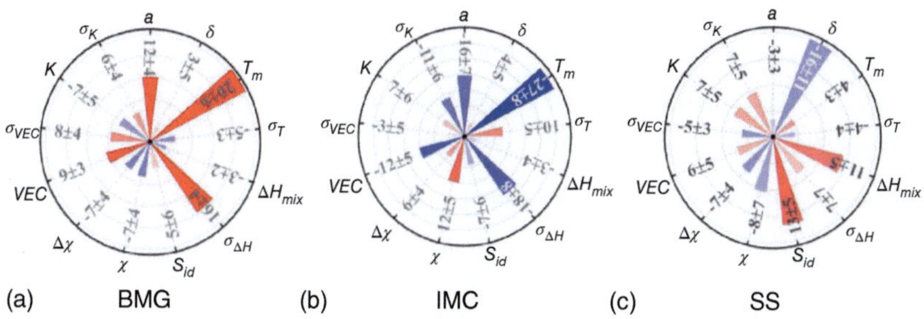


**Figure 3.6** Correlations between descriptors and the appearance of (a) bulk metallic glasses (BMG), intermetallic compounds (IMC), and solid solutions (SS) phases. Source: Zhou et al. [30]/Springer Nature/CC BY-4.0.

opposite sign. Large values in mean melting temperature, in standard deviation of binary mixing enthalpy, and in lattice parameter promote the formation of BMGs; in contrast, they suppress the formation of IMCs. For a SS, large $\Delta H_{mix}$ (should not be substantial negative) and $\Delta S_{mix}$ promote its formation, and large $\delta$ suppresses its formation, which is similar to the criteria of empirical models.

The descriptors in ML models can not only be properties of elements or their interactions, but they can also be any type of data that can be consistently obtained. For example, Qi et al. [31] trained an RF model to perform classification task with descriptors generated from binary phase diagrams. In their model, more detailed phase information is provided in comparison to other models. For example, they can determine what a NSS state is constituted of, e.g. ordered B2 structure + SS or IMCs (Sigma or Laves phase) + SS, etc.

### 3.2.2 Thermodynamic Computations

In principle, if the thermodynamic information of a system, e.g. Gibbs free energy $G(P, T, c_i)$, can be accurately obtained, then the phase equilibria can be fully predicted with no need for any of the above descriptors. The CALPHAD approach is the fundamental means to deal with thermodynamic computations. Unfortunately, CALPHAD parametrizations of multicomponent systems, e.g. parametrizations of $G(P, T, c_i)$, are almost unavailable, which restricts the quantitative design of HEMs. In addition, even if there is a parametrization for a given system, it could not be guaranteed that the parametrization can describe the whole system correctly, since it might just be optimized for limited situations. DFT-based methods serve as an alternative choice to estimate $G(P, T, c_i)$, which have been widely adopted in literatures [32–38]. Ikeda et al. [37] published a thorough review paper on the application of DFT-based methods on HEAs. Following their work, we will briefly introduce how $G(P, T, c_i)$ is estimated using DFT-based methods.

For a given phase $\alpha$, its Gibbs free energy $G^\alpha\left(P,T,c_i^\alpha\right)$ is:

$$G^\alpha\left(P,T,c_i^\alpha\right)=F^\alpha\left(V,T,c_i^\alpha\right)+PV \tag{3.1}$$

where $F^{\alpha}\left(V,T,c_i^{\alpha}\right)$ is the Helmholtz free energy, $P$ is pressure, and $V$ is volume. $F^{\alpha}\left(V,T,c_i^{\alpha}\right)$ is calculated by DFT-based methods under Born–Oppenheimer approximation:

$$F^{\alpha}\left(V,T,c_i^{\alpha}\right)=E^{\alpha}\left(V,c_i^{\alpha}\right)-TS_{conf}^{\alpha}\left(c_i^{\alpha}\right)+F_{el}^{\alpha}\left(V,T,c_i^{\alpha}\right)$$
$$+F_{mag}^{\alpha}\left(V,T,c_i^{\alpha}\right)+F_{vib}^{\alpha}\left(V,T,c_i^{\alpha}\right) \tag{3.2}$$

where $E^{\alpha}\left(V,c_i^{\alpha}\right)$ is the total electronic energy for a given atomic configuration of $\alpha$ phase with all atoms relaxed to their equilibrium positions at 0 K; $T$ is the absolute temperature; $S_{conf}^{\alpha}\left(c_i^{\alpha}\right)$ is the chemical configurational entropy; and $F_{el}^{\alpha}\left(V,T,c_i^{\alpha}\right)$, $F_{mag}^{\alpha}\left(V,T,c_i^{\alpha}\right)$, and $F_{vib}^{\alpha}\left(V,T,c_i^{\alpha}\right)$ are respectively the free energy from electronic, magnetic, and vibrational contributions, which can be expressed as follows:

$$F_{\lambda}^{\alpha}\left(V,T\right)=U_{\lambda}^{\alpha}\left(V,T\right)-TS_{\lambda}^{\alpha} \tag{3.3}$$

where $\lambda$ is *el*, *mag*, or *vib* in Eq. (3.2).

Usually, it is assumed that the electron and phonon density of states (DOS) are fully determined by the chemical configuration and cell volume. Then, $F_{el}^{\alpha}\left(V,T,c_i^{\alpha}\right)$ and $F_{vib}^{\alpha}\left(V,T,c_i^{\alpha}\right)$ are approximated by integration of electron and phonon DOS. $U_{el}^{\alpha}\left(V,T\right)$ and $S_{el}^{\alpha}$ can be expressed as follows:

$$U_{el}^{\alpha}\left(V,T\right)=\int_{-\infty}^{\infty}n^{\alpha}\left(\varepsilon\right)f\left(T\right)\varepsilon d\varepsilon-\int_{-\infty}^{\varepsilon_F}n^{\alpha}\left(\varepsilon\right)\varepsilon d\varepsilon \tag{3.4}$$

$$S_{el}^{\alpha}\left(V,T\right)=-k_B\int_{-\infty}^{\infty}n^{\alpha}\left(\varepsilon\right)\left[f\ln f+\left(1-f\right)\ln\left(1-f\right)\right]d\varepsilon \tag{3.5}$$

where $f = f(T)$ is the Fermi–Dirac distribution function, $n^{\alpha}(\varepsilon)$ is the electron DOS, $\varepsilon_F$ is the Fermi level at 0 K, $k_B$ is the Boltzmann constant, and $\varepsilon$ is electronic energy.

Since the calculation of phonon DOS is computationally demanding, $F_{vib}^{\alpha}\left(V,T\right)$ is usually estimated according to the Debye–Grüneisen approximation:

$$U_{vib}^{\alpha}\left(V,T\right)=\frac{9}{8}k_B\Theta_D^{\alpha}\left(V\right)+3k_BTD\left(y^{\alpha}\right) \tag{3.6}$$

$$S_{vib}^{\alpha}\left(V,T\right)=3k_B\left[\frac{4}{3}D\left(y^{\alpha}\right)-\ln\left(1-e^{-y^{\alpha}}\right)\right] \tag{3.7}$$

where $\Theta_D^{\alpha}\left(V\right)$ is the Debye temperature and $D(y^{\alpha})$ is the Debye function with $y^{\alpha}=\Theta_D^{\alpha}\left(V\right)/T$. According to Moruzzi et al. [39], the ground state Debye temperature $\Theta_D^{\alpha}\left(V_0\right)$ is evaluated as follows:

$$\Theta_D^{\alpha}\left(V_0\right)=C\sqrt{\frac{r_0B_0}{\bar{M}}} \tag{3.8}$$

where $C$ is an empirical constant, $\bar{M}$ is the averaged atomic mass $1/\left(\sum_i c_i / M_i\right)$, $r_0$ is the equilibrium Wigner–Seitz radius, and $B_0$ is the bulk modulus at 0 K. The volume-dependent Debye temperature $\Theta_D^\alpha(V)$ is calculated as follows:

$$\Theta_D^\alpha(V) = \Theta_D^\alpha(V_0)(V_0 / V)^\gamma \tag{3.9}$$

where $\gamma$ is the Grüneisen parameter given as follows:

$$\gamma = -g + \frac{1}{2}(1 + B_0') \tag{3.10}$$

where $B_0'$ is the derivative of bulk modulus with respect to pressure and g is a constant that depends on the choice of the approximation, which is 2/3 for high temperatures [40] and 1 for low temperatures [41]. If phonon DOS is calculated, then $F_{vib}^\alpha(V,T)$ can be more accurately estimated by quasi-harmonic approximation:

$$F_{vib}^\alpha(V,T) = k_B T \int_0^\infty \left[ \frac{\hbar\omega}{2k_B T} + \ln\left(1 - e^{-\frac{\hbar\omega}{k_B T}}\right) g(\omega) \right] d\omega \tag{3.11}$$

where $\hbar$ is the reduced Planck constant, $\omega$ is phonon frequency, and $g(\omega)$ is phonon DOS.

Different from electronic and vibrational contributions, the magnetic contribution of free energy is usually assumed to be independent of the chemical configuration, and it should be treated in a similar way to chemical configurations. For example, any atom in a system is defined by its element symbol and its spin state, e.g. spin up (↑) or spin down (↓). At high temperature ($T \gg T_C$), the magnetic free energy can be approximated as follows:

$$F_{mag}^\alpha(V, T \gg T_C) = U_{mag}^\alpha(V,T) - TS_{mag}^\alpha(V) \approx -TS_{mag}^\alpha(V) \tag{3.12}$$

$$S_{mag}^\alpha(V) = -k_B \sum_{i=1}^{n} c_i^\alpha \ln\left(\left|m_i^\alpha(V)\right| + 1\right) \tag{3.13}$$

where $T_C$ is the Curie temperature for ferromagnetic systems (or Néel temperature for antiferromagnetic systems) and $m_i^\alpha(V)$ is the magnetic moment of the $i$'th element.

With Eqs. (3.1) to (3.13), the free energy of a given phase $\alpha$ can be calculated by DFT-based methods. If $\alpha$ phase is an ordered crystal, the unit cell of $\alpha$ is a unified representation of $\alpha$ phase. However, for a disordered SS phase, we cannot define a period cell to represent it. The state-of-the-art methods within DFT to deal with disorder can be categorized into two classes: one is based on the coherent-potential approximation (CPA) [42]; the other is based on supercell models, usually with the supercell generated under the concept of special quasi-random structures (SQS) [43]. Ikeda et al. [37] have discussed the advantages and shortcomings of these two methods in detail, which is abstracted here.

In CPA, the chemical disorder is approximated by a mean field approach, where each element is assumed to be embedded into an effective medium that is determined self-consistently from the mutual effect of all elements together. Due to the mean field nature of CPA, perfect chemical disorder and any chemical composition can be treated within a single primitive cell, which guarantees its low computational cost. Therefore, efficient exploration of a large compositional space is possible within the CPA scenario. In addition, similar to chemical disorder, magnetic disorder can also be comparably simulated by the disordered-local-moment (DLM) approach. Furthermore, efficient extraction of chemical interaction can be achieved with the generalized perturbation method. The shortcomings of CPA include the following: (i) lattice distortions cannot be treated in the standard CPA, which might be important for HEMs and (ii) interatomic forces cannot be calculated; consequently, phonon dispersion calculation and MD simulation are impossible. Therefore, the vibrational free energy can be calculated only under Debye–Grüneisen approximation.

In supercell-based methods, a disordered configuration is simulated by a supercell with its sites randomly occupied by alloying elements. The advantage of supercell-based methods is that energy and atomic force can be precisely calculated, which guarantees better vibrational free energy estimation, MD simulation, and evaluation of lattice distortions. Due to the computational demanding nature of DFT-based methods, a supercell is usually small (~100 atoms). Such a small cell may not be big enough to represent perfect chemical randomness due to the long-range order imposed by periodical boundary condition. Therefore, the SQS concept [43] is adopted to generate supercells by minimizing the atomic pair correlations during construction, which provides a good representation of chemical randomness in small supercells. In a supercell, the concentration of an element can only vary with interval of $1/N$, where $N$ is the number of sites that can be occupied by alloying elements. It may result in a real concentration deviating from the ideal one due to the small size of the supercell, especially for ceramics, where only part of the sites is allowed for alloying. In addition, a single small supercell cannot represent the real system, where different configurations may exhibit different energies. To overcome the problem, many configurations are generated, and their energies and sometimes also interatomic forces are evaluated to generate a dataset. Then, other methods are used to train a model that predicts configurational energy based on the dataset. Traditionally, the typical method is cluster expansion (CE) [44], and ML methods are becoming more popular nowadays.

In the following, four typical examples will be given to illustrate how DFT-based methods have been applied in HEMs. In the first example, phase stability versus composition is predicted in $CoCrFeNiAl_x$HEA. Similar to the first example, the second example predicts phase stability versus composition in $Co_{20}Cr_{20}Fe_{40-x}Mn_{20}Ni_x$ ($x$ = 0–20 at.%) HEA, which is adopted to design TRIP (transformation-induced plasticity) HEA. In the third example, transformation between ferromagnetic and paramagnetic in CoCrFeNi-based HEAs is predicted, which shows how DFT-based methods can be used as a design tool for magnetic HEAs. In the fourth example, order–disorder transformation in refractory HEAs, e.g. NbTaMoW and VNbTaMoW, is studied, where many different methods have been applied. The third and fourth examples share some similarities, which are second-order phase transformation that are important to materials' properties but have not been covered by empirical models.

As mentioned above, the crystal structure of a SS phase is strongly related to its mechanical properties. In $CoCrFeNiAl_x$HEA, experimental measurements reveal that the crystal structure is FCC for $x \leq 0.5$, FCC/BCC duplex phase for $0.5 < x < 0.9$, and BCC for $x \geq 0.9$. Tian et al. [32] employed the exact muffin-tin orbitals (EMTOs) method in combination with CPA to estimate the free energy of $(CoCrFeNi)_{1-y}Al_y$ ($y = 0.0-0.5$) for BCC and FCC phases. In their work, the Gibbs free energy is approximated as follows:

$$G^{\alpha}(y) \approx E^{\alpha}(y) - TS^{\alpha}_{mix}(y) - TS^{\alpha}_{mag}(y) \tag{3.14}$$

where $E^{\alpha}(y)$ is the total electronic energy per atom and $S^{\alpha}_{mix}(y)$ and $S^{\alpha}_{mag}(y)$ are ideal mixing entropy and magnetic entropy, respectively. With the calculated free energy, they predicted the phase stability at different temperatures by the common tangent construction under the CALPHAD approach, as shown in Figure 3.7. They found that $CoCrFeNiAl_x$ has single FCC phase for $x \leq 0.579$ ($y \leq 0.130$), single BCC phase for $x \geq 1.229$ ($y \geq 0.235$), and duplex phases between the two limits. They also pointed out that the two limits are corresponding to $\overline{VEC} \geq 7.57$ and $\overline{VEC} \leq 7.04$, which is comparable to the criteria proposed by Guo et al. [15] (see Figure 3.2c).

TRIP effect is a phenomenon used for designing high-performance alloy with both high strength and plasticity. Li et al. [34] employed a similar method as that of Tian et al. [32] to predict phase stability versus composition in $Co_{20}Cr_{20}Fe_{40-x}Mn_{20}Ni_x$ ($x = 0$–$20$ at.%) HEA. The FCC + hexagonal close-packed (HCP) duplex phase $Fe_{50}Mn_{30}Co_{10}Cr_{10}$ alloy was taken as a reference. They predicted that the crossing points between the reference alloy and the quinary HEAs at around 12 at.% Ni concentration (6 at.% if magnetic excitations were not included), as shown in Figure 3.8a. It can be seen from Figure 3.8a that the stability of HCP phase decreases with the increase in Ni. Therefore, decreasing the content of Ni of CoCrFeMnNi alloy may result in HCP + FCC duplex phase TRIP alloy. Figure 3.8b shows the experimental phase structures versus $x$ for $Co_{20}Cr_{20}Fe_{40-x}Mn_{20}Ni_x$ alloys, which is consistent with the prediction. In their work, the TRIP effect was indeed obtained for the design alloys.

Design materials with desired magnetic properties, e.g. the Curie temperature $T_C$, is crucial in applications. HEAs are promising candidate magnetic materials, where their magnetic properties may be finely tuned by adjusting the alloying elements and their concentrations. However, exploring the immense configurational space of HEAs to search for a desired magnetic material is still a non-trivial task. As mentioned above, the CPA is suitable for such tasks due to its low computational cost. Körmann et al. [33] employed the EMTO-CPA method to evaluate energies of ferromagnetic and paramagnetic states of a HEA, where the paramagnetic state is approximated by the DLM approach. Then, the Curie temperature $T_C$ versus compositions was determined according to:

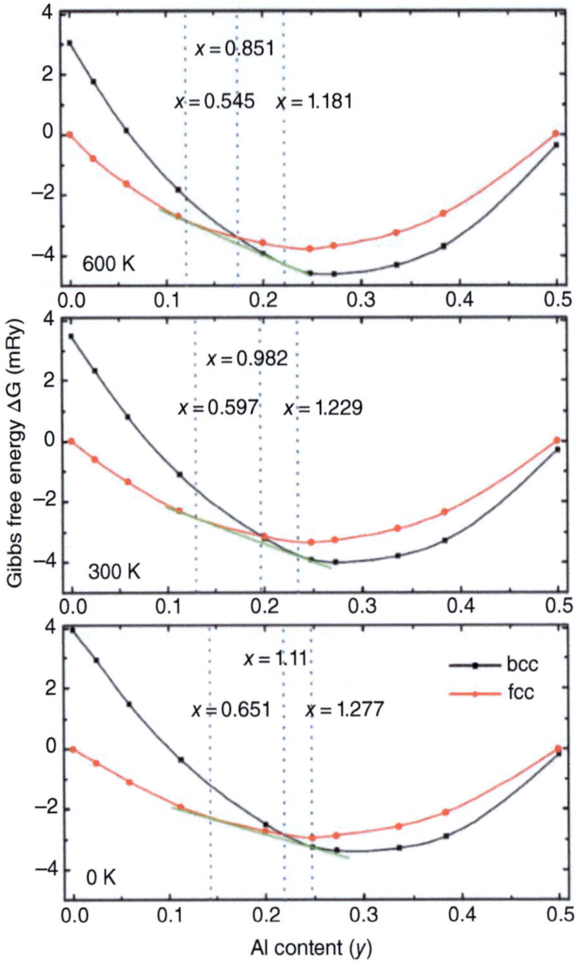


**Figure 3.7** Comparison of the Gibbs free energies of BCC and FCC $(CoCrFeNi)_{1-y}Al_y$ ($y = 0.0-0.5$) for $T = 0$, 300, and 600 K. Note that $y = x/(4+x)$, where $x$ is the atomic fraction of Al in $CoCrFeNiAl_x$. Source: Reproduced with permission from Tian et al. [32]/American Physical Society.

$$k_B T_C = \frac{2}{3(1-c)}(E_{FM} - E_{PM}) \tag{3.15}$$

where $c$ is the concentration of nonmagnetic elements and $E_{FM}$ and $E_{PM}$ are the energy of the ferromagnetic and paramagnetic state, respectively. Figure 3.9 shows the Treasure Maps (the Curie temperature maps and the corresponding saturation magnetization per site maps at $T = 0\,K$) for CoCrFeNiM (M = Ag, Au, Pd, Cu) HEAs, which are useful to explore and design alloys with well-defined magnetic properties. Their predictions agree well with experiments, indicating that DFT-based methods, especially CPA, are particularly useful for the design of HEMs.

Refractory HEAs have attracted great attention due to their ability to retain high strength at elevated temperatures, and they are believed to be promising candidates

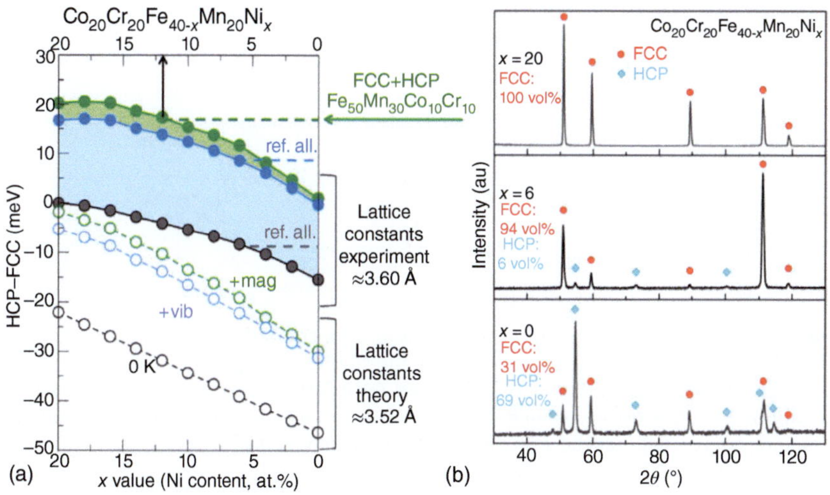


**Figure 3.8** (a) DFT-derived HCP–FCC energy differences for $Co_{20}Cr_{20}Fe_{40-x}Mn_{20}Ni_x$ ($x$ = 0–20 at.%) alloy at 0 K (black), including lattice vibrations (blue) and magnetic entropy effects (green) at 300 K. The dash and solid lines are corresponding to lattice parameters of ≈3.52 Å (theoretically predicted at 0 K from DFT) and ≈3.60 Å (experimental value), respectively. The reference energies are derived for the FCC + HCP duplex phase $Fe_{50}Mn_{30}Co_{10}Cr_{10}$ alloy. (b) X-ray diffraction (XRD) patterns of homogenized $Co_{20}Cr_{20}Fe_{40-x}Mn_{20}Ni_x$ ($x$ = 0, 6, 20 at.%) HEAs. Source: Reproduced with permission from Li et al. [34]/Elsevier.

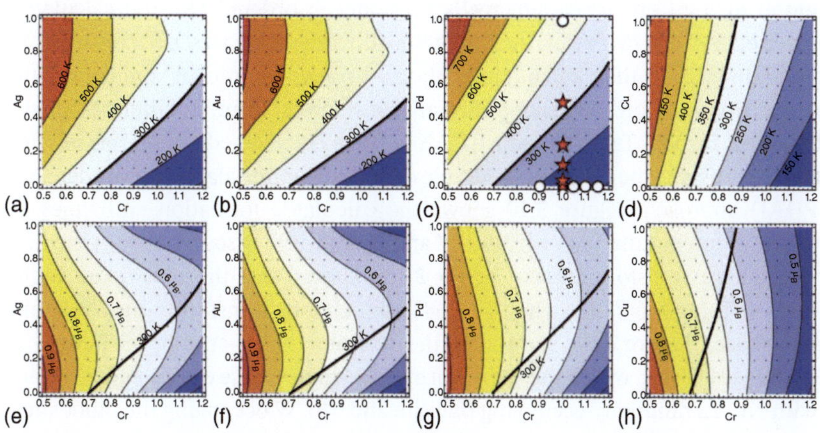


**Figure 3.9** Upper row (a)–(d): the Curie temperature maps and lower row (e)–(h): the corresponding saturation magnetization per site at $T$ = 0 K for CoCrFeNiM (M = Ag, Au, Pd, Cu) HEAs. The $x$ and $y$ axis are the content of Cr and M, respectively. Markers are experimental results. Thick lines indicate compositions with $T_C$ close to room temperature. Source: Reproduced with permission from Körmann et al. [33]/AIP Publishing.

for applications in aerospace, nuclear energy, and national defense. An important problem in refractory HEAs is the order–disorder transformation or the possible existence of short-range order (SRO), where plenty of work have been undertaken to study it. The SRO is usually described by the Warren–Cowley SRO parameter:

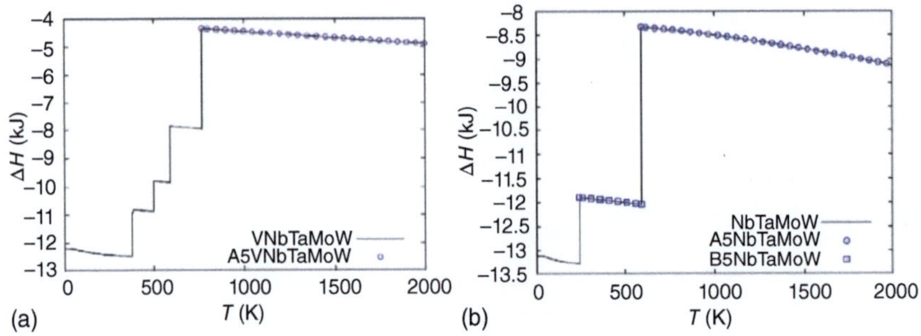


**Figure 3.10** The calculated formation enthalpies as a function of temperature for (a) VNbTaMoW and (b) NbTaMoW. Symbols illustrate the stable single phases. A5VNbTaMoW means a disordered solid solution in BCC structure. A4NbTaMoW also means a disordered solid solution in BCC structure, while B4NbTaMoW is a partially disordered solid solution in B2 structure. Source: Reproduced with permission from Wang et al. [36]/Elsevier.

$$\alpha_{ij}^{m} = 1 - \frac{c_{ij}^{m}}{c_i c_j} \tag{3.16}$$

where $c_{ij}^{m}$ is the probability to find the combination of the $i$'th and the $j$'th elements in the $m$'th pair cluster. If there is no SRO, $c_{ij}^{m}$ equals to the product $c_i c_j$ and $\alpha_{ij}^{m}$ vanishes. The two typical refractory HEAs that were mostly studied are NbTaMoW and VNbTaMoW.

Wang et al. [36] employed supercells to model disordered SSs and calculated Gibbs free energy of 178 phases, including fully disordered phases, partially ordered phases, and IMCs. In their calculations, the electronic, vibrational, and configurational contributions on free energy are all considered. Then, equilibrium phase combinations for VNbTaMoW and NbTaMoW were determined for a series of temperature by using the Dantzig's simplex minimization algorithm under the CALPHAD approach. Figure 3.10 shows the calculated formation enthalpies as functions of temperature for VNbTaMoW and NbTaMoW. Discontinuity in the line indicates phase transformations. Figure 3.10a shows that the disordered high-entropy VNbTaMoW is stable when $T > 768$ K, which then transforms into a mixture of multi-phases. Figure 3.10b shows that the disordered high-entropy NbTaMoW is stable when $T > 594$ K, which then transforms to a single-phase disordered B2 structure with Nb, Ta randomly occupying one site and Mo, W occupying the other site.

Körmann et al. [35] employed generalized perturbation method under the EMTO-CPA scenario to evaluate the configurational-dependent energy in NbTaMoW and simulated the order–disorder transformation by MC simulations. In their work, the configurational-dependent energy is expressed by effective pair interactions (EPIs) with the following equation:

$$\Delta E_{conf} = \frac{1}{2} \sum_{\mu\nu} \sum_{p=1}^{p_{max}} E_{\mu\nu}^{p} \sum_{n,m \in p} \delta c_{\mu}^{n} \delta c_{\nu}^{m} \tag{3.17}$$

where the second sum runs over all pair sp, and $E_{\mu\nu}^{p}$ are the EPIs between different elements $\mu$ and $\nu$. The first sum runs over the $N-1$ species $\mu$,$\nu$, and $\delta c_{\mu}^{n}$, $\delta c_{\nu}^{m}$ are the

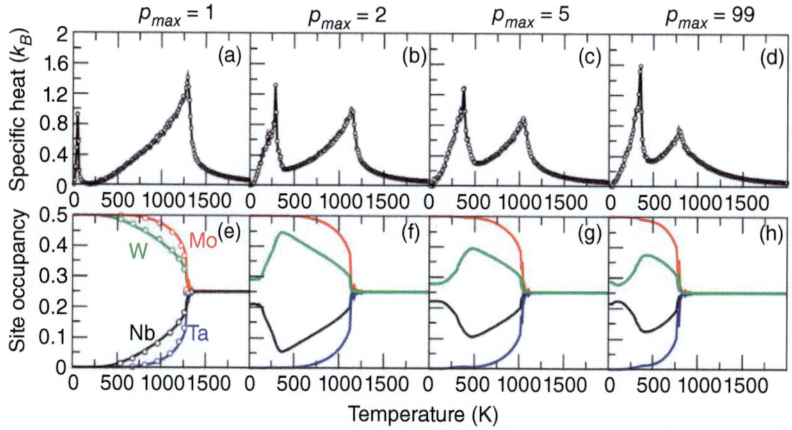


**Figure 3.11** Specific heat capacity (a)–(d) and site occupation (e)–(h) for different scenarios of included pair interactions. Source: Körmann et al. [35]/Taylor & Francis/ CC BY-4.0.

concentration variables at lattice sites *n* and *m*. Their results reveal that the B2 ordering temperature depends strongly on the long-range interactions. The transformation temperature is ≈ 1300 K when only the first neighbor interaction is included, which decreases to ≈ 1000 K for $p_{max}$ = 5 and further decreases to ≈ 750 K for $p_{max}$ = 99, as shown in Figure 3.11.

As stated above, lattice distortions are neglected in CPA scenario, which may cause some errors in the simulated results for HEMs. Later, Körmann et al. [38] employed a ML approach to estimate configurational-dependent energy by fitting a low-rank potential (LRP). When fitting the LRP, they used two types of DFT datasets: one is rigid lattice (no distortions in the lattice), and the other is relaxed lattice (all atoms are relaxed to their equilibrium position). For the rigid lattice approach, the transformation temperature is similar to their prediction. However, when lattice distortions were taken into consideration, they found that the order–disorder transformation temperature further reduced to ≈ 240 K. Recently, Liu et al. [45] did a similar work that used ML methods to extract EPIs and simulated order–disorder transformations in refractory HEAs by MC. The order–disorder transformation temperature of NbTaMoW is predicted to be slightly higher than 800 K (see Figure 3.12a), while it is higher than 2000 K for VNbTaMoW (see Figure 3.12d). Figure 3.12b,c,e,f shows how the Warren–Cowley SRO parameters vary with temperature from their simulation, which show how the SSs decompose to different compounds during cooling. Except for ML methods to fit Hamilton, a traditional method called cluster expansion (CE) is usually adopted to fit chemical configuration-dependent Hamilton. Fernández-Caballero et al. [46] employed CE in combination with MC simulations to study order–disorder transformations in VNbTaMoW HEA and predicted that the transformation temperature is ~1000 K. Due to the exponentially increasing nature of expanding clusters in CE with the number of elements, CE may not be a good choice for HEMs. Therefore, ML methods are more popular in HEMs.

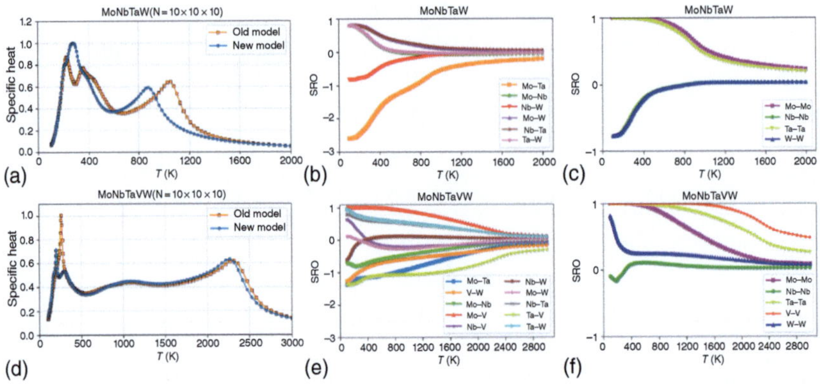


**Figure 3.12** (a) The specific heats of NbTaMoW. (b and c) The nearest-neighbor short-range order parameters of NbTaMoW. (d) The specific heats of VNbTaMoW. (e and f) The nearest-neighbor short-range order parameters of VNbTaMoW. New model means the EPI model is fitted by using both random and MC configurations, while the old model represents using only the random configurations. Source: Reproduced with permission from Liu et al. [45]/ Elsevier.

From these results, it can be seen that there are still large inconsistencies in predictions by different methods. In the works of Körmann et al. [35, 38] and Liu et al. [45], vibrational contributions and thermal expansion effects are all neglected. Wang et al. [36] took the vibrational and electronic contributions into consideration. However, only one chemical configuration was considered for each disordered structure, which might not be sufficient to describe the disordered structure. In recent years, ML methods have been adopted to fit interatomic potentials, which can be used to do MD simulations, instead of just Hamilton and MC simulations as the works of Körmann et al. [38] and Liu et al. [45]. Such ML potentials include deep potential (DP) [47, 48], spectral neighbor analysis potential (SNAP) [49], etc. Then, the combination of MD and MC may provide better predictions with accurate ML potentials.

As stated above, $G(P, T, c_i)$ can be directly estimated by CALPHAD parametrizations. Using the parametrized thermodynamic properties, phase equilibria for a given system at different temperatures can be predicted. As a matter of fact, most commercial thermodynamic databases are developed based on conventional alloys with one or two principal elements. Therefore, the accuracy of the databases in describing the central region of high-dimensional phase diagrams is questionable. For example, Senkov et al. [50] showed that discrepancies were found between the CALPHAD calculated results and experimental results in $CrMo_{0.5}NbTa_{0.5}TiZr$ HEA. They believed that the main reason for the discrepancies is the limitation of the thermodynamic database, which is developed for Ti-rich corner instead of the entire composition space. Nevertheless, the database can still be adopted as a fast screen tool to predict phase equilibria of multicomponent systems and provide useful information for materials design. For example, Senkov et al. [51] screened over 130 000 equimolar alloy systems by the CALPHAD method and pointed out that SS alloys become less likely as the number of alloy elements increases. It contradicts the major premise of HEAs that increasing configurational entropy enhances disordered SSs. As suggested by them, the configurational entropy rises slowly, while the probability of at least one pair of elements favoring the formation of IMCs increases more rapidly with the increase in element number, which explains the contradiction.

## 3.3 Properties Prediction

Exploring the "composition–structure–property" relations has long been the central problem in materials science. In Section 3.2, it is hoped that the microstructure of a material can be predicted based on its composition and treatment history. In this section, on the one hand, it is expected that the mechanical, physical, and chemical properties of the material can be predicted based on its composition and microstructure; on the other hand, it is hoped that simulations tools can help us to explore the underlying mechanism that controls materials properties. However, there is still a long way to go, and only limited properties can be predicted nowadays. Here, how computational methods are applied in properties prediction is briefly introduced. In Figure 3.9 and its related contents, the prediction of magnetic properties has been introduced. Therefore, magnetic property prediction will not be repeated later.

### 3.3.1 Lattice Distortions

Due to atomic radius mismatch and chemical bond strength mismatch, it is believed that the lattice of a HEM is severely distorted, which is suggested as one of the core effects in HEMs. Lattice distortions are related to many mechanical and physical properties, e.g. SS strengthening, thermal conductivity reduction, electrical conductivity reduction, or even some other interesting phenomena. For example, Jahn–Teller distortion is believed to related to superconductivity in cuprates with transition temperatures over 35 K and colossal magnetoresistance in manganese-based perovskites.

Lattice distortions in a HEM are usually characterized by two types of information. One is the distribution of chemical bond length [52]; the other is the distribution of atom displacements from their ideal lattice positions [53–55]. Additionally, local volume strain fluctuations may also be used to characterize lattice distortions [54, 55]. Usually, atomistic simulations are adopted to simulate intrinsic lattice distortions in HEMs [52–55].

In HEAs, lattice distortion induced SS strengthening may belong to one of the most concerned problems, since SS strengthening is one of the key strengthening mechanisms in alloys. Oh et al. [52] calculated lattice distortions in CrMnFeCoNi HEA by DFT-based methods with SQS supercells. They characterized lattice distortions by element-resolved bond lengths and pointed out that lattice distortions in HEMs come from two contributions: the bond length fluctuations (standard deviation in bond length) and the mean bond length deviations (the deviations of mean

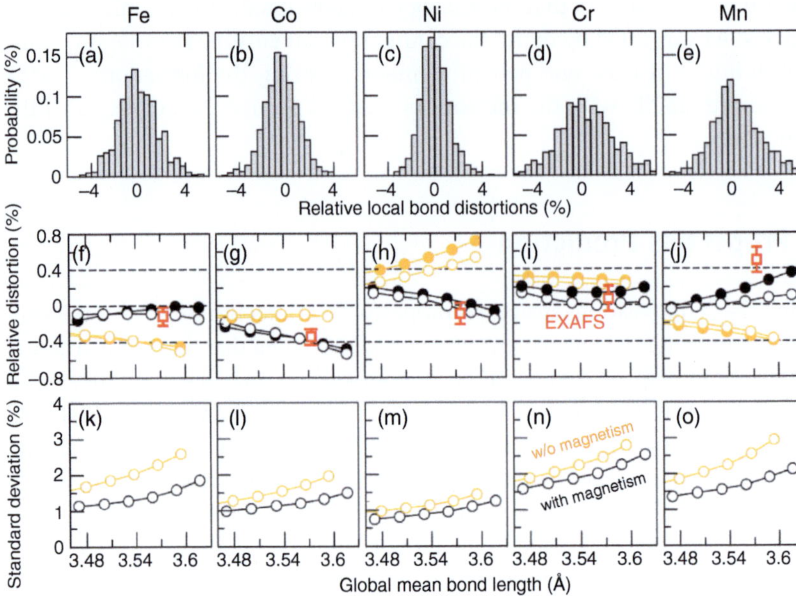


**Figure 3.13** (a–e) The distribution of element-resolved bond length deviations for Fe, Co, Ni, Cr, and Mn. The standard deviations of the bond distortions are an order of magnitude larger than the mean bond length deviation. (f–j) Comparison between experimental EXAFS (extended X-ray absorption fine structure) data (red open squares) and the predicted mean bond length deviation (black filled circles) as well as the mean value derived from a Gaussian fit (black open circles) for different lattice parameters. Non-spin polarized calculations are shown for comparison (open and filled orange circles), which reveals that calculations with spin-polarization give better predictions in comparison to experiments. (k–o) Standard deviation of the Gaussian fits revealing the much larger local bond distortions compared to mean bond length deviations. Source: Oh et al. [52]/MDPI/ CC BY-4.0.

bond lengths from the value determined from lattice parameter of the alloy). They expressed the local bond distortion as follows:

$$\Delta b_i = \Delta \bar{b}_i + \delta b_i \tag{3.18}$$

where $\Delta b_i$ is the overall bond length distortion of element $i$, $\Delta \bar{b}_i$ is the mean bond length deviation, and $\delta b_i$ is the bond length fluctuation. They found that the bond length fluctuations are one magnitude larger than the mean bond length deviations, as shown in Figure 3.13. According to this finding, it is suggested that local lattice distortions due to fluctuations in local chemical environments may significantly impact the SS strengthening in HEAs.

Okamoto et al. [53] characterized lattice distortions by mean-square atomic displacements (MSADs) from ideal lattice positions and suggested that the strength at 0 K of HEAs can be approximated linearly from the root of averaged MSADs, as shown in Figure 3.14. This reveals that SS strengthening can be quantitatively predicted with appropriate description on lattice distortions. In principle, averaged MSAD can be obtained from experiments from broadening of peaks in X-ray

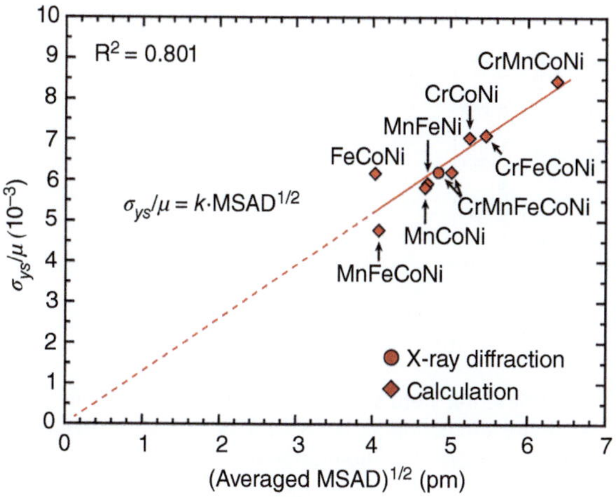


**Figure 3.14** Scaled yield strength ($\sigma_{YS}/\mu$) extrapolated to 0 K versus the square root of the mean MSAD value. $\sigma_{YS}$ is yield strength at 0 K, and $\mu$ is shear modulus. Source: Okamoto et al. [53]/AIP Publishing/CC BY-4.0.

diffraction. Due to the finite temperature effects in experiments, the experimental averaged MSAD contains two contributions: intrinsic lattice distortion due to disordered chemical configurations and thermal effect due to lattice vibration. It is usually assumed that these two contributions are independent of each other and follow Gaussian distribution. Then, the total distortion is as follows:

$$\sigma_{tot}^2 = \sigma_{thermal}^2 + \sigma_{chemical}^2 \tag{3.18}$$

where $\sigma$ is the standard deviation. Intrinsic lattice distortions may be deduced from measurements at different temperatures or from direct measurement at low temperature where thermal contribution can be neglected.

Even though lattice distortions can be evaluated from atomistic simulations, predictive models on how lattice distortions influence SS strengthening is more essential for materials' design. Varvenne et al. [56] developed a parameter-free SS strengthening model for FCC HEAs based on the mean field theory with fluctuations, which can predict the temperature-, composition-, and strain rate-dependence of the plastic yield strength. In this model, the interaction energy between elements and dislocation cores are necessary, which may not be easily obtained. A simplified model by approximating the interaction energy as $-p\Delta V_i$ is then derived. In the simplified model, the key quantity is the elemental-resolved misfit volume. Similar to Oh et al. [52], the misfit volume is separated to the mean misfit volume of element $i(\Delta\overline{V}_i)$ and its standard deviation due to local fluctuations ($\sigma_{\Delta V_i}$), but with a different formula in the quadradic form instead of the linear form.

$$\Delta V_i^2 = \Delta\overline{V}_i^2 + \sigma_{\Delta V_i}^2 \tag{3.19}$$

Toda-Caraballo et al. [20, 57, 58] have made great efforts to develop simplified models based on elastic theory to evaluate lattice distortions and to predict SS strengthening in HEAs. They [58] modified and extended models developed for binary alloys

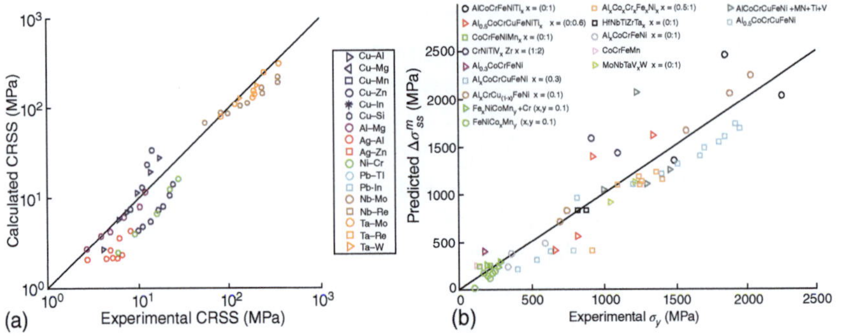


**Figure 3.15** Comparison between experimental yield strength and computed $\Delta\sigma_{ss}^{m}$ by model of Toda-Caraballo et al. for (a) binary alloys, and (b) HEAs. CRSS = critical resolved shear stress. Source: Reproduced with permission from Toda-Caraballo [58]/Elsevier.

suitable for multicomponent systems and checked the generality of the model by applying it to both binary and multicomponent alloys, which agree well with experiments, as shown in Figure 3.15. The central idea in their model is to calculate the variation in the local lattice parameter by means of variation in the local concentration of the elements, which can be viewed as the configurational dependence of lattice distortions.

Lattice distortions in HEMs may be more complicated in comparison to HEAs due to different sites, e.g. cations and anions. Usually, cations and anions exhibit different magnitudes of mean distortions. For example, atomistic simulations indicate that anion distortions in high-entropy carbides, diborides, and rock salt oxides are larger because of their relative smaller radii [54, 55, 59]. Lattice distortions in HEAs are usually symmetry conserving, which means that the distribution of the distortions is Gaussian. However, symmetry breaking distortions, Jahn–Teller distortions, have been reported in $MgCoNiCuZnO_5$ high-entropy oxide. As stated above, the Jahn–Teller distortion is associated with many intriguing functional properties of materials. Therefore, studying symmetry breaking distortions is crucial to further expand the functional properties of HEMs. The Jahn–Teller distortion is usually related to octahedral complexes of transition metals, where the ligand-field splitting of the *d*-orbitals gives rise to a degenerate ground state [60]. In an octahedral field, the five *d*-orbitals split into $e_g(d_{z^2}$ and $d_{x^2-y^2})$ and $t_{2g}(d_{xy}, d_{yz},$ and $d_{zx})$ orbitals. For $Cu^{2+}$ ion with nine *d* electrons, the $t_{2g}$ orbitals are fully filled, while $e_g$ orbitals are partially filled. The partially filled $e_g$ orbitals give rise to a doubly degenerate ground state, which then may proceed to symmetry-breaking Jahn–Teller distortion along one of the fourfold axes, as shown in Figure 3.16. Rák et al. [60] evaluated lattice distortions in $MgCoNiCuZnO_5$ by DFT-based calculations and checked the distortions of Cu—O bonds. They found that a majority of the bonds clearly reveal the elongated distortions, and a minority of the bonds display compressed distortions. In the future, the dependence of functional properties on lattice distortions, especially symmetry-breaking Jahn–Teller distortions, needs to be explored.

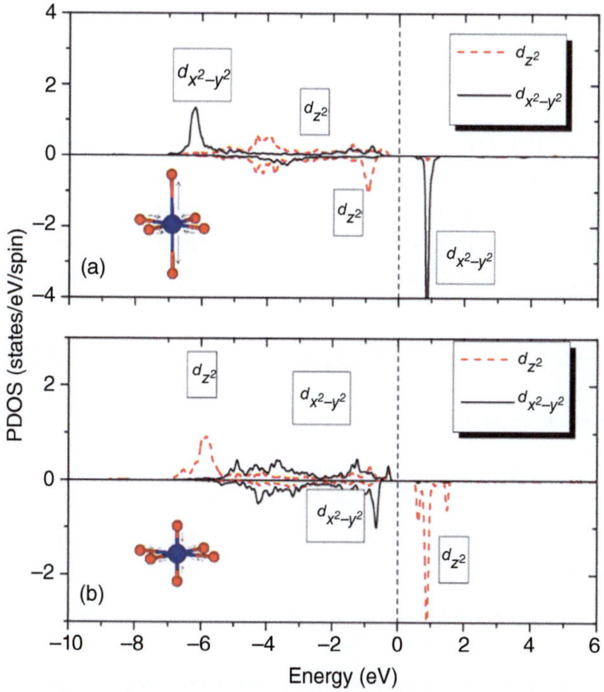


**Figure 3.16** Partial DOS associated with the Cu $d_{z^2}$ and $d_{x^2-y^2}$ orbitals in $MgCoNiCuZnO_5$ high-entropy oxide. (a) A half-filled $d_{x^2-y^2}$ orbital corresponding to a Jahn–Teller elongated octahedron; (b) A half-filled $d_{z^2}$ orbital corresponding to a Jahn–Teller compressed octahedron. PDOS = Partial density of state. Source: Reproduced with permission from Rák et al. [60]/Elsevier.

### 3.3.2 Elastic Properties

Elastic properties are fundamental for the application of a material, which can be accurately calculated by atomistic simulations, especially by DFT-based methods. The elastic property of a single crystal is anisotropic, which can be fully characterized by its elastic stiffness tensor $\mathbb{C}$, and the relation between stress and strain is as follows:

$$\boldsymbol{\sigma} = \mathbb{C}\,\boldsymbol{\varepsilon} \text{ or } \sigma_{ij} = \mathbb{C}_{ijkl}\varepsilon_{kl}\left(i,j,k,l=1,2,3\right) \tag{3.20}$$

where $\boldsymbol{\sigma}$ is stress tensor and $\boldsymbol{\varepsilon}$ is strain tensor. In the right equation, Einstein's summation convention is adopted. Einstein's summation role will also be adopted in the following equations. The inverse of Eq. (3.20) is as follows:

$$\boldsymbol{\varepsilon} = \mathbb{S}\ \boldsymbol{\sigma} \text{ or } \varepsilon_{ij} = \mathbb{S}_{ijkl}\sigma_{kl}\left(i,j,k,l=1,2,3\right) \tag{3.21}$$

where $\mathbb{S}$ is elastic compliance tensor. Usually, these tensors are expressed in Voigt notation as follows:

$$\sigma_I = \mathbb{C}_{IJ}\varepsilon_J\left(I,J=1,2,3,4,5,6\right) \tag{3.22}$$

$$\varepsilon_I = \mathbb{S}_{IJ}\sigma_J\left(I,J = 1,2,3,4,5,6\right) \tag{3.23}$$

$$\mathbb{C}_{IJ}\mathbb{S}_{JK} = \delta_{IK} \tag{3.24}$$

where $\delta_{IK}$ is an identity matrix. For a stable crystal, $\mathbb{C}_{IJ}$ should be a positive definite matrix following Born–Huang's criteria [61], e.g. all its six eigen values are positive. Here, the number of subindices is adopted to distinguish what the notation is. For example, stress and strain symbols with one subindex are in Voigt notation, while they are in tensor notation with two subindices. The relations between these two notations are as follows:

$$\sigma_1 = \sigma_{11}, \sigma_2 = \sigma_{22}, \sigma_3 = \sigma_{33}, \sigma_4 = \sigma_{23}, \sigma_5 = \sigma_{31}, \sigma_6 = \sigma_{12} \tag{3.25}$$

$$\varepsilon_1 = \varepsilon_{11}, \varepsilon_2 = \varepsilon_{22}, \varepsilon_3 = \varepsilon_{33}, \varepsilon_4 = 2\varepsilon_{23}, \varepsilon_5 = 2\varepsilon_{31}, \varepsilon_6 = 2\varepsilon_{12} \tag{3.26}$$

The values of $\mathbb{C}_{IJ}$ and $\mathbb{C}_{ijkl}$ can be directly replaced by each other following the role of subindices in Eqs. (3.25) and (3.26) with the correspondence of $I \leftrightarrow ij$, $J \leftrightarrow kl$. Transformation between $\mathbb{S}_{IJ}$ and $\mathbb{S}_{ijkl}$ needs some prefactors. Following the correspondence $I \leftrightarrow ij$, $J \leftrightarrow kl$ to find prefactors $p_{I\leftrightarrow ij}$ and $p_{J\leftrightarrow kl}$ in Eq. (3.26), $\mathbb{S}_{IJ} = p_{I\leftrightarrow ij}p_{J\leftrightarrow kl}\mathbb{S}_{ijkl}$ or inversely $\mathbb{S}_{ijkl} = 1/\left(p_{I\leftrightarrow ij}p_{J\leftrightarrow kl}\right)\mathbb{S}_{IJ}$. Usually, $\mathbb{C}_{IJ}$ is calculated, and $\mathbb{S}_{IJ}$ can be calculated from Eq. (3.24). Then, their tensor form can also be determined. Either the Voigt notation or the tensor notation is more convenience in the calculation of other elastic properties. So, both of them are introduced here. Usually, we need to know the anisotropy in Young's modulus, shear modulus, or Poisson's ratio. In this situation, the tensor notation is more convenient. The Young's modulus for the direction $\boldsymbol{d}$ = [$d_1$, $d_2$, $d_3$] (in this section, all the directional vectors are unit vector) is:

$$E_d = \frac{1}{\mathbb{S}_{ijkl}d_i d_j d_k d_l}\left(i,j,k,l = 1,2,3\right) \tag{3.27}$$

The shear modulus of a shear system with shear plane normal $\boldsymbol{n}$ = [$n_1$, $n_2$, $n_3$] and shear direction $\boldsymbol{b}$ = [$b_1$, $b_2$, $b_3$] is:

$$\mathrm{G}_{n,b} = \frac{1}{4\mathbb{S}_{ijkl}b_i n_j b_k n_l}\left(i,j,k,l = 1,2,3\right) \tag{3.28}$$

The Poisson's ratio for a system defined two directions $\boldsymbol{d}$ = [$d_1$, $d_2$, $d_3$] and $\boldsymbol{p}$ = [$p_1$, $p_2$, $p_3$] is:

$$\nu_{d,p} = -\frac{\mathbb{S}_{ijkl}d_i d_j p_k p_l}{\mathbb{S}_{ijkl}d_i d_j d_k d_l}\left(i,j,k,l = 1,2,3\right) \tag{3.29}$$

where $\boldsymbol{d}$ is the tensile or compression direction, and $\boldsymbol{p}$ is a direction that is perpendicular to $\boldsymbol{d}$.

Elastic properties of a polycrystalline is isotropic, which can be estimated from single crystal elastic properties following Voigt–Reuss–Hill approximations [62–64]:

$$9B_V = \left(\mathbb{C}_{11} + \mathbb{C}_{22} + \mathbb{C}_{22}\right) + 2\left(\mathbb{C}_{12} + \mathbb{C}_{23} + \mathbb{C}_{31}\right) \tag{3.30}$$

$$15G_V = \left(\mathbb{C}_{11} + \mathbb{C}_{22} + \mathbb{C}_{22}\right) - \left(\mathbb{C}_{12} + \mathbb{C}_{23} + \mathbb{C}_{31}\right) + 3\left(\mathbb{C}_{44} + \mathbb{C}_{55} + \mathbb{C}_{66}\right) \tag{3.31}$$

$$1/B_R = \left(\mathbb{S}_{11} + \mathbb{S}_{22} + \mathbb{S}_{22}\right) + 2\left(\mathbb{S}_{12} + \mathbb{S}_{23} + \mathbb{S}_{31}\right) \tag{3.32}$$

$$15/G_R = 4\left(\mathbb{S}_{11} + \mathbb{S}_{22} + \mathbb{S}_{22}\right) - 4\left(\mathbb{S}_{12} + \mathbb{S}_{23} + \mathbb{S}_{31}\right) + 3\left(\mathbb{S}_{44} + \mathbb{S}_{55} + \mathbb{S}_{66}\right) \tag{3.33}$$

The Voigt approximation and the Reuss approximation are upper and lower bounds for $B$ and G. Hill [64] suggested to use their arithmetic means to estimate $B$ and G:

$$B_H = \left(B_V + B_R\right)/2 \tag{3.34}$$

$$G_H = \left(G_V + G_R\right)/2 \tag{3.35}$$

Then, the Young's modulus $E_H$ and Poisson's ratio $\nu_H$ can be evaluated as follows:

$$E_H = \frac{9B_H G_H}{3B_H + G_H} \tag{3.36}$$

$$\nu_H = \frac{3B_H - 2G_H}{2\left(3B_H + G_H\right)} \tag{3.37}$$

The ratio of bulk to shear modulus ($B/G$ or Pugh's ratio [65]) is often employed to estimate the brittle–ductile transition. A value of $B/G > 1.75$ indicates ductility, otherwise it is brittle. With Eq. (3.37), the Pugh's ratio is also related to Poisson's ratio. Then, it is assumed that a material with $\nu > 0.26$ is generally ductile, or it is brittle if $\nu < 0.26$. Moreover, for a material with cubical symmetry, the Cauchy pressure $C_{12} - C_{44}$ is usually taken as a brittle–ductile indicator, where a negative Cauchy pressure means directional bonding and brittle nature of a material. For a brittle material, $B$ and $G$ can also be used to estimate its hardness according to Chen's model [66]:

$$H = 2\left(G^3 / B^2\right)^{0.585} - 3 \tag{3.38}$$

The polycrystalline shear moduli can also be used to measure the elastic anisotropy via:

$$A_{VR} = \left(G_V - G_R\right)/\left(G_V + G_R\right) \tag{3.39}$$

where the isotropic system exhibits $A_{VR} = 0$.

In the DFT-based method, elastic properties are calculated at 0 K. To calculate temperature-dependent elastic properties, one can refer to the following steps:

1) Calculate elastic stiffness matrix $\mathbb{C}_{IJ}(V)$ as a function of volume at 0 K;
2) Determine temperature-dependent isothermal elastic stiffness matrix $\mathbb{C}_{IJ}^T\left(V\right)$ according to $\mathbb{C}_{IJ}(V)$ and $V(T)$;
3) Convert isothermal elastic stiffness matrix $\mathbb{C}_{IJ}^T\left(T\right)$ to isentropic elastic stiffness matrix $\mathbb{C}_{IJ}^s\left(T\right)$ based on the equation given by Davies [67]:

$$\mathbb{C}_{IJ}^{s}(T)=\mathbb{C}_{IJ}^{T}(V)+\frac{TV\xi_I\xi_J}{C_V} \tag{3.40}$$

where $C_V$ is the heat capacity at constant volume and $\xi_I$ is calculated by

$$\xi_I=-\alpha_J\mathbb{C}_{IJ}^{T}(V),J=1,2,\cdots 6 \tag{3.41}$$

where $\alpha_J$ is directional-dependent thermal expansion coefficient, which is:

$$\mathbf{A}=\mathbf{R}^{-1}\left(\frac{\partial \mathbf{R}}{\partial V}\right)_{T=0}\frac{\partial V}{\partial T} \tag{3.42}$$

with $\alpha_1 = A_{11}$, $\alpha_2 = A_{22}$, $\alpha_3 = A_{33}$, $\alpha_4 = A_{23}+A_{32}$, $\alpha_5 = A_{31}+A_{13}$, $\alpha_6 = A_{12}+A_{12}$. In Eq. (3.42), **R** is the matrix representation of a lattice with its row vectors being ***a***, ***b***, and **c** of the lattice cell. Super-scripter −1 means matrix inversion.

As one of the easily accessible properties, elastic properties of many HEMs are predicted by DFT-based calculations or MD simulations [32, 54, 55, 68, 69]. In DFT-based calculations, either the EMTO-CPA method or the supercell method is adopted. Most of the calculations show that lattice distortions have negligible effect on elastic properties, which means that fast screening of elastic properties by the EMTO-CPA method is possible. Therefore, a series of compositional-dependent elastic properties have been reported [32, 69]. For example, Tian et al. [32] studied the influence of Al on elastic properties of $CoCrFeNiAl_x$ HEA, as shown in Figure 3.17. Their results reveal that elastic properties vary nearly linearly with respect to Al concentration for both FCC and BCC phases. Linear-like dependence of elastic properties on element concentration was also found by Niu et al. [69], who systematically studied over 2700 compositions of the NiFeCrCo alloy. Whether

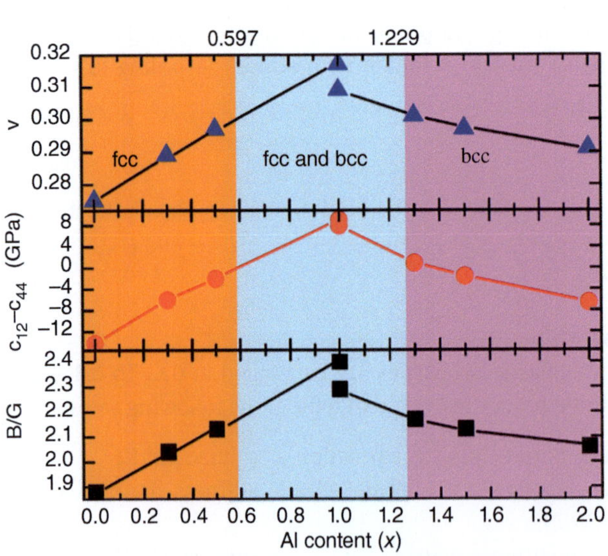


**Figure 3.17** Elastic properties versus Al fraction, (top) Poisson's ratio, (middle) Cauchy pressure, and (lower) Pugh's ratio for $CoCrFeNiAl_x$ HEA. Source: Reproduced with permission from Tian et al. [32]/American Physical Society.

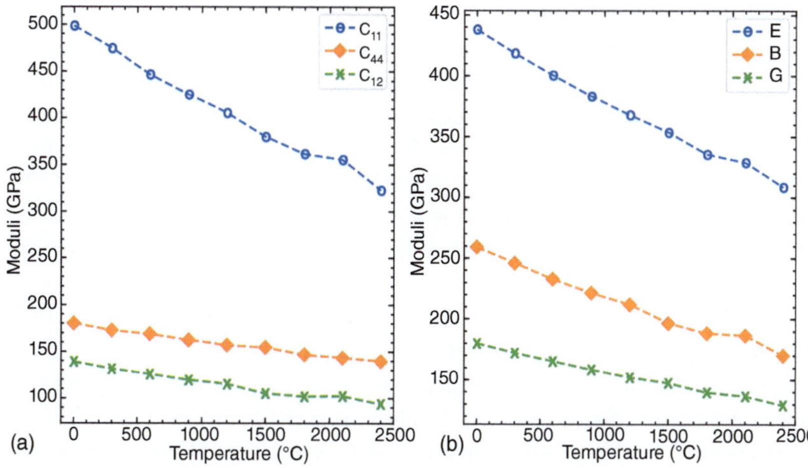


**Figure 3.18** Temperature-dependent elastic properties. (a) $C_{11}$, $C_{12}$, and $C_{44}$;(b) $B, G, E$ based on Voigt–Reuss–Hill approximations. Source: Reproduced with permission from Dai et al. [54]/Elsevier.

lattice distortion is important or not in determining elastic properties may depend on the lattice stability. When the lattice stability is low, lattice distortion may play a dominant role. For example, Zheng et al. [70] reported that substantial differences in elastic properties are obtained with and without consideration of lattice distortions for TaNbHfZrTi. Results with lattice distortions agree better with experiments. The reason for the discrepancies is believed due to the instability of BCC Ti, Zr, and Hf at 0 K. By the way, large discrepancy in the two values may provide some guidelines for materials' design, e.g. searching for materials with structural transformations.

Most elastic properties predicted by DFT-based methods are values at 0 K, and only limited work reported temperature-dependent elastic properties. Until now, only temperature-dependent isothermal elastic modulus $\mathbb{C}_{IJ}^{T}\left(V\right)$ was reported [36, 68, 71]. As discussed above, the full set of isentropic elastic properties can be calculated when temperature-dependent free energies are determined with Eqs. (3.2)–(3.11). We believe that more temperature-dependent elastic properties will be reported in the future. MD simulations provide an alternative choice to predict temperature-dependent elastic properties. However, the accuracy of MD simulations depends strongly on the accuracy of interatomic potentials. Dai et al. [54, 55] trained high accurate ML potentials for high-entropy carbides and diborides with DFT accuracy and predicted temperature-dependent elastic properties for high-entropy carbides and diborides. Figure 3.18 shows the results for high-entropy $TiZrHfNbTaC_5$, which exhibit a typical linear decrease tendency with respect to temperature.

### 3.3.3 Stacking Fault Energy

Stacking fault energy (SFE) is one of the most important parameters, which links tightly to mechanical properties of alloys. Especially in FCC alloys, the dominating plastic deformation mechanism relies strongly on SFE. It is well established that plastic deformation is dominated by dislocation slip when SFE is high, by twinning when SFE is low, and possibly by transformation to the HCP phase when SFE is further lower. TRIP and twinning induced plasticity (TWIP) are fundamental aspects in the design of high-performance alloys with both high toughness and high strength. For example, Gludovatz et al. [1] reported a fracture-resistant CrMnFeCoNi HEA for cryogenic applications, which is a TWIP HEA at cryogenic temperature. Therefore, temperature- and composition-dependent SFEs are essential for the design of high-performance HEA.

Figure 3.19 shows a schematic diagram of intrinsic stacking fault in an FCC lattice. The corresponding shear vector of intrinsic stacking fault in an FCC lattice is $<\bar{2}11>/6$ along a {111} plane. In principle, a stacking fault can be generated by a shear along any plane with any in-plane vector for any crystal, which is called generalized stacking fault. When calculating intrinsic SFE in an FCC lattice, there are three methods. The first is calculating SFE according to the following formula by constructing supercells with and without stacking fault (see Figure 3.19):

$$\gamma_{SF} = \frac{E_{SF} - E_{FCC}}{A} \tag{3.43}$$

where $E_{SF}$ and $E_{FCC}$ are the energies of supercells with and without stacking fault, respectively. $A$ is the area of the faulted region. This method is general, which is also applicable to other structure types or shear models corresponding to a generalized stacking fault. The second and third methods are approximation methods for FCC lattice only. The stacking fault in an FCC lattice can be viewed as a local transformation from FCC to HCP, as shown in Figure 3.19. When considering interaction of the {111} layers up to the nearest neighbor layers or up to the next-nearest neighbor layers, the following two approximations [37] can be derived:

$$\gamma_{SF} \approx 2\frac{E_{HCP} - E_{FCC}}{A} \tag{3.44}$$

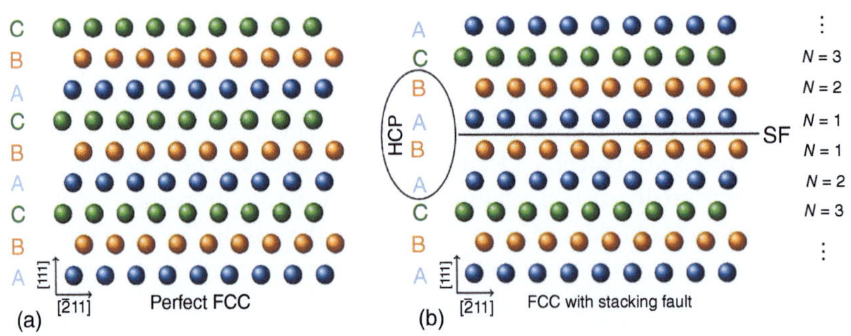


**Figure 3.19** Schematic diagram of intrinsic stacking fault in an FCC lattice: (a) perfect FCC stacking sequence with ABCABCABC, (b) FCC with a stacking fault with sequence ABC"AB|AB"CA. The vertical line illustrates the position of stacking fault, where a local HCP stacking region with an ABAB sequence can be observed. Source: Reproduced with permission from Ikeda et al. [37]/Elsevier.

$$\gamma_{SF} \approx \frac{E_{HCP} + 2E_{DHCP} - 3E_{FCC}}{A} \tag{3.45}$$

where $E_{HCP}$ and $E_{DHCP}$ are the energies of the HCP and double hexagonal close-packed (DHCP) structures, respectively. The stacking sequence of DHCP is ABAC. The approximation based on Eq. (3.44) is called the axial-nearest-neighbor-Ising (ANNI) model, and the one based on Eq. (3.45) is the axial-next-nearest-neighbor-Ising (ANNNI) model. All these equations have been used in literatures [71–73].

SFE of CrMnFeCoNi HEA has been studied by many works due to the intriguing TWIP effect of CrMnFeCoNi HEA. Huang et al. [71] employed EMTO-CPA DFT-based methods to calculate temperature-dependent SFEs of CrMnFeCoNi HEA and separated the total SFEs $\gamma_{SF}$ to chemical, magnetic, and strain contributions ($\gamma_{chem}$, $\gamma_{mag}$, and $\gamma_{strain}$), as shown in Figure 3.20a,b. According to their results, $\gamma_{SF}$ increases with temperature, $\gamma_{chem}$ also increases with temperature, $\gamma_{mag}$ first increases with temperature and then becomes saturated when $T > 300$ K, and $\gamma_{strain}$ shows negligible temperature dependence. The temperature-dependent $\gamma_{chem}$ results from the temperature-dependent formation energy difference between HCP and FCC phases. The relative stability of FCC phase increases with temperature, which results in increasing $\gamma_{SF}$ with temperature. In addition, they also show that the magnetic state of Fe and Mn atoms around the faulted plane exhibits substantially different magnetic moment in comparison to the FCC bulk (see Figure 3.20c,d), which then contributes significantly to SFE.

Zhao et al. [73] employed supercells to determine both temperature- and composition-dependent SFEs of CrMnFeCoNi and CrFeCoNiPd and subsystems from these two HEAs. All results calculated from Eqs. (3.43)–(3.45) were compared, indicating that all results are comparable, as shown in Figure 3.21. Additionally, they also discussed the electronic and vibrational contributions to SFEs, which

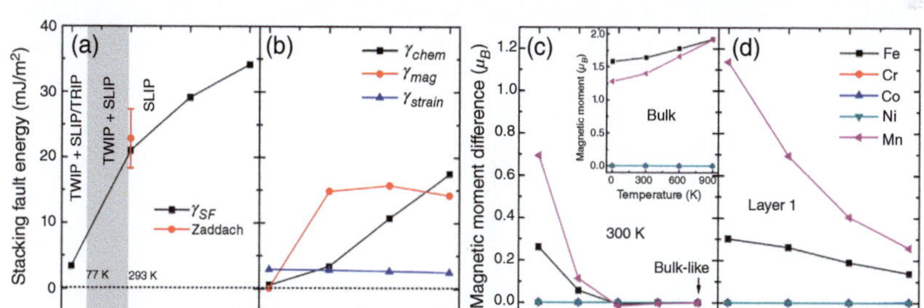


**Figure 3.20** Temperature-dependent SFE of CrMnFeCoNi HEA: (a) Total SFE; (b) contributions from chemical part $\gamma_{chem}$, magnetic part $\gamma_{mag}$, and strain part $\gamma_{strain}$. In (a), Zaddach means results reported by Zaddach et al. [72], and rough boundaries between different plastic mechanisms (SLIP → SLIP + TWIP → TWIP + SLIP/TRIP) are illustrated. Magnetic moment difference between the perfect and faulted lattice in paramagnetic CrMnFeCoNi HEA as a function of (c) number of layers N at room temperature, and (d) temperature for atoms in layer 1. The inset shows the temperature dependence of the magnetic moments in the perfect FCC lattice. Source: Reproduced with permission from Zaddach et al. [72]/Elsevier.

reveals that temperature dependence of SFE is dominated by vibrational contributions for most of the alloys. The results are consistent with the results reported by Huang et al. [71]. Evident discrepancy between SFE of CrMnFeCoNi HEA reported by Huang et al. [71] (small positive) and Zhao et al. [73] (significant negative) can be seen. The discrepancy may result from the difference in their methods, where lattice distortions are neglected in EMTO-CPA adopted by Huang et al. [71]. In DFT-based calculations, energy difference of 1 meV/atom is usually adopted as the convergency criterion. This value corresponds to an error of 5.7 $mJ/m^2$ for SFE by assuming a typical FCC lattice parameter of 3.6 Å. Therefore, the accuracy of the DFT calculations has strong influence on SFE, especially when SFE is low. Therefore, supercell methods may be a better choice for quantitative SFE calculations.

### 3.3.4 Thermal Properties

Thermal properties of a material mainly include thermal expansion coefficient and thermal conductivity, which are practically important for the design of materials' processing and for materials service at high temperatures, materials service in a board temperature range, or materials service in thermal management. In HEMs, chemical disorder and chemical bond disorder results in severe lattice distortions, which will induce significant scattering for phonons and electrons, and finally reduce the thermal conductivity and electrical conductivity. Mechanical properties are the main concern in HEAs, while thermal properties are not. Thermal properties attracted research interests only after HECs were proposed. Therefore, only limited theoretical works on thermal properties of HEMs have been reported.

Both DFT-based methods and MD simulations can be used to predict thermal properties. The calculation of thermal expansion is highly correlated with the calculation of temperature-dependent thermodynamic properties in DFT-based scenario. Equations (3.1)–(3.13) show how to calculate temperature-dependent free energies. Equilibrium lattice volume at each temperature can be determined by minimizing

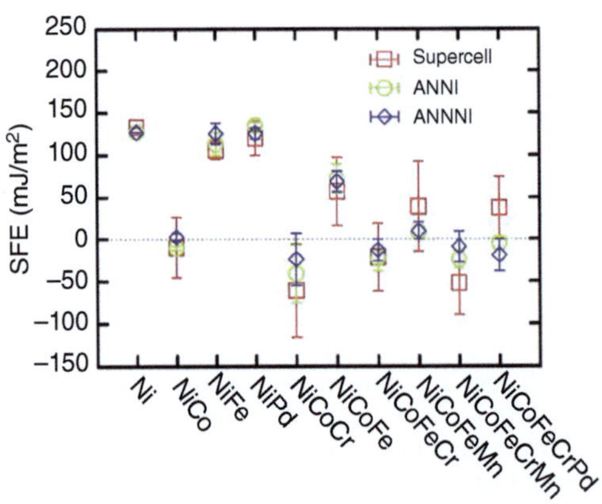


**Figure 3.21** SFEs calculated by the supercell method as well as ANNI and ANNNI approximations.

the free energy, which can then be used to determine thermal expansion coefficient. For example, Ma et al. [68] calculated thermal expansion coefficient of CrMnFeCoNi HEA by employing the EMTO-CPA method, which agrees well with experiments.

In comparison to DFT-based methods, MD simulations are more suitable for predicting thermal properties. For example, temperature-dependent lattice parameters can be obtained by equilibrating a lattice under NPT ensemble at different temperatures. Then, thermal expansion coefficient can be calculated. Phonon thermal conductivity can be calculated according to the Green–Kubo method:

$$\kappa_{\mu\nu} = \frac{1}{\text{©} k_B T^2} \int_0^{\infty} \langle J_\mu(t) J_\nu(0) \rangle dt \tag{3.46}$$

where $\Omega$ is the system volume and the angular brackets denote time average of heat current auto-correlation function, $J$ is heat current, and $\mu, \nu = x, y, z$. Dai et al. [54, 55] predicted thermal expansions and phonon thermal conductivities of high-entropy $(Ti_{0.2}Zr_{0.2}Hf_{0.2}Nb_{0.2}Ta_{0.2})C$ and $(Ti_{0.2}Zr_{0.2}Hf_{0.2}Nb_{0.2}Ta_{0.2})B_2$ with a ML potential and MD simulations. Figure 3.22 shows the predicted results for high-entropy $(Ti_{0.2}Zr_{0.2}Hf_{0.2}Nb_{0.2}Ta_{0.2})C$, which reveals that the thermal conductivity is really very low. The low phonon thermal conductivity may be approximated by using Clarke's model [74]:

$$\kappa_{\min} = k_B v_m \left( \frac{M}{n \rho N_A} \right)^{-2/3} \tag{3.47}$$

where $v_m$ is the average sound velocity, $\rho$ is the density, $N_A$ is the Avogadro's constant, and $M$ is the molecular weight. Braun et al. [6] and Lim et al. [75] studied the amorphous-like thermal conductivities in MgCoNiCuZnX-O family oxides with MD simulations. By designing empirical potentials, they revealed that charge nonuniformity (chemical bond distortions or atomic force disorder) is more important in reducing thermal conductivity in comparison to atomic mass disorder.

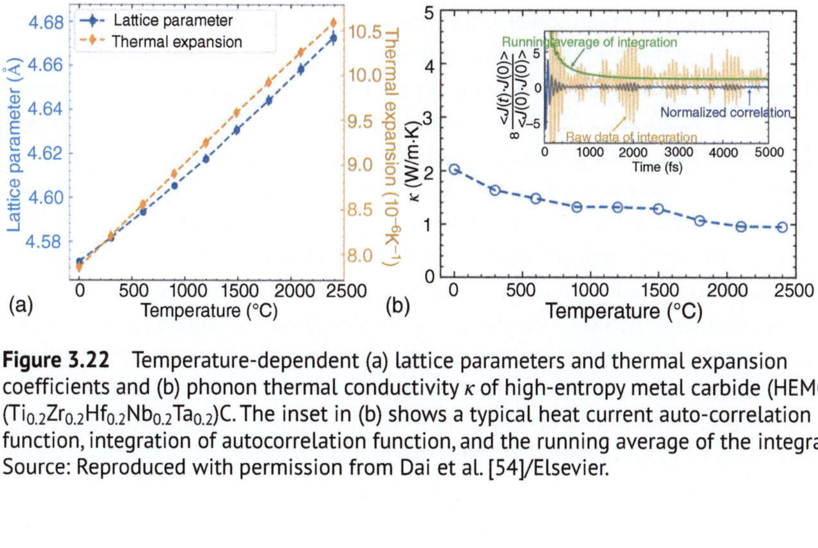


**Figure 3.22** Temperature-dependent (a) lattice parameters and thermal expansion coefficients and (b) phonon thermal conductivity $\kappa$ of high-entropy metal carbide (HEMC) $(Ti_{0.2}Zr_{0.2}Hf_{0.2}Nb_{0.2}Ta_{0.2})C$. The inset in (b) shows a typical heat current auto-correlation function, integration of autocorrelation function, and the running average of the integration. Source: Reproduced with permission from Dai et al. [54]/Elsevier.

### 3.3.5 Simulation on Defects

Previous works are focused on the properties of ideal lattice without defects, except for SFE. Defects, e.g. dislocations, and surface and grain boundaries are relevant to materials properties, especially mechanical properties. However, simulation on the properties of these defects is in its initial phase, and only limited works have been reported [76–82].

Dislocation glide dominates plastic deformations in most alloys and plastic deformations in some ceramics at elevated temperatures. Understanding the mechanisms of dislocation glide plays key roles in tailoring mechanical properties of alloys. Atomistic simulations provide us atomistic-level details of dislocation core structures and dynamic behavior of dislocation glide, and it also allow us to find out the underlying mechanisms during dislocation glide. Dislocation core structures, critical shear stress, temperature- and stress-dependent dislocation glide velocities, and the underlying strengthening mechanisms of different HEAs have been simulated by MD with either conventional atomic potentials or with ML potentials [76–78, 80, 81] and compared with conventional alloys. The most striking feature of dislocation core structure in HEAs is the irregular and waving shape, e.g. non-uniform stacking fault spacing when dissociating into partials [81], automatically kinked structure [77], which result from local fluctuations in HEAs. The rate-limiting mechanisms in different HEAs may be different and may also be different from conventional alloys. In conventional BCC alloys, it is believed that kink pair nucleation is the rate-limiting step for screw dislocation motion, whereas the kink spread on the dislocation line is very fast with low barrier. However, Chen et al. [77] simulated dislocation glide in $Co_{16.67}Fe_{36.67}Ni_{16.67}Ti_{30}$ BCC-HEA and found that the controlling step for screw dislocation motion changes to kink spread on dislocation lines. They employed the climbing image-nudged elastic band method to study energy landscape by confronting the gliding dislocations, which shows that the highest barrier relates to propagation of kinks along the dislocations, as shown in Figure 3.23. Another finding of their simulation is that edge and screw dislocations experience a comparable barrier for motion in HEA, and hence, their contribution to strength becomes comparable. In contrast, critical shear stress of edge dislocation is much lower than that of screw dislocation in conventional BCC alloys, where the strength is governed solely by screw dislocations. The mechanism may vary from system to system, depending on what the alloy is. For example, Yin et al. [78] simulated dislocation glide in NbTaMoW BCC-HEA by MD and found that the rate-limiting process is kink pair nucleation, which is similar to conventional BCC alloys. Furthermore, they revealed that cross-slip locking mechanism, which results from intrinsic fluctuation of energy along the dislocation line, may also contribute to high strength of HEA. They also investigated the impact of SRO on dislocation glide and found a striking phenomenon that edge dislocation glide was enhanced by SRO. Simulation on dislocation mechanisms in HEAs are still limited, where dislocation cross-slip, dislocation-grain boundary interaction, and dislocation reactions have not been investigated.

The structure, composition, energy, migration, and sliding of grain boundaries and the interaction of grain boundaries with other defects play key roles in determining

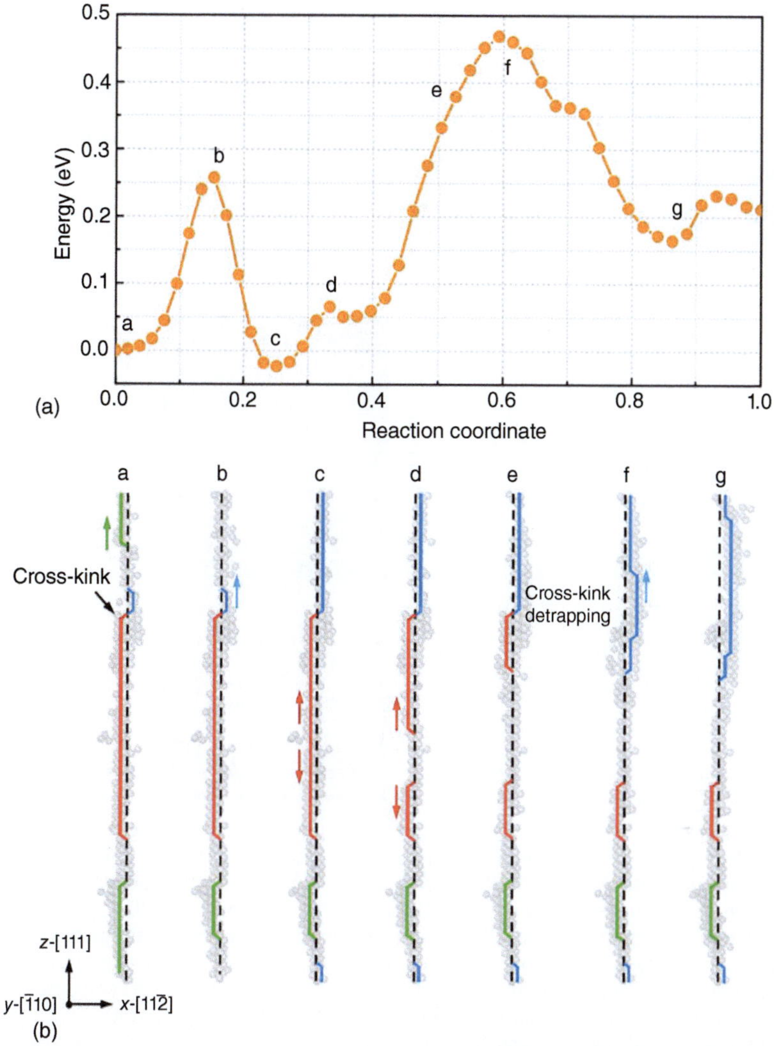


**Figure 3.23** The minimum energy pathway for screw dislocation glide in $Co_{16.67}Fe_{36.67}Ni_{16.67}Ti_{30}$ BCC-HEA: (a) the energy landscape, and (b) snapshots of dislocation cores with their corresponding indices in (a). In (a), the landscape is quite rugged due to fluctuations of HEA, and the energy barriers indicate the activated events that control the dislocation movement, i.e. the rate-limiting processes. In (b), it can be seen that the initial screw dislocation contains a high density of kinks. With reference to the initially created straight dislocation (the dashed black line), the trapped dislocation segments on three {110} planes are marked by three different colors (red, blue, and green solid lines). A cross-kink is formed when two differently oriented kinks meet, as indicated by the black arrow in position a. The movement of the screw dislocation requires de-trapping of the kinks and cross-kinks from lower-energy local environments. The red, blue, and green arrows indicate the propagation directions of the corresponding kinks. As an example, from position a to c, the kink propagation requires de-trapping to overcome barrier (~0.25 eV, see saddle point in position b). As another example, from position d to g, the activation barrier for de-trapping the cross-kink is ~0.4 eV. Source: Chen et al. [77].

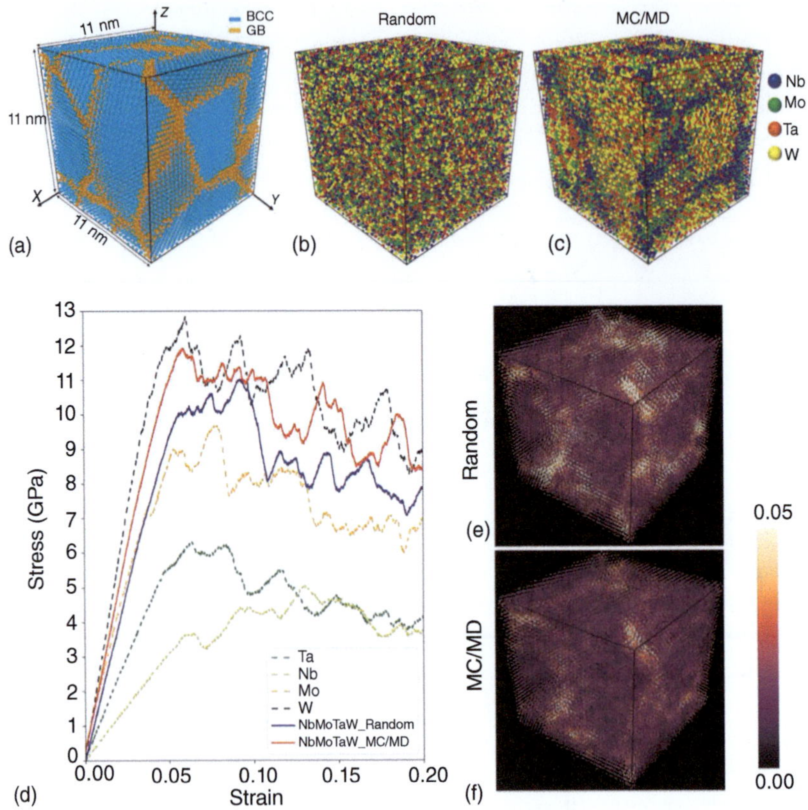


**Figure 3.24** A polycrystalline model for the equimolar quaternary NbTaMoW HEA: (a) atoms colored by local lattice structure (BCC or GB: grain boundary), (b and c) atoms colored by element type, (c) is a snapshot of a polycrystalline model after hybrid MC/MD simulations, where Nb segregation into GBs is clear. (d) Polycrystalline stress–strain curves under compression for element substrates Nb, Ta, Mo, and W, purely random HEA and HEA after MC/MD simulation. Evident improvement in strength is obtained when GB segregates with Nb. Atomic von Mises strain map for (e) random HEA and (f) HEA after MC/MD simulation. Source: Li et al. [82]/Springer Nature/CC BY 4.0.

grain coarsening, creep, strength, and other mechanical properties of a material. For example, Li et al. [82] fitted a ML potential and applied it to study NbTaMoW HEA. They found that Nb tends to segregate into grain boundaries, which reduces the von Mises strain, and finally enhances the compressive strength, as shown in Figure 3.24. In the future, the segregation of grain boundaries, migration and sliding of grain boundaries, and interaction of grain boundaries with other defects need to be studied.

## 3.4 Conclusions and Perspectives

In this chapter, theoretical aspects on the design of HEMs have been briefly reviewed. The contents are separated into two major parts: (i) prediction on the formability and

(ii) prediction on properties and exploring the underlying mechanisms of properties. Efforts have been made to propose new models, to predict composition-dependent formability and properties, and to explore new mechanisms. Intriguing new phenomena are emerging, e.g. decreasing SFE with the decrease in temperature that induces TWIP HEA with an exceptional combination of high strength and toughness for cryogenic applications; cross-slip locking and rate-limiting process transfer from kink nucleation to kink propagation of screw dislocations are studied in some BCC HEAs. There are also some pioneer works on tailoring the properties of HEA under inverse design scenario, e.g. Li et al. [34] designed a duplex-phase TRIP $Co_{20}Cr_{20}Fe_{40-x}Mn_{20}Ni_x$ ($x = 0$–20 at.%) HEA assisted by DFT-based calculations, Choi et al. [79] improved the strength of CoCrFeMnNi-based HEA by MD simulation of critical resolved shear stress, Feng et al. [83] tailored the microstructure of AlCrFeMnTi-based HEA by high-throughput CALPHAD computation to avoid precipitation of brittle C14 Laves phase. Although many works have been published, this field is still young, and more studies need to be done in the future, e.g. to develop physical-based or ML-assisted models for HEMs, to modify the accuracy of CALPHAD databases, to build "composition–structure–property" databases, to fit accurate ML potentials for atomistic simulations, and to explore distinct phenomena that result from the intrinsic fluctuations.

A fundamental complexity for the design of HEMs is their infinite combination of elements and variation in their concentrations. Up to now, most theoretical investigations have focused on DFT calculations and MD simulations, and empirical property prediction models derived from physical assumptions or trained from data with ML methods are still rare. Though DFT calculations and MD simulations can provide in-depth understanding of HEMs, empirical models are still essential for property tailoring and materials design due to their ability of fast scanning on a large compositional and structural space. With the development of computational capabilities, ML methods, and enrichment of databases either from theoretical calculations or experimental results, computational-assisted approaches would play key roles in the design of HEMs in the future.

## References


1 Gludovatz, B., Hohenwarter, A., Catoor, D. et al. (2014). A fracture-resistant high-entropy alloy for cryogenic applications. *Science* 345: 1153–1158. https://doi.org/10.1126/science.1254581.

2 Senkov, O.N., Wilks, G.B., Scott, J.M., and Miracle, D.B. (2011). Mechanical properties of $Nb_{25}Mo_{25}Ta_{25}W_{25}$ and $V_{20}Nb_{20}Mo_{20}Ta_{20}W_{20}$ refractory high entropy alloys. *Intermetallics* 19: 698–706. https://doi.org/10.1016/j.intermet.2011.01.004.

3 Zhang, Y., Stocks, G.M., Jin, K. et al. (2015). Influence of chemical disorder on energy dissipation and defect evolution in concentrated solid solution alloys. *Nat. Commun.* 6: 8736. https://doi.org/10.1038/ncomms9736.

4 Bérardan, D., Franger, S., Dragoe, D. et al. (2016). Colossal dielectric constant in high entropy oxides. *Phys. Status Solidi RRL* 10: 328–333. https://doi.org/10.1002/pssr.201600043.

**5** Bérardan, D., Franger, S., Meena, A.K., and Dragoe, N. (2016). Room temperature lithium superionic conductivity in high entropy oxides. *J. Mater. Chem. A.* 4: 9536–9541. https://doi.org/10.1039/C6TA03249D.

**6** Braun, J.L., Rost, C.M., Lim, M. et al. (2018). Charge-induced disorder controls the thermal conductivity of entropy-stabilized oxides. *Adv. Mater.* 30: 1805004. https://doi.org/10.1002/adma.201805004.

**7** Zhang, W., Liaw, P.K., and Zhang, Y. (2018). Science and technology in high-entropy alloys. *Sci. China Mater.* 61: 2–22. https://doi.org/10.1007/s40843-017-9195-8.

**8** Chang, X., Zeng, M., Liu, K., and Fu, L. (2020). Phase engineering of high-entropy alloys. *Adv. Mater.* 32: 1907226. https://doi.org/10.1002/adma.201907226.

**9** Zhang, R., Zhao, S., Ding, J. et al. (2020). Short-range order and its impact on the CrCoNi medium-entropy alloy. *Nature* 581: 283–287. https://doi.org/10.1038/s41586-020-2275-z.

**10** Wright, A.J. and Luo, J. (2020). A step forward from high-entropy ceramics to compositionally complex ceramics: a new perspective. *J. Mater. Sci.* 55: 9812–9827. https://doi.org/10.1007/s10853-020-04583-w.

**11** Xiang, H., Xing, Y., Dai, F. et al. (2021). High-entropy ceramics: present status, challenges, and a look forward. *J. Adv. Ceram.* 10: 385–441. https://doi.org/10.1007/s40145-021-0477-y.

**12** Hume-Rothery, W. and Coles, B.R. (1954). The transition metals and their alloys. *Adv. Phys.* 3: https://doi.org/10.1080/00018735400101193.

**13** Yang, X. and Zhang, Y. (2012). Prediction of high-entropy stabilized solid-solution in multi-component alloys. *Mater. Chem. Phys.* 132: 233–238. https://doi.org/10.1016/j.matchemphys.2011.11.021.

**14** Guo, S., Hu, Q., Ng, C., and Liu, C.T. (2013). More than entropy in high-entropy alloys: forming solid solutions or amorphous phase. *Intermetallics* 41: 96–103. https://doi.org/10.1016/j.intermet.2013.05.002.

**15** Guo, S., Ng, C., Lu, J., and Liu, C.T. (2011). Effect of valence electron concentration on stability of fcc or bcc phase in high entropy alloys. *J. Appl. Phys.* 109: 103505. https://doi.org/10.1063/1.3587228.

**16** Troparevsky, M.C., Morris, J.R., Kent, P.R.C. et al. (2015). Criteria for predicting the formation of single-phase high-entropy alloys. *Phys. Rev. X* 5: 011041. https://doi.org/10.1103/PhysRevX.5.011041.

**17** King, D.J.M., Middleburgh, S.C., McGregor, A.G., and Cortie, M.B. (2016). Predicting the formation and stability of single phase high-entropy alloys. *Acta Mater.* 104: 172–179. https://doi.org/10.1016/j.actamat.2015.11.040.

**18** Ye, Y.F., Liu, C.T., and Yang, Y. (2015). A geometric model for intrinsic residual strain and phase stability in high entropy alloys. *Acta Mater.* 94: 152–161. https://doi.org/10.1016/j.actamat.2015.04.051.

**19** Wang, Z., Huang, Y., Yang, Y. et al. (2015). Atomic-size effect and solid solubility of multicomponent alloys. *Scr. Mater.* 94: 28–31. https://doi.org/10.1016/j.scriptamat.2014.09.010.

**20** Toda-Caraballo, I. and Rivera-Díaz-del-Castillo, P.E.J. (2016). A criterion for the formation of high entropy alloys based on lattice distortion. *Intermetallics* 71: 76–87. https://doi.org/10.1016/j.intermet.2015.12.011.

**21** Sarker, P., Harrington, T., Toher, C. et al. (2018). High-entropy high-hardness metal carbides discovered by entropy descriptors. *Nat. Commun.* 9: https://doi.org/10.1038/s41467-018-07160-7.

**22** Pitike, K.C., Kc, S., Eisenbach, M. et al. (2020). Predicting the phase stability of multicomponent high-entropy compounds. *Chem. Mater.* 32: 7507–7515. https://doi.org/10.1021/acs.chemmater.0c02702.

**23** Wang, Y., Csanádi, T., Zhang, H. et al. (2020). Enhanced hardness in high-entropy carbides through atomic randomness. *Adv. Theo. Simul.* 3: 2000111.

**24** Lun, Z., Ouyang, B., Kwon, D.-H. et al. (2020). Cation-disordered rocksalt-type high-entropy cathodes for Li-ion batteries. *Nat. Mater.* https://doi.org/10.1038/s41563-020-00816-0.

**25** de Boer, F.R., Mattens, W.C.M., Boom, R., et al. (1988). Cohesion in metals. Transition metal alloys, 1. https://www.osti.gov/etdeweb/biblio/5556327 (accessed 9 August 2022).

**26** Lindemann, F.A. (1910). The calculation of molecular vibration frequencies. *Phys. Z.* 11: 609–612.

**27** Agarwal, A. and Prasada Rao, A.K. (2019). Artificial intelligence predicts body-centered-cubic and face-centered-cubic phases in high-entropy alloys. *JOM* 71: 3424–3432. https://doi.org/10.1007/s11837-019-03712-4.

**28** Huang, W., Martin, P., and Zhuang, H.L. (2019). Machine-learning phase prediction of high-entropy alloys. *Acta Mater.* 169: 225–236. https://doi.org/10.1016/j.actamat.2019.03.012.

**29** Zhang, Y., Wen, C., Wang, C. et al. (2020). Phase prediction in high entropy alloys with a rational selection of materials descriptors and machine learning models. *Acta Mater.* 185: 528–539. https://doi.org/10.1016/j.actamat.2019.11.067.

**30** Zhou, Z., Zhou, Y., He, Q. et al. (2019). Machine learning guided appraisal and exploration of phase design for high entropy alloys. *npj Comput. Mater.* 5: 128. https://doi.org/10.1038/s41524-019-0265-1.

**31** Qi, J., Cheung, A.M., and Poon, S.J. (2019). High entropy alloys mined from binary phase diagrams. *Sci. Rep.* 9: 15501. https://doi.org/10.1038/s41598-019-50015-4.

**32** Tian, F., Delczeg, L., Chen, N. et al. (2013). Structural stability of $NiCoFeCrAl_x$ high-entropy alloy from ab initio theory. *Phys. Rev. B: Condens. Matter* 88: https://doi.org/10.1103/PhysRevB.88.085128.

**33** Körmann, F., Ma, D., Belyea, D.D. et al. (2015). "Treasure maps" for magnetic high-entropy-alloys from theory and experiment. *Appl. Phys. Lett.* 107: 142404. https://doi.org/10.1063/1.4932571.

**34** Li, Z., Körmann, F., Grabowski, B. et al. (2017). Ab initio assisted design of quinary dual-phase high-entropy alloys with transformation-induced plasticity. *Acta Mater.* 136: 262–270. https://doi.org/10.1016/j.actamat.2017.07.023.

**35** Körmann, F., Ruban, A.V., and Sluiter, M.H.F. (2017). Long-ranged interactions in bcc NbMoTaW high-entropy alloys. *Mater. Res. Lett.* 5: 35–40. https://doi.org/10.1080/21663831.2016.1198837.

**36** Wang, Y., Yan, M., Zhu, Q. et al. (2018). Computation of entropies and phase equilibria in refractory V–Nb–Mo–Ta–W high-entropy alloys. *Acta Mater.* 143: 88–101. https://doi.org/10.1016/j.actamat.2017.10.017.

**37** Ikeda, Y., Grabowski, B., and Körmann, F. (2019). Ab initio phase stabilities and mechanical properties of multicomponent alloys: a comprehensive review for high entropy alloys and compositionally complex alloys. *Mater. Charact.* 147: 464–511. https://doi.org/10.1016/j.matchar.2018.06.019.

**38** Kostiuchenko, T., Körmann, F., Neugebauer, J., and Shapeev, A. (2019). Impact of lattice relaxations on phase transitions in a high-entropy alloy studied by machine-learning potentials. *npj Comput. Mater.* 5: 55. https://doi.org/10.1038/s41524-019-0195-y.

**39** Moruzzi, V.L., Janak, J.F., and Schwarz, K. (1988). Calculated thermal properties of metals. *Phys. Rev. B: Condens. Matter* 37: 790–799. https://doi.org/10.1103/PhysRevB.37.790.

**40** Slater, J.C. (1939). *Introduction to Chemical Physics*, 1e. McGraw-Hill.

**41** Dugdale, J.S. and MacDonald, D.K.C. (1953). The thermal expansion of solids. *Physiol. Rev.* 89: 832–834. https://doi.org/10.1103/PhysRev.89.832.

**42** Vitos, L. (2001). Total-energy method based on the exact muffin-tin orbitals theory. *Phys. Rev. B: Condens. Matter* 64: 014107. https://doi.org/10.1103/PhysRevB.64.014107.

**43** Zunger, A., Wei, S.-H., Ferreira, L.G., and Bernard, J.E. (1990). Special quasirandom structures. *Phys. Rev. Lett.* 65: 353–356. https://doi.org/10.1103/PhysRevLett.65.353.

**44** Sanchez, J.M., Ducastelle, F., and Gratias, D. (1984). Generalized cluster description of multicomponent systems. *Physica A* 128: 334–350. https://doi.org/10.1016/0378-4371(84)90096-7.

**45** Liu, X., Zhang, J., Yin, J. et al. (2021). Monte Carlo simulation of order–disorder transition in refractory high entropy alloys: a data-driven approach. *Comput. Mater. Sci.* 187: 110135. https://doi.org/10.1016/j.commatsci.2020.110135.

**46** Fernández-Caballero, A., Wróbel, J.S., Mummery, P.M., and Nguyen-Manh, D. (2017). Short-range order in high entropy alloys: theoretical formulation and application to Mo–Nb–Ta–V–W system. *J. Phase Equilibria Diffus.* 38: 391–403. https://doi.org/10.1007/s11669-017-0582-3.

**47** Han, J., Zhang, L., and Car, R. (n.d.). Deep potential: a general representation of a many-body potential energy surface. *Commun. Comput. Phys.* 11.

**48** Zhang, L., Han, J., Wang, H. et al. (2018). Deep potential molecular dynamics: a scalable model with the accuracy of quantum mechanics. *Phys. Rev. Lett.* 120: 143001. https://doi.org/10.1103/PhysRevLett.120.143001.

**49** Trott, C.R., Hammond, S.D., and Thompson, A.P. (2014). SNAP: strong scaling high fidelity molecular dynamics simulations on leadership-class computing platforms. In: *Supercomputing* (ed. J.M. Kunkel, T. Ludwig and H.W. Meuer), 19–34. Springer International Publishing.

**50** Senkov, O., Zhang, F., and Miller, J. (2013). Phase composition of a $CrMo_{0.5}NbTa_{0.5}TiZr$ high entropy alloy: comparison of experimental and simulated data. *Entropy* 15: 3796–3809. https://doi.org/10.3390/e15093796.

**51** Senkov, O.N., Miller, J.D., Miracle, D.B., and Woodward, C. (2015). Accelerated exploration of multi-principal element alloys with solid solution phases. *Nat. Commun.* 6: 6529. https://doi.org/10.1038/ncomms7529.

52 Oh, H., Ma, D., Leyson, G. et al. (2016). Lattice distortions in the FeCoNiCrMn high entropy alloy studied by theory and experiment. *Entropy* 18: 321. https://doi.org/10.3390/e18090321.

53 Okamoto, N.L., Yuge, K., Tanaka, K. et al. (2016). Atomic displacement in the CrMnFeCoNi high-entropy alloy – a scaling factor to predict solid solution strengthening. *AIP Adv.* 6: 125008. https://doi.org/10.1063/1.4971371.

54 Dai, F.-Z., Wen, B., Sun, Y. et al. (2020). Theoretical prediction on thermal and mechanical properties of high entropy $(Zr_{0.2}Hf_{0.2}Ti_{0.2}Nb_{0.2}Ta_{0.2})C$ by deep learning potential. *J. Mater. Sci. Technol.* S1005030220300050. https://doi.org/10.1016/j.jmst.2020.01.005.

55 Dai, F.-Z., Sun, Y., Wen, B. et al. (2021). Temperature dependent thermal and elastic properties of high entropy $(Ti_{0.2}Zr_{0.2}Hf_{0.2}Nb_{0.2}Ta_{0.2})B_2$: molecular dynamics simulation by deep learning potential. *J. Mater. Sci. Technol.* 72: 8–15. https://doi.org/10.1016/j.jmst.2020.07.014.

56 Varvenne, C., Luque, A., and Curtin, W.A. (2016). Theory of strengthening in fcc high entropy alloys. *Acta Mater.* 118: 164–176. https://doi.org/10.1016/j.actamat.2016.07.040.

57 Toda-Caraballo, I., Wróbel, J.S., Dudarev, S.L. et al. (2015). Interatomic spacing distribution in multicomponent alloys. *Acta Mater.* 97: 156–169. https://doi.org/10.1016/j.actamat.2015.07.010.

58 Toda-Caraballo, I. (2017). A general formulation for solid solution hardening effect in multicomponent alloys. *Scr. Mater.* 127: 113–117. https://doi.org/10.1016/j.scriptamat.2016.09.009.

59 Zs.Rak, Rost, C.M., Lim, M. et al. (2016). Charge compensation and electrostatic transferability in three entropy-stabilized oxides: results from density functional theory calculations. *J. Appl. Phys.* 120: 095105. https://doi.org/10.1063/1.4962135.

60 Zs.Rák, Maria, J.-P., and Brenner, D.W. (2018). Evidence for Jahn-Teller compression in the (Mg, Co, Ni, Cu, Zn)O entropy-stabilized oxide: a DFT study. *Mater. Lett.* 217: 300–303. https://doi.org/10.1016/j.matlet.2018.01.111.

61 Born, M. and Huang, K. (1954). *Dynamical Theory of Crystal Lattices*. London: Oxford University Press.

62 Voigt, W. (1928). *Lehrbuch der KrystallPhysik*. Leipzig: Teubner.

63 Reuss, A. and Angew, Z. (1929). Berechnung der fliegrenze von mischkristallen auf grund der plastizittsbedingung für einkristalle. *Z. Angew. Meath. Mech.* 49–58.

64 Hill, R. (1952). The elastic behavior of a crystalline aggregate. *Proc. Phys. Soc. A* 349–354.

65 Pugh, S.F. and XCII. (1954). Relations between the elastic moduli and the plastic properties of polycrystalline pure metals. *Lond. Edinb. Dublin Philos. Mag. J. Sci.* 45: 823–843. https://doi.org/10.1080/14786440808520496.

66 Chen, X.-Q., Niu, H., Li, D., and Li, Y. (2011). Modeling hardness of polycrystalline materials and bulk metallic glasses. *Intermetallics* 19: 1275–1281. https://doi.org/10.1016/j.intermet.2011.03.026.

67 Davies, G.F. (1974). Effective elastic moduli under hydrostatic stress – I. quasi-harmonic theory. *J. Phys. Chem. Solids* 35: 1513–1520.

68 Ma, D., Grabowski, B., Körmann, F. et al. (2015). Ab initio thermodynamics of the CoCrFeMnNi high entropy alloy: Importance of entropy contributions beyond the configurational one. *Acta Mater.* 100: 90–97. https://doi.org/10.1016/j.actamat.2015.08.050.

69 Niu, C., Zaddach, A.J., Koch, C.C., and Irving, D.L. (2016). First principles exploration of near-equiatomic NiFeCrCo high entropy alloys. *J. Alloys Compd.* 672: 510–520. https://doi.org/10.1016/j.jallcom.2016.02.108.

70 Zheng, S., Feng, W., and Wang, S. (2018). Elastic properties of high entropy alloys by MaxEnt approach. *Comput. Mater. Sci.* 142: 332–337. https://doi.org/10.1016/j.commatsci.2017.09.060.

71 Huang, S., Li, W., Lu, S. et al. (2015). Temperature dependent stacking fault energy of FeCrCoNiMn high entropy alloy. *Scr. Mater.* 108: 44–47. https://doi.org/10.1016/j.scriptamat.2015.05.041.

72 Zaddach, A.J., Niu, C., Koch, C.C., and Irving, D.L. (2013). Mechanical properties and stacking fault energies of NiFeCrCoMn high-entropy alloy. *JOM* 65: 1780–1789. https://doi.org/10.1007/s11837-013-0771-4.

73 Zhao, S., Stocks, G.M., and Zhang, Y. (2017). Stacking fault energies of face-centered cubic concentrated solid solution alloys. *Acta Mater.* 134: 334–345. https://doi.org/10.1016/j.actamat.2017.05.001.

74 Clarke, D.R. and Levi, C.G. (2003). Materials design for the next generation thermal barrier coatings. *Annu. Rev. Mater. Res.* 33: 383–417. https://doi.org/10.1146/annurev.matsci.33.011403.113718.

75 Lim, M., Zs.Rak, Braun, J.L. et al. (2019). Influence of mass and charge disorder on the phonon thermal conductivity of entropy stabilized oxides determined by molecular dynamics simulations. *J. Appl. Phys.* 125: 055105. https://doi.org/10.1063/1.5080419.

76 Varvenne, C., Leyson, G.P.M., Ghazisaeidi, M., and Curtin, W.A. (2017). Solute strengthening in random alloys. *Acta Mater.* 124: 660–683. https://doi.org/10.1016/j.actamat.2016.09.046.

77 Chen, B., Li, S., Zong, H. et al. (2020). Unusual activated processes controlling dislocation motion in body-centered-cubic high-entropy alloys. *Proc. Natl. Acad. Sci. U.S.A.* 117: 16199–16206. https://doi.org/10.1073/pnas.1919136117.

78 Yin, S., Zuo, Y., Abu-Odeh, A. et al. (2021). Atomistic simulations of dislocation mobility in refractory high-entropy alloys and the effect of chemical short-range order. *Nat. Commun.* 12: 4873. https://doi.org/10.1038/s41467-021-25134-0.

79 Choi, W.-M., Jo, Y.H., Sohn, S.S. et al. (2018). Understanding the physical metallurgy of the CoCrFeMnNi high-entropy alloy: an atomistic simulation study. *npj Comput. Mater.* 4: 1. https://doi.org/10.1038/s41524-017-0060-9.

80 Rao, S.I., Varvenne, C., Woodward, C. et al. (2017). Atomistic simulations of dislocations in a model BCC multicomponent concentrated solid solution alloy. *Acta Mater.* 125: 311–320. https://doi.org/10.1016/j.actamat.2016.12.011.

81 Rao, S.I., Woodward, C., Parthasarathy, T.A., and Senkov, O. (2017). Atomistic simulations of dislocation behavior in a model FCC multicomponent concentrated solid solution alloy. *Acta Mater.* 134: 188–194. https://doi.org/10.1016/j.actamat.2017.05.071.

**82** Li, X.-G., Chen, C., Zheng, H. et al. (2020). Complex strengthening mechanisms in the NbMoTaW multi-principal element alloy. *npj Comput. Mater.* 6: 70. https://doi.org/10.1038/s41524-020-0339-0.

**83** Feng, R., Zhang, C., Gao, M.C. et al. (2021). High-throughput design of high-performance lightweight high-entropy alloys. *Nat. Commun.* 12: 4329. https://doi.org/10.1038/s41467-021-24523-9.

# 4

# Synthesis and Processing of High-Entropy Materials

From practical point of view, the synthesis and processing routes of materials should be selected based on the phase stability, product shape, product size, desirable property and application, consumption, and complexity of the process. For high-entropy alloys (HEAs) and high-entropy ceramics (HECs), the choice of the synthesis methods is largely different due to their distinguished distinction in properties. For most alloys, they are easily fused, intrinsically plastic, and vulnerable to oxidation, while for most ceramics, they are refractory, brittle, and resistant to oxidation (especially for oxides). Thus, properly choosing synthesis methods for different forms of high entropy materials (HEMs) is important. In this chapter, the synthesis and processing routes for different forms of HEMs will be comprehensively reviewed.

## 4.1 Powders

### 4.1.1 Powders of HEAs

#### 4.1.1.1 Mechanical Alloying

There are limited methods for powder synthesis of HEAs since it is more sophisticated to produce bulk alloys from their melts. Most of the HEA powders are synthesized by mechanical alloying (MA). MA is normally a dry, high-energy ball milling (HEBM) technique and has been employed to produce a variety of materials such as intermetallics, ordered compounds, solid solution alloys, amorphous structures, quasi-crystalline phases, and nanocomposites, as demonstrated in Figure 4.1 [1]. MA process is described by the repeated welding and fracturing of powder particles entrapped between milling media, the extent of which depends on the mechanical attributes of powder constituents [1]. MA was first used to synthesize nanocrystalline AlFeTiCrZnCu alloy by Varalakshmi et al. [2]. Al, Fe, Ti, Cr, Zn, and Cu powders with purity higher than 99.5% and particle size of ≤45 μm (~325 mesh) were mechanically alloyed from binary Al–Fe to hexanary Al–Fe–Ti–Cr–Zn–Cu systems in an equiatomic ratio in a high energy planetary ball mill. They found that, in all compositions, the nanostructured solid solutions crystallized in the body-centered cubic (BCC) structure as determined by X-ray diffraction. The HEA prepared by MA

*High-Entropy Materials: From Basics to Applications,* First Edition.
Huimin Xiang, Fu-Zhi Dai, and Yanchun Zhou.

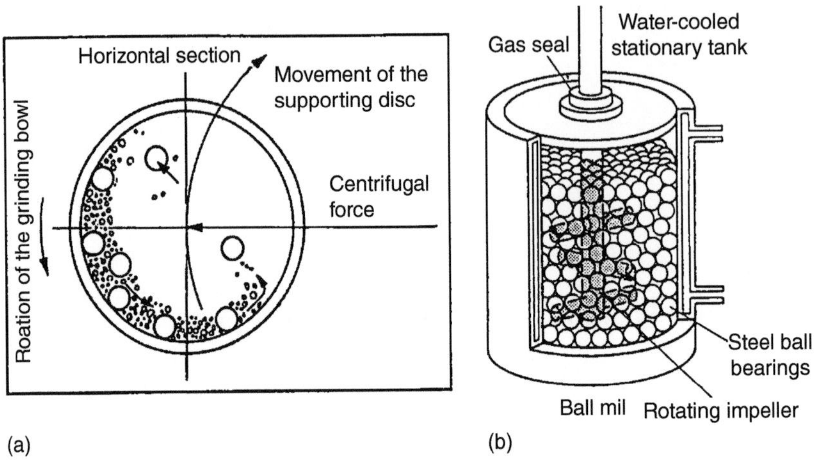


**Figure 4.1** Schematic depicting (a) the ball motion inside the ball mill and (b) arrangement of the rotating arms in the attrition ball mill. Source: Reprinted with permission from Suryanarayana. Source: Reproduced with permission from Suryanarayana [1]/Elsevier.

had a crystallite size of about 10 nm and was stable after treatment at 800 °C for one hour. Similar results have also been obtained in their work on the Cu–Ni–Co–Zn–Al–Ti system, where single nano-crystallized solid solution with BCC structure was prepared by MA [3]. A slight difference was that the heat-treated equiatomic CuNiCoZnAlTi alloy at 800 °C consisted of a BCC phase and two minor face-centered cubic (FCC) phases, indicating the metastable nature of MA powders of HEAs. Shortly, Chen et al. extended the system to Cu–Ni–Al–Co–Cr–Fe–Ti–Mo octonary alloys [4]. They found that binary and ternary alloys formed FCC or BCC phases, respectively, and remain crystalline even after long-time milling. While for quaternary and alloys with more components, the FCC solution phase was first formed and then transformed into the amorphous phase after prolonged milling. The wider range of atomic size difference provided by more constituent elements was believed to be responsible for the formation of amorphous structure. Further, they determined the alloying sequence for these eight elements to be Al → Cu → Co → Ni → Fe → Ti → Cr → Mo in the order of decreasing alloying rate [5], and this order correlated best with the melting point among metallurgical factors. This empirical relationship between the alloying rate and the melting point is useful in the prediction of alloying competition and phase evolution for a wide range of alloys during solid-state alloying.

In the process of MA, several process variables matter in determining the product phase and/or microstructure, such as type of mill, milling speed, milling time, grinding medium, and ball-to-powder weight ratio [6]. In the preparation of HEAs, planetary mills are the most common choice. With regard to the milling speed, Salemi et al. have investigated the effects of ball diameter and rotation speed on the phase formation of CuNiCoZnAl alloy [7]. They kept the ball-to-powder weight ratio to 10 : 1. They found that the alloying was not complete even after 50 hours of milling when the ball diameter was 20 mm solely and the rotation speed was 300 rpm, while single-phase FCC CuNiCoZnAl alloy was obtained when the milling speed increased to 350 rpm and balls with varying diameter (10, 16, and 20 mm) were used. The total energy transferred to the powders during ball milling was estimated to be 1.7 times larger in the second route and hence accounts for the solid solution formation. Tan et al. synthesized the powders of $Al_2NbTi_3V_2Zr$ alloy by MA in low-energy (150 rpm) and high-energy (350 rpm) modes, respectively [8]. They found that HEBM not only reduced the particle size (also observed in the low-energy ball milling process) but also resulted in dissolution of Al into other elements or lattices, which was beneficial for alloying and achieving nano-crystallinity of the alloy system.

For the synthesis of HEAs through MA, sufficient milling time is needed. In most reports, the solid solution phases in HEAs form within 15–40 hours [6]. Prolonged milling time often leads to the formation of an amorphous solid solution. As we have stated above, the quaternary and alloys with more components in the Cu–Ni–Al–Co–Cr–Fe–Ti–Mo system transformed into an amorphous phase after 36 hours ball milling [4]. For FeSiBAlNi HEA, fully amorphous powders were obtained after 240 hours of milling [9]. It seems that the powders are more homogenous with extended milling time; however, it also increases the likely contamination induced during milling. Thus, rational control of milling time is vital to the formation of phase-pure, nano-crystallized HEA powders. Another interesting phenomenon regarding the phase formation during MA in HEAs is the addition sequence of constituent elements. The conventional way to fabricate HEAs is to mix the constituent elements in a single step in required proportion. Vaidya et al. found that the phase formation in HEAs was governed by inherent thermodynamic and kinetic factors of the system [10]. In the formation of nanocrystalline AlCoCrFeNi by MA, they proposed 13 different addition sequences of constituent elements. The choice of the initial binary and the order of mixing of other constituents affect the phase evolution during MA. For example, sequence AlNiCoFeCr showed a single-phase BCC structure, while FeNiCoCrAl had the maximum amount (25%) of FCC phase present.

MA is proven to be an efficient and convenient method to produce nanocrystalline, mechanically activated HEA powders with exciting properties and promising applications. However, several drawbacks still exist. The first drawback is the contamination from milling media or oxidation of the milled materials. Second, the milled powders are rarely used directly and often required to be either consolidated into bulk form or sprayed as coatings. Considering the metastable nature of mechanical-alloyed powders, suppression of the structure, and phase transformation during the treatment requires delicate selection of the subsequent processing methods and parameters [6]. Nevertheless, MA is still the main choice when HEA powders are required.

#### 4.1.1.2 Atomization

Atomization has been established as a promising technology for the fabrication of spherical metal powders. Two typical methods, i.e. gas atomization and water atomization, have been used to prepare HEA powders. To manufacture the HEA AlCoCrFeNi alloy in the form of a thin-walled element on a stainless-steel substrate using the laser engineered net shaping (LENS) method, Kunce et al. employed inert-gas atomization to produce high-entropy particles from the cast alloy [11]. The particles of the powder were spherical or nearly spherical and had relatively smooth surfaces. A few aggregates of the powder particles, called "satellite particles," formed because of the turbulent flow in the atomization chamber and owing to the collisions of the solidifying droplets [11]. The same technique has also been used in the preparation of other HEA powders such as CrMnFeCoNi [12], $AlCoCrFeNi_{2.1}$ [13], AlCoCrFeNiCu [14], and $Al_{0.5}CoCrFeNi$ [15].

The morphologies and phase selections of atomized HEA powders are controlled by several process parameters. In the preparation of $Al_{0.6}CoCrFeNi$ HEA powder, Zhou et al. discovered that gas-atomized powders with size larger than 75 μm consisted of FCC + BCC structure, while those with size smaller than 75 μm crystallized in BCC structure only [16]. The high cooling rates in small particles was responsible for the phase change by simultaneously inhibiting eutectic solidification and increasing the solubility of Al in CoCrFeNi. They also observed smoothening of powder surface and variation in grain structure as the particle size decreased. Yang et al. found that the addition of Si caused the phase transformation of $Al_{0.5}CoCrCuFeNiSi$ powders from FCC to BCC structure [17]. Due to the high cooling rate of the gas atomization process, the intermetallic impurities in the as-casted bulk materials disappeared in the gas-atomized powders.

#### 4.1.1.3 Wet Chemistry

Wet chemical synthesis is a newly developed method to prepare nanoparticles of HEAs for their catalysis applications. Singh and Srivastava used the chlorides of Cr, Co, Fe, Ni, and Cu as precursors to synthesize the NiFeCrCuCo nanoparticles by a wet chemical synthesis-based method [18]. The synthesis condition was mild, and the as-synthesized nanoparticles were spherical with an average size of 26.7 ± 3.3 nm. Their work clearly demonstrated the possibility of the use of the wet chemical method in the synthesis of HEA nanoparticles. Soon, Niu et al. demonstrated a two-step sol–gel autocombustion synthesis of CoCrCuNiAl HEA nanoparticles [19]. They described the processing routes as different metal oxides form complex multi-component oxides, followed by *in situ* reduction to HEAs by the reducing gases. The produced solid solutions with an FCC and BCC structure and an average grain size of 14 nm exhibited a distinctive magnetic behavior similar to a superparamagnetic behavior. Bondesgaard et al. employed a solvothermal synthesis method to produce high-entropy noble-metal alloys PtPdRhRu and PtPdIrRhRu [20]. They carefully chose the acetone–ethanol system with the ratio of 50 : 50 as the optimal solvent, and the metal acetylacetonates were dissolved in this optimal solvent, where metal acetylacetonates were dissolved in. They found that the average crystallite size of the synthesized nanoparticles depended systematically on the enthalpy of fusion/melting point of the constituent metals, which could allow prediction of crystallite sizes of alloys with varying stoichiometries. Yang et al. used a one-step aerosol spray pyrolysis technique to produce FeCoNiCuPt, FeCoNiPdPt, and FeCoNiCuPd HEA nanoparticles [21]. In their setup, the precursor solution containing an equimolar amount of metal nitrates was nebulized into small aerosol droplets using a 10% $H_2$/90% Ar mixture gas with a pressure of 30 psi. After fast heating in a high-temperature tubular furnace reactor, the metal salts were rapidly decomposed and reduced to metal nanoparticles. The use of wet chemical methods to synthesize HEA powders is still in its nascent state and has not been utilized extensively. Further investigations are still needed to introduce more sophisticated and practical wet chemical routes.

#### 4.1.1.4 Hydrogenation–Dehydrogenation

For brittle materials, grinding is always a choice to fabricate powders. However, the HEA ingot is not brittle enough to be crushed. To make HEAs brittle, Lee et al. employed hydrogen embrittlement (HE) [22]. Hydrogen embrittlement is a good approach to obtain the required brittleness, which is simply achieved by exposure of metals to hydrogen atmosphere at an adequate temperature, resulting in the hydrogen-triggered ductile-to-brittle phenomenon [23]. They prepared the ingot of VNbMoTaW HEA first and exposed the ingot to a 100% hydrogen atmosphere at 600 °C for one hour. The hydrogen-annealed ingot was crushed into particles and ball-milled for five hours at the rotation speed of 200 rpm. Lastly, to eliminate hydrogen, they annealed the powder at 600 °C for one hour in a high-vacuum (~$10^{-5}$ Torr) atmosphere. Afterward, the powder was spheroidized using inductively coupled thermal plasma. The resulting powder showed a spherical shape and its size was revealed to be 15.0, 28.0, and 43.8 μm in terms of $d_{10}$, $d_{50}$, and $d_{90}$. Using the same method, Park et al. prepared TaNbHfZrTi HEA powders where powder sizes for $d_{10}$, $d_{50}$, and $d_{90}$ were measured to be 7.4, 18.1, and 31.6 μm, respectively [24]. Since the reaction between hydrogen and metal is common, this hydrogenation–dehydrogenation (HDH) reaction is expected to be applicable to various compositions of other HEAs.

### 4.1.2 Powders of HECs

#### 4.1.2.1 Mechanical Alloy

In the synthesis of HECs, MA is not an ideal choice, since there are plenty of other methods capable of preparing ultrafine and homogenous ceramic powders. The main usage of milling in the processing of ceramic powders is the ball milling in the homogenization of raw materials. Nevertheless, MA is an alternative in HEC powder preparation. Moskovskikh et al. used Hf, Ta, Ti, Nb, Zr, and Mo powders and graphite particles as precursors to fabricate (HfTaTiNbZr)C and (HfTaTiNbMo)C high-entropy carbides [25]. The single-phase solid solutions were obtained at room temperature after 60 minutes of milling. The particle size of the produced powders distributed in the range of 0.1–20 μm. After sintering at 2000 °C, fine grain size was preserved. Similarly, six five-components high-entropy carbides were synthesized by Chicardi et al. using transition metals from the IVB (Ti, Zr, Hf), VB (V, Nb, Ta) groups and graphite powder as raw materials [26]. In their work, the minimum milling time was 50 hours to ensure the complete formation of the expected HECs. The crystalline domain size of the final products was several nanometers with slight Fe contamination. They have also investigated the effect of ball ratio on the synthesis of high-entropy carbide [27]. Two kinds of balls (10 and 15 mm in diameter) were used in their work. They found that low energetic milling conditions (10 mm diameter ball used) was crucial to the formation of HECs by simultaneous metal alloying and carbon diffusion into the alloy structure. To avoid the contamination, Guan et al. employed silicon nitride vials and grinding balls in the synthesis of high-entropy metal boron carbonitride ceramic powders [28]. They adopted a two-step process in MA, where the HEC powders were obtained by ball milling the mixture of metal powders and pre-milled amorphous BCN by MA the mixture of graphite and h-BN. The nanosized HEC powders with FCC rock-salt structure were obtained after 24 hours of HEBM. Good thermal stability was observed after annealing the as-synthesized metal boron carbonitride at different temperatures.

Another variation of MA is also utilized, i.e. mechanochemical synthesis. It can be defined as chemical reaction that is induced by the direct absorption of mechanical energy, where shearing, stretching, and grinding are typical methods for the mechano-chemical generation of reactive sites in polymer chains that undergo mechanochemical reactions [29]. It has attracted increasing attention since it can be used to react solids quickly, quantitatively, and with little-to-no solvent use. Jin et al. used metal chlorides and urea for synthesizing high-entropy transition metal nitride (VCrNbMoZr)N; the process is demonstrated in Figure 4.2 [30]. The solvent-free mixture was added to a commercially available zirconia vial reactor along with three zirconia balls, and the vial was placed in a high-speed vibrating ball miller for 30 minutes. The resulted gel-like product was amorphous and subsequently crystallized into rock salt structure after pyrolyzing at high temperatures under an inert $N_2$ atmosphere. Chen et al. utilized the metastable feature of materials being milled to

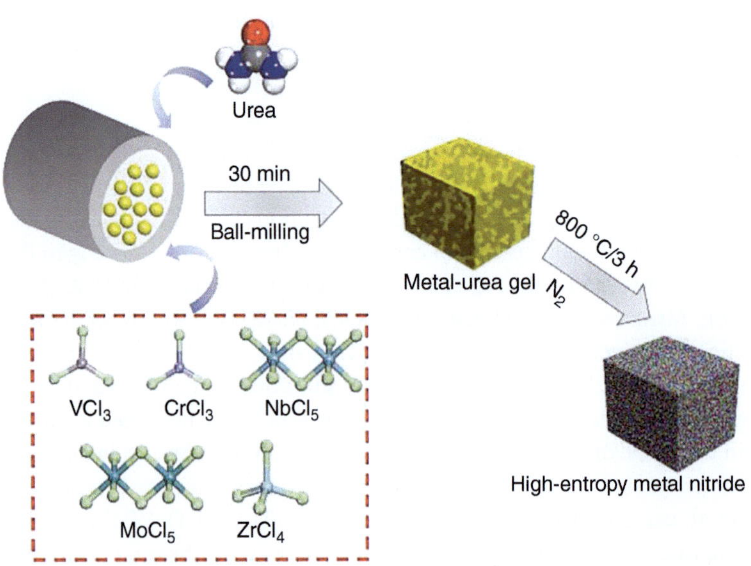


**Figure 4.2** Schematic illustration of the two-step synthetic strategy for high-entropy metal nitride. Source: Reproduced with permission from Jin et al. [30]/John Wiley & Sons.

bear catalytically active Pt/Ru in high-entropy oxide (NiMgCuZnCo)O without post-heat treatment [31]. In their work, an equimolar ratio of NiO, MgO, CuO, ZnO, and CoO was mixed with $PtO_2$ or $RuO_2$ with desired loading. Mixed powders were then put into a stainless-steel milling pot and ball-milled with three steel balls (10 mm diameter) for two hours at 30 Hz. They observed remarkable thermal stability of milled Pt/Ru-(NiMgCuZnCo)O powder even after annealing at 700 °C. The ability of mechanochemical synthesis to convert a high-temperature process into a readily scalable solid-state low-temperature synthesis and retain the microstructure of high surface area is fascinating and appealing for the development of novel catalysts.

Although MA is efficient in producing nano-sized, high-sintering active HEC powders under mild conditions, the effect of process parameters on the quality of final products is far from being fully understood.

#### 4.1.2.2 Wet Chemistry

The wet chemical methods are common routes in the preparation of fine and controllable ceramic powders, such as co-precipitation, sol–gel method, spray pyrolysis, and hydrothermal. Several processes will be reviewed in the following text. The co-precipitation technique is one of the most used wet chemical methods in modern chemical synthesis. The aim in co-precipitation is to prepare multicomponent materials through the formation of intermediate precipitates, usually hydrous oxides or oxalates, so that an intimate mixture of components is formed during precipitation and chemical homogeneity is maintained on calcination. In the typical process of co-precipitation, aqueous metal salts are mixed at sufficient temperatures with a base, which acts as a precipitating agent. After filtering, the precipitate is dried, lightly ground in an agate mortar, and finally calcined at a high temperature to obtain the desired crystalline oxide phase. Sarkar et al. used nitrates as precursors to produce (Co, Cu, Mg, Ni, Zn)O and (Co, Mg, Ni, Zn)O powders [32]. Ammonia solution was added to maintain the pH level of the mixed solution at 10. After calcination at 1000 °C for one hour, powders with particle size ranging from 50 to 150 nm were obtained. The pH, rate of addition of precursors' solution into the base, and temperature of calcination are the factors affecting the crystal structure, morphology, and crystallite size of the powder synthesized [32]. A similar process was utilized by Biesuz et al. in the synthesis of (CoCuMgNiZn)O powders except that potassium carbonate dissolved in deionized water was used as the base solution [33]. More complex oxides such as $(LaCeNdSmEu)_2Zr_2O_7$ [34], $(NdSmEuYYb)_4Al_2O_9$ [35], and $(LaCeNdSmEu)PO_4$ [36] were also successfully synthesized with ammonia solution and $H_3PO_4$ as precipitants, respectively. The co-precipitation technique is a perfect method to achieve chemical homogeneity in the powder, and the crystal structure, morphology, and crystallite size are easily adjustable. However, post-treatment at high temperature always leads to the coarsening and agglomerate of the particles.

Spray pyrolysis is a technique used to synthesize a wide variety of high-purity, homogeneous ceramic powders with very uniform size and spherical morphology [37]. In the pursuit of HECs powders, nebulized spray pyrolysis (NSP) and flame spray pyrolysis (FSP) are the most commonly used pyrolysis methods. A scheme of NSP and FSP is shown in Figure 4.3. In NSP, the precursor solution is nebulized in a chamber using an ultrasonic generator and carried by flowing gas into the hot zone of a tubular furnace. At elevated temperature, the precursor solution transforms into the desired crystalline oxide. Sarkar et al. used NSP to produce (Co, Cu, Mg, Ni, Zn)O and (Co, Mg, Ni, Zn)O powders [32]. They found that at lower temperature of 1150 °C, the 5-cationic system was single-phase, while the 4-cationic system was multi-phase with the presence of wurtzite phase (2.51 wt%). They believed that relatively lower configurational entropy in the 4-cationic system was responsible for the phase composition difference since an increase in synthesis temperature from 1150 to 1250 °C led to the formation of single-phase rock-salt structure even in the 4-cationic system. Due to the long particle residence time in the reactor (around 0.1 seconds), the obtained powder was homogeneous in structure but with large crystallite sizes of around 40 nm. The same procedure is also used in their preparation of high entropy perovskite type oxides [39], Li-containing spinel-type oxides [40], and high-entropy rare-earth oxides [41]. For FSP, the atomized precursor solution is sprayed into a flame zone where the metal oxide is produced by combustion. Sarkar et al. used liquefied petroleum gas (LPG) and oxygen as the fuel and oxidizer, respectively, to maintain the flame [32]. The residence time of the particles formed in the flame was very short (typically around 0.01 seconds [32]). Asymmetric X-ray diffraction (XRD) peaks were observed in the powders produced using the FSP method, which could be attributed to the convolution of peaks from two rock-salt phases, resulting from the uneven temperature field in the flame, with lattice parameters close to each other and similar compositions. Beneficial from the short residence time, the crystallite size was smaller than those from NSP. For both methods, the quality of the final powders is dependent on the concentration of

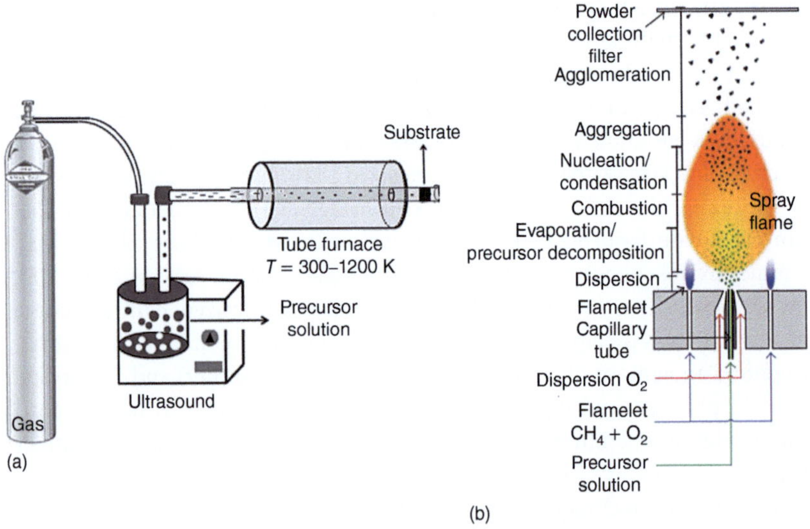


**Figure 4.3** A schematic representation of (a) nebulized spray pyrolysis (NSP) and (b) flame spray pyrolysis (FSP). Source: Reproduced with permission from Nair and Dhoble [37]/ Elsevier; (b) Karhunen et al. [38]/Hindawi/CC BY-3.0.

reactants in the precursor solution, nature of additives, flow rate of droplets, and other preparatory conditions [37], which have not yet been investigated. Nevertheless, the spray pyrolysis technique is still a low-cost and efficient method for obtaining ultrafine powders with small particle size (<1 μm), narrow size distribution, high purity, and large surface area.

Polymer-derived synthesis of HEC powders is also appealing for its ability to achieve chemical homogeneousness. The basic idea is to create a polymeric steric matrix where all cation atoms distribute homogeneously. To do so, Tseng et al. used polyvinyl alcohol solution as polymerization agent to entrap the five cations in a homogeneous, amorphous organic–inorganic structure [42]. After calcination at 1149 °C, the high-entropy, lanthanide sesquioxide became a single phase solid solution with the bixbyite structure. Yoon et al. employed the same scheme [43]. They mixed stoichiometric amounts of metal nitrates and chloride with a citric acid solution to chelate the metal atoms. Ethylene glycol (EG) was then added to this mixture to promote polymerization. After annealing this precursor at 550 °C for five hours, the organic compounds were removed, and multi-phase powders were obtained. This technique can also be used to prepare non-oxide powders with the addition of a proper carbon source and high temperature treatment. Du et al. mixed the commercially available metal chloride ($HfCl_4$, $NbCl_5$, $ZrCl_4$, and $TiCl_4$) with acetyl methane and butanol to yield a homogeneous solution under an argon atmosphere [44]. By annealing the polymerized precursor to 2200 °C for one hour, phase-pure high entropy carbide powders with the average particle size of ~800 nm were obtained. Sun et al. used the commercially available yttrium(III) acetylacetonate, lanthanum(III) acetylacetonate hydrate, titanium *n*-propoxide, zirconium *n*-propoxide, and homemade metal alkoxide from the corresponding metal chloride, and *n*-propanol as metal-containing monomers [45]. Acetylacetone and *n*-propanol were used as solvents. Controlled co-hydrolysis and subsequent polycondensation reaction were carried out by adding deionized water drop wise, followed by distillation to generate a clear reddish-brown viscous solution. The allyl-functional novolac (AN) resin was mixed as the carbon source. Single-phase solid solution with uniform dispersed elements could be achieved at temperature as low as 1800 °C. Due to the existence of oxygen atoms in the precursors, it seems inevitable to detect the oxygen from the high entropy carbide powders in both works (0.78 at.% in Du's work [44], and 1.2 wt% in Sun's work [45]).

Some other wet chemical methods are also employed in the synthesis of HEC powders. Mao et al. chose the modified solution combustion synthesis (SCS) method to prepare (CoCuMgNiZn)O high-entropy oxide (HEO) nano-crystalline powder [46]. In the precursors, the metal nitrates were used as oxidizers and glycine as a fuel. They found that full conversion to single-phase rock-salt structure occurred only at the synthesis temperatures higher than 1123 K. Taking the advantage of the acoustic cavitation phenomenon in the ultrasonication process, single-phase $BaSr(ZrHfTi)O_3$, $BaSrBi(ZrHfTiFe)O_3$, and $Ru/BaSrBi(ZrHfTiFe)O_3$ nanoparticles with an average crystallite size of approximately 5.9 nm were crystallized through ultrasonication exposure without calcinations [47]. With the excellent dispersion of Ru in the perovskite phase, Ru bearing high-entropy nanoparticles exhibited superior catalytic activity over the conventional perovskite counterparts ($BaRuO_3$). Dong et al. synthesized $(YbYLuScGd)_2Si_2O_7$ powders using a sol–gel method [48]. In this process, tetraethoxysilane (TEOS) was dissolved in distilled water and ethanol. Afterward, the TEOS solution was stirred with the addition of $RE(NO_3)_3$ solution until the solution was fully hydrolyzed. Subsequently, the solution was treated at 70 °C for 24 hours and 120 °C for 24 hours to obtain the dry gel. After calcining at 800 °C for two hours and 1500 °C for two hours in air, submicron β-type $(YbYLuScGd)_2Si_2O_7$ powders were obtained. The solvothermal synthesis route is a crystallization process directly from organic solutions, where ultrafine nanoparticles could be produced. Wang et al. loaded five transition-metal acetates and hexamethylenetetramine into the mixed solution of polyethylene oxide–polypropylene oxide–polyethylene oxide ($PEO_{20}$–$PPO_{70}$–$PEO_{20}$, Pluronic P123) surfactant, ethylene glycol (EG) co-surfactant, and ethanol to make a homogeneous solution [49]. The solution was transferred into a 100 ml stainless steel Teflon-lined autoclave and heated at 170 °C for 15 hours. After cooled to room temperature, the HEO nanoparticles were obtained by calcinating the precursor at 400 °C for two hours without any post-treatment. Ultrafine particle size (5 nm) was observed due to the suppressing of particle size and spontaneous agglomeration from surfactant and co-surfactant.

#### 4.1.2.3 Solid-State Reaction

So far, the wet chemical methods are mostly used to synthesize high-entropy oxide powders. Solid-state reaction synthesis is also a commonly chosen method for the preparation of HECs, which is suitable not only for oxides but also for silicates, borides, carbides, silicide carbides, and sulfides. Rost et al. used the solid-state synthesis route to prepare the first entropy-stabilized (Mg, Ni, Co, Cu, Zn)O ceramic [50]. Liu et al. used boro/carbothermal reduction (BCTR) to produce $(HfTaNbTi)B_2$ powders from commercially available $HfO_2$, $Ta_2O_5$, $Nb_2O_5$, $TiO_2$, and $B_4C$ powders [51]. Chen et al. [52] synthesized high-entropy rare-earth silicide carbides $(TmYPrGdDy)_3Si_2C_2$ and $(TmYPrGdTb)_3Si_2C_2$ powders at 1600 °C with dwell time of one hour in Ar using rare-earth elements and SiC powders as the starting materials. Although the concept of solid-state reaction is simple, several processing parameters matter in powder preparation, i.e. raw materials, ratio of raw materials, heat treatment temperature, etc.

Gild et al. synthesized seven high-entropy metal diborides using HEBM of individual commercial diboride powders followed by spark plasma sintering (SPS) at 2000 °C [53]. They observed that a single, high-entropy phase of the hexagonal $AlB_2$ structure formed after SPS at 2000 °C with the presence of minor secondary (Zr, Hf) $O_2$ phases, which came from the intrinsic or extrinsic (from HEBM) surface oxides. Liu et al. used commercially available $HfO_2$, $Nb_2O_5$, $ZrO_2$, $TiO_2$, $Ta_2O_5$, and amorphous B powders to synthesize high-entropy metal diboride $(HfZrTaNbTi)B_2$ powders by borothermal reduction (BTR) at 1700 °C for one hour in flowing argon [54]. The as-synthesized powders exhibited sphere-like microstructure with an average particle size of ~310 nm. An amorphous $B_2O_3$ layer of 2–3 nm can be observed on the surface of the as-synthesized powders. Similarly, Zhang et al. used BTR to produce $(HfZrTaCrTi)B_2$, $(HfMoZrNbTi)B_2$, and $(HfMoTaNbTi)B_2$ [55]. The high-entropy borides formed at 1600 °C for one hour under vacuum with some oxide impurities due to incomplete reaction and B loss during heating. Chen et al. used $B_4C$ as a reductor in the synthesis of $(ZrHfNbTaTi)B_2$ HEC using BCTR [56]. By examining the shrinkage behavior of the columnar-shaped green bodies during heating, they recognized that the reaction process for high-entropy $(ZrHfNbTaTi)B_2$ formation started at 870 °C and finished at 1700 °C. The final products showed sphere-like microstructure with small grain size of 400–800 nm. An alternative method has also been proposed by Tallarita et al. to synthesize high-entropy borides: self-propagating high-temperature synthesis (SHS) [57]. They mixed Hf, Mo, Nb, Ta, and Ti metal powders with excess amorphous B powders and activated the mixture at one pellet end using an electrically heated tungsten coil. Upon ignition, the reaction front displayed a self-sustaining character with a measured maximum temperature of about 2000 °C and average velocity of $4.75 \pm 0.25$ mm/s. Minor amounts of other diborides ($(Ta_{0.5}Ti_{0.5})B_2$, $(Hf_{0.5}Ti_{0.5})B_2$, and $HfB_2$) and traces of $HfO_2$ had been detected and could be eliminated after further heat treatment. These reports reveal that using different raw materials leads to different synthesis routes, and the thermodynamical barrier needs to overcome is also different. In these four synthesis routes for high entropy borides using solid state reaction, the minimum temperature for BTR/BCTR to go to completion seems to be the lowest, due to their more negative standard reaction Gibbs free energy than reaction of individual metal diborides [54]. The SHS is fast, but hard to achieve completely homogeneous products. Thus, post-treatment is needed to complete the solid solution formation. The same phenomenon has also been observed in the synthesis of high-entropy carbides [58–60].

Another interesting phenomenon regarding the effect of raw materials on the solid-state reaction of HEC powders is the introduction of nonstoichiometric compound resulting in a dramatic decrease in synthesis temperature. Peng et al. reported that by replacing the stoichiometric TiC and TiN in the raw materials with non-stoichiometric $TiC_{0.4}$ and $TiN_{0.3}$ [61], single-phase solid solution in ultrahigh temperature carbide systems ($TiN_{0.3}$/VC/NbC/TiC, $TiN_{0.3}$/VC/NbC/TiC/TaC, $TiC_{0.4}$/VC/NbC/TiN/$Mo_2C$/WC, $TiN_{0.3}$/VC/NbC/TiC/TaC/$Mo_2C$/WC) could form at temperature as low as 1400–1500 °C, which is much lower than other reported ones, where solid solution formation started at least at 1700 °C and completed by 2000 °C [58, 59]. They attributed this to the carbon vacancy-induced acceleration of mass transfer [62].

The microstructure and properties of final products are also connected to the ratio of raw materials. Slight deviation from equal atomic ratios of raw materials leads to the decrease in configuration entropy, as stated in Chapter 2. By introducing the non-equimolar ratio of cations intentionally, Wright et al. extend the concept of HECs to compositionally complex ceramics (CCCs) [63]. They synthesized $(Hf_{0.284}Zr_{0.284}Ce_{0.284}Y_{0.074}Yb_{0.074})O_{2-\delta}$ medium-entropy ceramics by solid state reaction from commercially purchased nano powders. These medium-entropy nonequimolar ceramics outperformed their equimolar high-entropy counterparts with lower thermal conductivity ($\kappa$) and higher $E/\kappa$ ratios. This has extensively expanded the chemical space for design of desirable properties and microstructures. Another example of effect of stoichiometric ratio of raw materials on the HEC powders by solid state reaction is the synthesis of ultrahigh temperature borides and carbides using BTR/BCTR and carbothermal reaction. Liu et al. demonstrated the synthesis of (HfTaNbTi)$B_2$ using $HfO_2$, $TiO_2$, $Nb_2O_5$, $Ta_2O_5$, $B_4C$, and carbon black as raw materials [51]. When using stoichiometric ratio of raw materials, phase pure solid solution could not be achieved even when the mixture had been treated at 1800 °C. A minor rock-salt structural transition metal carbide coexisted with the target borides. By examining the Gibbs free energy change of all possible reactions, they argued that the direct reaction between metal oxides and $B_4C$ consumed the boron source by releasing $B_2O_3$ from the system. Thus, extra $B_4C$ was needed to compensate the consumption of the boron source. When 40% extra $B_4C$ was added, single-phase (HfTaNbTi)$B_2$ ceramic was successfully synthesized. A similar result was also reported by Gu et al. [64]. They used metal oxides and $B_4C$ to synthesize (ZrHfTaNbTi)$B_2$ ceramic, and the optimized excess amount of $B_4C$ to completely remove $MO_x$ was found to be in the range from 5 to 10 wt%, according to the X-ray diffraction patterns. Very recently, Song et al. investigated the effect of carbon non-stoichiometry on the formation of high-entropy carbides by solid state reaction [65]. They synthesized (ZrHfTaNb)C and (ZrHfTaTi)C powders from metal oxides and carbon black at 2000 °C and found that (ZrHfTaTi)C was dual-phase while (ZrHfTaNb)C exhibited a single-phase rock-salt structure. When the carbon stoichiometry in the raw materials was adjusted to 0.78, i.e. the ratio of carbon to metal was 0.78, single-phase (ZrHfTaTi)C solid solution was formed under the same condition, while the synthesis temperature of (ZrHfTaNb)C could be reduced to 1800 °C. Thus, the rational choice of the raw materials and the stoichiometry of raw materials are critical for a successful and energy-saving solid state reaction of HEC powders.

## 4.2 Dense and Porous Bulks

For practical application of HEMs, dense and porous bulk samples are needed. The main choice on fabricating high quality bulk materials for HEAs and HECs is largely different due to their divergent properties.

### 4.2.1 HEAs

For the preparation of metal and alloys, arc melting and induction melting (IM) are the most common technique, by which all the constituent elements are fully mixed in the liquid state and then solidified in a crucible. Repetitive melting and solidification are often carried out to ensure the chemical homogeneity of the alloys. During the melting, vacuum or an inert environment is always needed which can significantly improve the purity of the alloy. In their very first published works, Yeh et al. and Cantor et al. used arc melting and IM to prepare the HEA ingots, respectively [66, 67]. The as-cast alloy exhibited a dendritic microstructure in both cases, which was typical in the solidified ingots, as shown in Figure 4.4. However, as pointed out by Zhang et al. [68], the solidification process in arc melting could not be easily controlled, which led to a microstructure change in the samples inwards, i.e. inhomogeneous steric distribution of the dendrites.

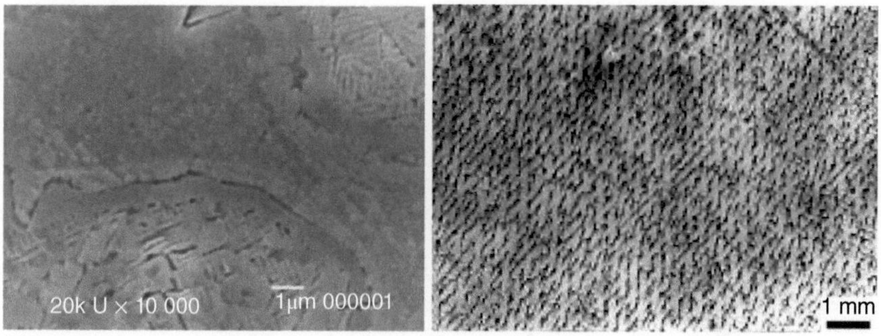


(a) (b)

**Figure 4.4** A typical dendritic microstructure of as-casting (a) CuCoNiCrAlFe and (b) FeCrMnNiCo high-entropy alloy. Source: (a) Reproduced with permission from Cantor et al. [66]/Elsevier; (b) Reproduced with permission from Yeh et al. [67]/John Wiley & Sons.

Three distinct areas could be observed from the surface to the center in the as-cast samples with diameters of 5 mm: the outermost were finery equiaxed grains originating from the large cooling rate, the intermediates were carrotlike columnar dendrites, which grew along a preferred orientation; and the innermost were thick equiaxed dendrites. Besides, as stated in Chapter 2.1.3, simulation and experimental results have demonstrated that chemical short-range order (CSRO) exists in the melts of multicomponent alloys [69, 70]. With the cooling of the melts, this structural feature remains in the as-cast ingot, resulting in element segregation. The situation may be worsened for those HEAs containing elements with variable melting points. Since the elements with high melting points tend to enrich the dendrite region, while those with low melting points distribute more in the interdendritic area [71]. In addition, a series of inevitable as-cast defects, including elemental segregation, suppression of equilibrium phases, microscopic and macroscopic residual stresses, cracks, and porosities, also exist and may have a negative effect on the mechanical properties of HEAs [68, 71]. Thus, other methods to synthesize bulk HEAs have also been introduced.

Bridgman solidification (BS) is introduced by Zhang et al. in the synthesis of bulk HEA [68]. The principle of the Bridgman technique is the directional solidification by translating a melt from the hot zone to the cold zone of the furnace, the cooling rate can be controlled by the withdrawal velocity and temperature gradient of the set. The scheme of BS is shown in Figure 4.5. After BS of AlCoCrFeNi HEA with withdrawal velocities of 200–1800 μm/s, apparent morphology transition from flowery dendrites by arc melting to equiaxed grains was observed [68]. The average grain size was about 100–150 μm, and independent of the withdrawal velocity. In the secondary electron images of the longitudinal sections of the solidified rods by BS, coherent and alternating platelets, with different sizes of ~150–400 nm in length and 100–200 nm in width was observed in the sample solidified at 200 μm/s, while honeycomb-like structure with a uniform dispersion of spherical particles formed in the sample solidified at 1800 μm/s, as shown in Figure 4.6. These nanoscale precipitations might originate from the spinodal decomposition, leading to a modulated structure composed of ordered (B2) and disordered (A2) phases, respectively.

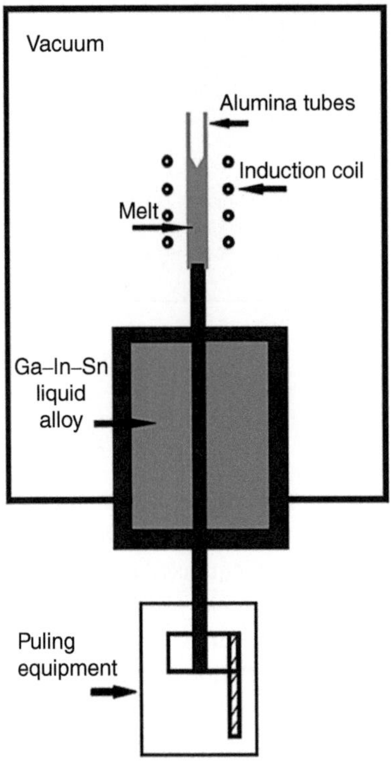


**Figure 4.5** A schematic representation of the Bridgman solidification technique. Source: Reproduced with permission from Hu et al. [72]/Elsevier.

Furthermore, they found that using BS, it was possible to grow single-crystal HEA. By reducing the withdrawal velocity of BS to an extremely slow value (5 μm/s), Ma et al. successfully synthesized FCC-structured single-crystal $CoCrFeNiAl_{0.3}$ HEA for the first time [73]. The single-crystal sample was of the order of 3 mm in diameter and about 50 mm in length, as shown in Figure 4.7. Using the same BS process, only columnar crystal was obtained for the HEA CoCrFeNiAl. The Al element in the alloys was mainly responsible for the different solidification behavior. To verify the single-crystal nature of the product, they conducted measurement on the crystallographic orientation of the single-crystal $CoCrFeNiAl_{0.3}$ HEA by electron backscattered diffraction (EBSD) mapping [74]. The results illustrated that the crystallographic orientation of the single-crystal sample mainly focused on the <001> direction. The misorientation level of the single-crystal product was very low and very low-angle grain boundaries basically within 2° were formed during solidification, indicating the successful synthesis of the single-crystal HEA.

Other than BS, SPS is also utilized to synthesize bulk HEAs from powders. Zhang et al. first combined the MA and SPS in consolidating elemental powders to synthesize CoCrFeNiTiAl HEA bulk material [75]. They found that the sample sintered at 800 °C was different from the powder mechanical milled for 30 hours in phase composition due to the formation of a metastable supersaturated solid solution during the

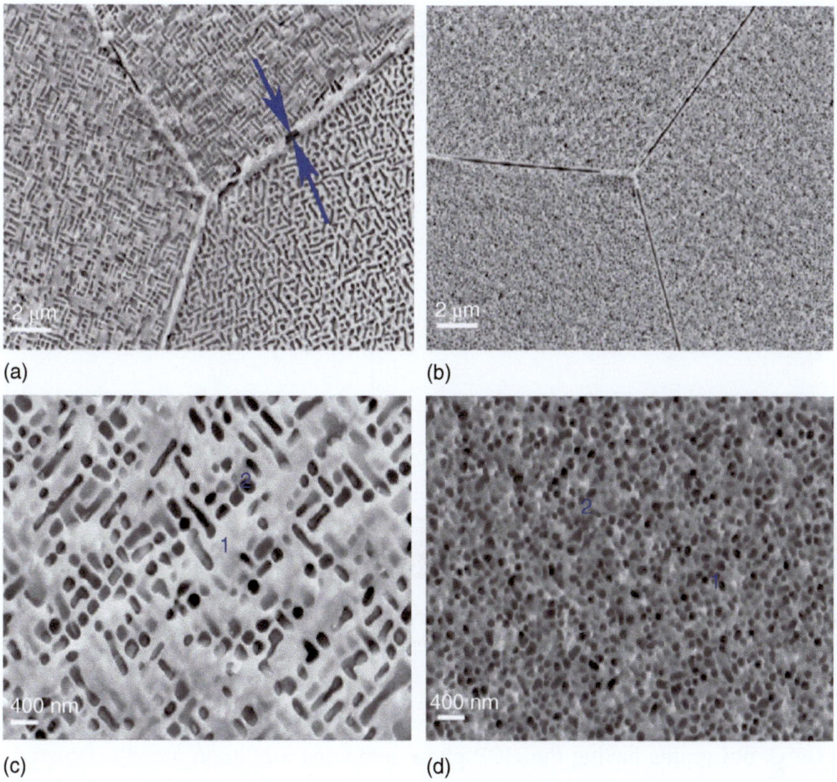


**Figure 4.6** SEM secondary electron images of AlCoCrFeNi HEAs by Bridgman solidification with withdrawal velocities (a) 200 μm/s and (b) 1800 μm/s. (c) and (d) are magnifications of (a) and (b), respectively. Source: Zhang et al. [68], Reproduced with permission from Springer Nature.

MA process. High relative density (98%) indicated the good sinterability of the mechanical alloyed powders. Using the same procedure, AlCoCrCuFe, NiCoCrCuFe, $Co_{0.5}FeNiCrTi_{0.5}$, CoCrFeNiMn, $Co_{25}Ni_{25}Fe_{25}Al_{7.5}Cu_{17.5}$, etc. [76–79] were prepared in bulk form. Although MA is a common choice in powder preparation for SPS, the long process time (usually more than 30 hours) and possible contamination from milling medium and the environment still hinder the large-scale application of this method. To tackle these two problems, Zhang et al. mixed high-purity element powders of Al, Co, Cr, Fe, and Ni in a mechanical mixer without solvent [80]. The mixed powder was directed sintered by SPS. The SPSed HEA consisted of an FCC phase and a duplex BCC structure composed of spinodally modulated disordered BCC and ordered BCC phases. Park et al. employed the hydrogenation–dehydrogenation reaction, as described above, to prepare TaNbHfZrTi HEA powders [24]. After dehydrogenation, the single-phase solid solution transformed into BCC with hexagonal-close packed (HCP) dual-phase. Examining the shrinkage curve demonstrated that the shrinkage started at about 600 °C and was completed at 1000 °C; thus, the sintering temperature of SPS should be above 1000 °C. The HEA powders were fully densified when sintered

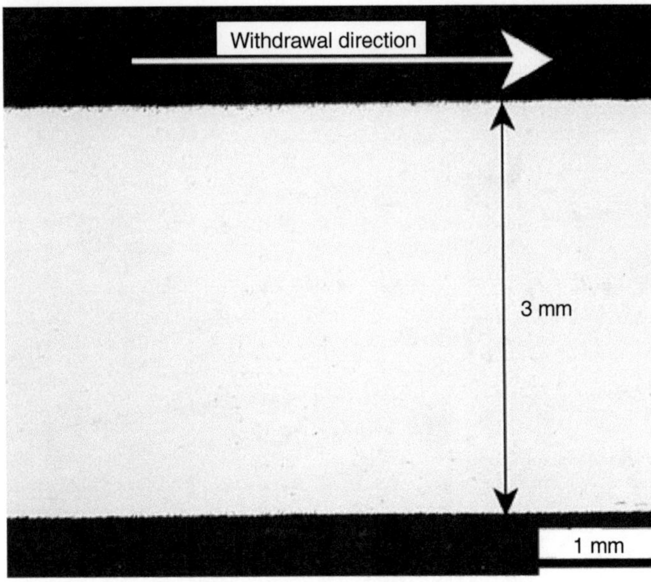


**Figure 4.7** The optical microphotos of single crystal $CoCrFeNiAl_{0.3}$ high-entropy alloy by Bridgman solidification. Source: Ma et al. [73], Reproduced with permission from Springer Nature.

at 1000 °C, but grains having the HCP phase remained and grain growth occurred non-uniformly. On the other hand, in the compact sintered at 1100 °C, all the HCP phase disappeared, and the grain growth occurred uniformly to form an equiaxed microstructure. The grain size of the compact sintered at 1100 °C was 22.5 μm, which was significantly reduced compared to that of the initial ingot (227.8 μm).

Porous structures are good for structural, catalytic, and biomedical applications. Up to now, the most used method for porous HEAs is the dealloying technique. Diverse solutions are used as a dealloying agent to etch metal elements from the matrix to generate pores. Qiu et al. synthesized senary AlNiCuPtPdAu, octonary AlNiCuPtPdAuCoFe, and senary all-non-noble metal AlNiCuMoCoFe nano-porous HEAs by dealloying the precursor alloys $Al_{97.5}Ni_{0.5}Cu_{0.5}Pt_{0.5}Pd_{0.5}Au_{0.5}$, $Al_{97.2}Ni_{0.4}Cu_{0.4}Pt_{0.4}Pd_{0.4}Au_{0.4}Co_{0.4}Fe_{0.4}$, and $Al_{95.5}Ni_{1.5}Cu_{1.5}Mo_{0.5}Co_{0.5}Fe_{0.5}$ in a 0.5 M NaOH solution [81]. The as-prepared nano-porous HEAs had uniformly distributed nano-ligaments which were less than 3 nm and coated with a thin spinel oxide layer and exhibited greatly enhanced high temperature stability (up to 600 °C) and CO oxidation activity. The low diffusion rate of Pt element suppressed the diffusion of Au and Cu elements, promoting the stability of HEAs as well as downsizing the nano-porous structure. Using the same procedure, they also synthesized a series of nano-porous Pt-based HEAs with four predetermining elements (Al, Cu, Ni, and Pt) and one alternative element, including Pd, V, Co, Mn [82], and a series of Ir-based nano-porous HEAs [83]. Uniform and ultrafine nanopores and nano-ligaments with typical sizes of ~2–3 nm were generated by chemical dealloying. Cai et al. expanded this facile synthetic route to the preparation of nano-porous ultra-HEA containing up to

14 elements (Al, Ag, Au, Co, Cu, Fe, Ir, Mo, Ni, Pd, Pt, Rh, Ru, and Ti) [84]. Besides dealloying Al elements in the base solution, etching Cu in the concentrated $HNO_3$ is also employed in the preparation of porous Cu–Au–Pt–Pd HEA [85]. In the above works, the contamination of O in the final products seems inevitable since an aqueous solution is used in the etching process. To tackle this problem, Joo and his coworkers improved the dealloying method by proposing liquid metal dealloying [86]. Metallic Ti, V, Nb, Mo, and Ta cannot be fused with metallic Mg, while metallic Ni melts in metallic Mg. Nano-porous TiVNbMoTa and TaMoNbVNi HEAs were prepared by removing the Ni elements from $(TiVNbMoTa)_{25}Ni_{75}$ and $(TaMoNbV)_{25}Ni_{75}$ precursor alloys in the Mg melt. The average ligament size achieved in a unimodal porous structure synthesized at 873 K was about 10 nm. The ultra-stable nanostructure of nano-porous HEA resulted from the high activation energy for coarsening and the minimized surface diffusivity. Tang et al. utilized the low melting point and easy evaporation of Mg to synthesize porous CoCrFeNiMo by microwave sintering at 950 °C for 20 minutes [87]. The grain size was fine with the average dimension of approximately 300 nm and the average pore diameter was 2.83 μm.

Apart from dealloying, additive manufacturing is also applicable to the fabrication of porous materials. Xu et al. employed binder jetting (BJ) to fabricate porous CoCrFeMnNi HEA [88]. The commercial CoCrFeMnNi HEA alloy powders were mixed with an aqueous binder with 60% binder saturation. The printed samples were cured at 200 °C in air for 12 hours, followed by debinding at 800 °C and sintering in a vacuum furnace. The structure of pores in the sintered samples was controlled by the sintering temperature and duration time. With the increase in temperature and time, the porosity decreased, while the pores openness index showed much weak connection with sintering conditions.

### 4.2.2 HECs

The conventional sintering process using pressureless sintering or a single step process combining synthesis and densification is a natural choice in the sintering of oxides, silicates, and aluminates [89–97] due to its advantage of being able to fabricate near-net-shape and complicated ceramic components. In the initial stage of HECs investigation, the conventional pressureless sintering was employed by Bérardan et al. to synthesize bulk $(MgCoNiCuZn)_{1-x}Li_xO$, $(MgCoNiCuZn)_{1-x}Na_xO$, $(MgCoNiCuZn)_{1-2x}Li_xGa_xO$, and $(MgCoNiCuZn)_{0.95}K_{0.05}O$ samples from binary oxides and carbonates at 1000 °C in air for 12 hours [98]. The geometrical density of the samples was in the range of 75–80%. The quality of the final products of conventional sintering is related to the microstructure of raw materials, sintering temperature and duration time. Biesuz et al. prepared two raw materials for the sintering of (MgCoNiCuZn)O ceramic, that is, co-precipitated and hydrothermal powders [33]. They found that the consolidation process for the co-precipitated powders started at about 400 °C and almost completed at 950 °C, while that for hydrothermal powders was 700 and 1100 °C, respectively. After sintering at 1050 °C for 1.5 hours, dense bulk materials with relative bulk density of 97% and 91% for co-precipitated and hydrothermal powders were obtained, respectively.

Due to strong covalent bonding and low self-diffusion coefficient, pressureless sintering is not an ideal choice to achieve dense ultrahigh temperature HECs including borides and carbides. Nevertheless, attempts on the pressureless sintering of high-entropy carbides have been conducted. Yu et al. used commercially available raw carbide powders and phenolic resin (PR) as raw materials to fabricate bulk (HfZrTaNbTi)C HECs [99]. They compressed different contents of PR into carbides billets using cold isostatically pressing and heated the green bodies to 2200 °C and then soaked for two hours without external pressure. The amount of free carbon from pyrolyzed PR needed delicate controlling to remove the surface oxides of the equimolar mixed raw materials and promote densification, and the optimal content of carbon could be achieved when 4 wt% PR was added. After dwelling at 2200 °C for two hours, HEC sample with relative density of 96.6% was obtained. The mean grain size and mean pore size was 7.0 and 0.9 μm, respectively. To avoid the grain coarsening-induced deterioration of mechanical properties due to high densification temperature, Chen et al. intended to introduce vanadium to effectively inhibit the grain growth and improve the mechanical properties of metallic carbide solid solution, as that in TaC–HfC system [100]. They used metal oxides and carbon black as raw materials to fabricate a series of bulk (TiZrHfVNbTa)C HECs with different vanadium contents by carbothermal reduction reaction at 2300–2500 °C [101]. All samples tended to densify with the increment in sintering temperature and vanadium content. For (TiZrHfNbTa)C, fully dense materials could not achieve even at 2500 °C, while for ceramics with 16.67 at.% vanadium, the relative density of sintered sample was 97.5% when the temperature was 2300 °C. The ability of V in inhibiting rapid grain growth was also notable.

Conventional sintering method is always connected to high sintering temperature, long processing time and undesired density of final products. To achieve high density in a short time and mild temperature, external pressure is needed to promote the densification process. Hot pressing (HP) and SPS are the most widely used methods to densify HECs because of enhanced densification attributed to high pressures and/or high heating rates. HP has already been used in the sintering of transparent fluoride laser ceramics, high-entropy carbides, borides, and nitride-carbides [102–105]. Depending on the material systems, the procedure requires at least 30 minutes at high temperature to achieve satisfactory density. As demonstrated in the sintering of ultrahigh temperature ceramics, the densification process can be promoted by reducing the particle size of the raw materials. Feng et al. used homemade (TiZrHfNbTa)C powders with an average particle size of ~100 nm in the sintering of dense bulk material by HP in the temperatures ranging from 1750 to 1900 °C [106]. They found that the density of HECs using fine powders was relatively higher than those using coarse powders, and the average grain size was considerably small due to the high driving force for densification of fine powders and elimination of oxide impurities. A variant of HP is also applied in the sintering of high-entropy borides, i.e. oscillatory pressure sintering (OPS). Li et al. hot pressed commercial $ZrB_2$, $HfB_2$, $TaB_2$, $NbB_2$, and $MoB_2$ powders with an oscillatory pressure of 25–35 MPa at 10 Hz for one hour under the vacuum at the desired temperature [107]. A relatively high density of 96.6% could be achieved at 1900 °C.

The duration time of SPS is shorter than that of HP, which is beneficial for the downsizing of particles in the sintered samples and leads to the enhancement of mechanical properties. Thus, SPS is quite popular in the sintering of HECs, including oxides, carbides, borides, and silicides. Gild et al. first employed SPS to densify high-entropy borides [53]. They fabricated six dense borides with ~92% relative density using commercial diboride powders and SPS conditions of 2000 °C and 30 MPa. Minor secondary (Zr, Hf)$O_2$ phases inhibited densification. Tallarita et al. used the metal powders and boron as raw materials. Two different synthesis routes were examined in the densification of high-entropy borides, i.e. the reactive SPS and SHS–SPS route [57, 108]. They found that the density of the final products seemed to be independent of the synthesis routes, while the purity of high-entropy borides was largely different. The sample prepared by SHS–SPS was pure, while a conversion of reactants to the desired high-entropy phase of about 67 wt%, at most for samples using reactive SPS. The densification of borides using SPS relies on reduction of the particle size and elimination of the oxygen impurity of the starting powders. Zhang et al. used submicron metal oxides, amorphous boron, boron carbide and graphite powders as raw materials and successfully fabricated near fully dense high-entropy (HfZrTaNbTi)$B_2$, (HfZrMoNbTi)$B_2$, and (HfMoTaNbTi)$B_2$ ceramics via borothermal and BCTR, respectively [109]. However, since no excess reducing agents were used in the reactions, the final products contained at least 2.4 wt% oxides impurity. To eliminate oxygen contamination, Gild et al. [110] and Gu et al. [64] used excess reductors (graphitic carbon and boron carbide) in the sintering of HECs, and high-density (>98%) borides free of observable oxide inclusions were obtained. Another processing parameter needs noticing is the sintering temperature, since high temperature leads to the coarsening of grain size. Both Feng et al. and Gu et al. observed significant grain growth at high densification temperatures [64, 106].

Other methods have also been employed in the sintering of HECs. Flash sintering (FS) can be used to densify and synthesize (reactive flash sintering) ceramics at low temperature and within a short period of time. Liu et al. successfully synthesized five-component entropy-stabilized oxide ceramic (MgNiCoCuZn)O by reactive flash sintering at room temperature [111]. Due to the limited duration time (60 seconds), the relative density was low (62%). By increasing the sintering temperature and time, Yoon et al. [43] and Kumar et al. [112] obtained (MgNiCoCuZn)O with high density of 74% and 95%, respectively. This technique has also been used in the sintering of high-entropy (BiNaKBaCa)$TiO_3$ ceramic at 840 °C in five minutes [113], and a dense ceramic with a density of 87.5% was obtained under the electric field of 129 kV/cm. Near dense $(LaNdSmEuGd)_2Zr_2O_7$ was prepared by FS at 1200 °C in seconds [114]. Arc melting is also an alternative in the densification of HECs. Failla et al. produced dense high-entropy metal diborides starting from $HfB_2$, $ZrB_2$, $TiB_2$, $TaB_2$, and $CrB_2$ powders [115]. It was difficult for $CrB_2$ to form a solid solution due to its low melting point contributed to phase separation, and its high vapor partial pressure induced almost complete sublimation.

Although it is arduous to accomplish fully dense ceramics via pressureless sintering, it can be utilized to prepare porous HECs. Zhou and his coworkers developed a novel *in situ* reaction/partial sintering method to fabricate highly porous bulk materials [116–118]. In their design, the porous materials are not prepared by pressureless sintering the target material powders. They carefully designed the reactions to ensure enormous gas release in the process of reduction. The released gas can not only increase the porosity but also block large shrinkage during reaction. Thus, no pore forming agent is needed, and the reaction/partial sintering of the porous HECs can be achieved in a single step. They applied this technique in the preparation of highly porous high-entropy carbides and borides with ultralow thermal conductivity [56, 60]. The whole process is illustrated in Figure 4.8. In the preparation of porous (ZrHfTiNbTa)C, carbothermal reduction was employed, and CO gas was used as the pore forming agent as described in the following Eq. [60]:

$$\begin{aligned} &2ZrO_2(s)+2HfO_2(s)+2TiO_2(s)+Nb_2O_5(s)+Ta_2O_5(s)+32C(s) \\ &=10(ZrHfTiNbTa)C(s)+22CO(g) \end{aligned} \tag{4.1}$$

After being reduced at 1500 °C and partial sintering at 1850 °C, porous high-entropy carbides with all element distributed homogeneously were obtained. The grain size was 100–500 nm, and the pore size was in the range of 0.2–1 μm. The

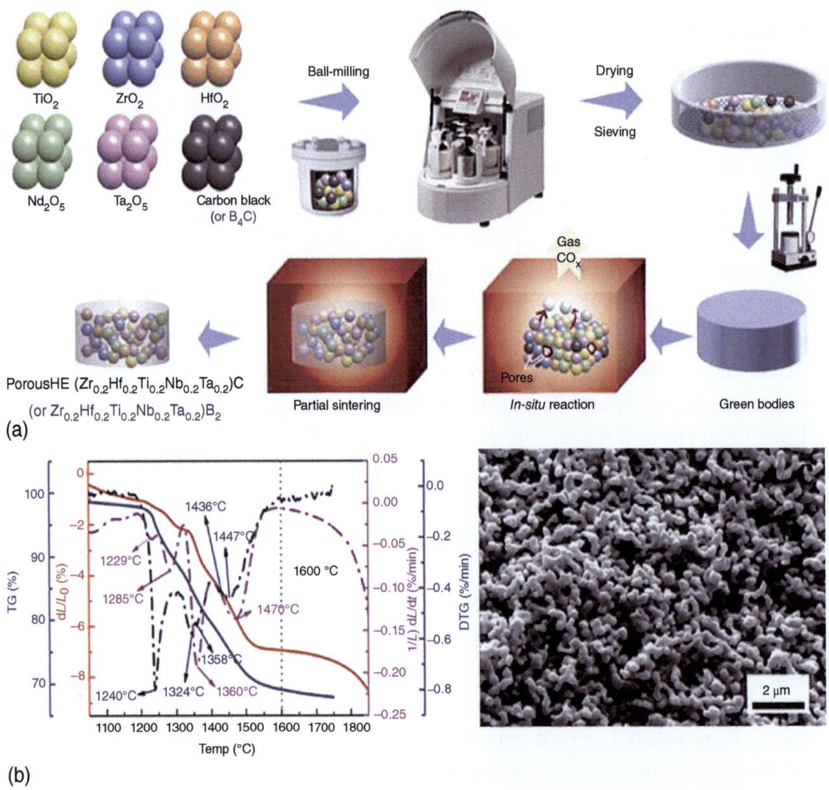


**Figure 4.8** (a) Schematic illustrartion of *in situ* synthesis/partial sintering for fabricating porous $(Zr_{0.2}Hf_{0.2}Ti_{0.2}Nb_{0.2}Ta_{0.2})C$. (b) The linear shrinkage, linear shrinkage rate, TG, DTG curves recorded during the heating of green body of porous HE carbide and the microstructure of HE carbide. Source: Reproduced with permission from Chen et al. [60]/Elsevier.

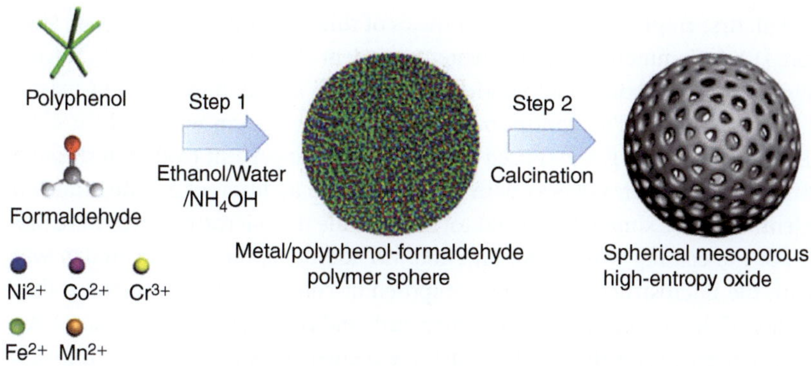


**Figure 4.9** Schematic illustration of the sol–gel synthesis of spherical mesoporous HEOs. Source: Reproduced with permission from Wang et al. [120]/American Chemical Society.

porosity is 80.99% with a compressive strength of 3.45 MPa and low room temperature thermal conductivity of 0.39 W/(m·K). Similar results are obtained for the porous $(ZrHfTiNbTa)B_2$, and $(YYbSmNdEu)B_6$ [56, 119]. Besides this partial sintering technique, a wet-chemistry sol–gel process was developed for the synthesis of spherical mesoporous HECs, where the metal precursors, tannic acid (TA) solution, and formaldehyde solution were adequately stirred followed by hydrothermal treatment [120], as shown in Figure 4.9. The metal/tannin-formaldehyde resin polymer spheres were exposed to heat treatment and a spinel structured spherical mesoporous HECs without any impure phase was successfully prepared. This porous material was characterized as high specific surface area (42–143 $m^2/g$), large pore size (5.5–8.3 nm), and unique spherical morphology with a solid or hollow structure. Zhang et al. employed cellulose nanocrystals as pore forming agents to prepare porous $(LaYNdGdSr)CrO_3$ ceramic [121]. A positive correlation was found between the porosity and conduction resistivity in electrons and phonons.

## 4.3 Films and Coatings

Owing to their superior properties, HEMs coating materials, deposited on the substrate surfaces in the form of thick coatings or thin films, can enhance the surface performance of structure components. Thus, a number of techniques have been used to prepare coatings and films of HEMs.

### 4.3.1 Laser Cladding

For HEAs, the most mature technique to prepare thick coatings is laser cladding (LC), in which the pre-selected coating material and the thin surface layer of the substrate are simultaneously melted and solidified rapidly under the heat source of laser. The surface coating with the thickness of hundreds of micrometers to several millimeters exhibits very low dilution and metallurgical bonding to the substrate.

Zhang et al. first employed LC in the synthesis of thick HEA coatings on Q235 steel substrate [122]. Commercially pure element powders of Fe, Co, Ni, Cr, and Cu were mechanically mixed. Additional Si, Mo, and Mn were added in the form of master alloys for comparison. The continuous-wave $CO_2$-laser system was used for cladding. High-purity argon gas was used as shielding gas to prevent oxidation through the coaxial nozzle. Both systems with or without the addition of Si, Mo, and Mn were identified to be simple FCC solid solution, while the microstructures were disparate. For the coating without Si, Mn, and Mo additions, the coating quality was poor with the microstructure mainly composed of columnar and equiaxed grains. The quality of the coating was greatly improved, and the microstructure was transformed to dendrite with the addition of three elements, as shown in Figure 4.10. They also examined the nucleation incubation time of various competing phases and concluded that this incubation time was strongly dependent on the atomic jump distance and, kinetically, rapid solidification process prevented the nucleation of the intermetallics [123]. Moreover, they pointed out that the addition of elements with a small atomic ratio aggravated the solute segregation due to the atomic size difference and high diffusion rate of small atoms, leading to the formation of secondary arms of dendrites. While sufficiently high solidification rate could suppress this process. Shu et al. also observed the composition-induced phase and

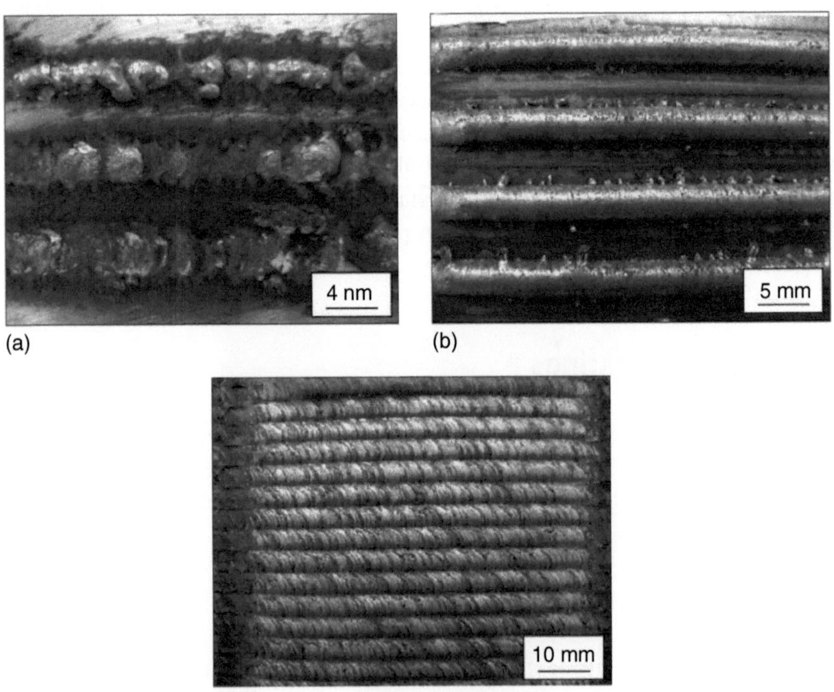


**Figure 4.10** Surface morphology of the FeCoNiCrCu coatings. (a) Single-track without Si, Mn, and Mo additives; (b) and (c) single-track and multi-tracks with Si, Mn, and Mo additions, respectively. Source: Reproduced with permission from Arif et al. [125]/Elsevier.

microstructure changing in laser-cladded FeCoCrBNiSi HEA coatings [124]. Their results showed that the increase in the Fe-to-Co ratio weakened the glass formation ability of the cladding coatings, thus decreasing the content of the amorphous phase.

Besides the composition of the feedstock material, various laser processing parameters in LC technology affect the characteristics of the claddings, as illustrated in Figure 4.11 [125]. Chao et al. comprehensively investigated the effect of key processing parameters on the microstructure and quality of $Al_{0.3}$CoCrFeNi claddings, such as laser power, laser scanning speed, laser beam diameter, powder mass feed rate, and hatch distance to establish the optimal parameters [126]. By examining the cross-sectional microstructure of a single-track $Al_{0.3}$CoCrFeNi deposition, the relationship between the dimensional characteristics of melt pool profile and processing parameters could be established, and the optimal parameters could be identified by balancing the efficient deposition and minimal dilution. Generally, their results indicated that the increase in powder feed rate decreased the clad depth and interdiffusion thickness and increased clad thickness. Furthermore, an increase in laser power increased clad height, width, depth, and interdiffusion thickness. Such conclusions have been verified by other researchers. Guo and Liu reported the influence of laser output power and scanning rate on the hardness and wear resistance of the AlMoFeCrTiWNb-based HEA coating and found that high laser power and low scanning rate led to thickening of the HEA coating [127]. They also found that delicate control of processing parameters was vital to a denser and more crack-free coating. Low or exorbitant laser powers and scanning rate made it difficult for HEA

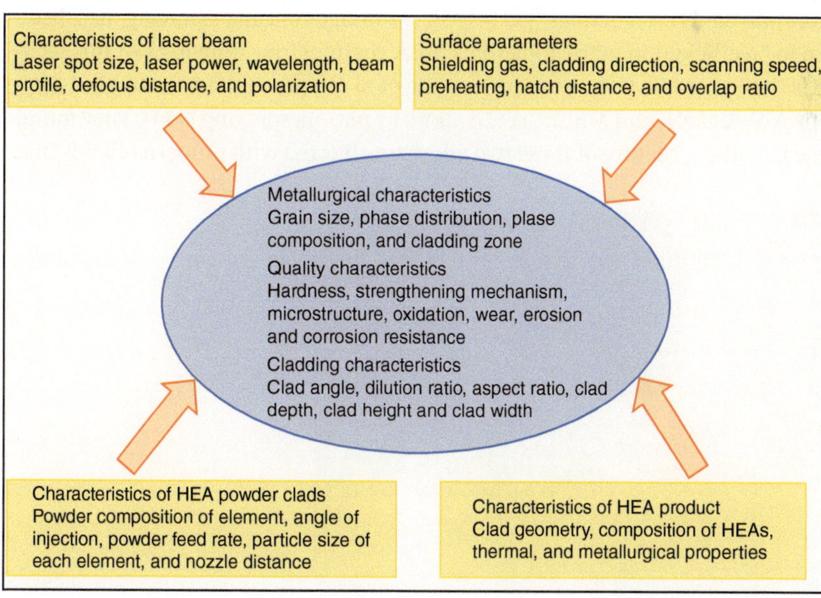


**Figure 4.11** Dependence of the metallurgical characteristics, quality characteristics, and cladding characteristics of the LC-HEACs on the processing parameters of the laser and HEA powder. Source: Reproduced with permission from Arif et al. [125]/Elsevier.

coatings to exhibited more homogeneous and denser microstructure. Shu et al. investigated the effect of laser power on the properties and microstructure of amorphous-containing $Co_{34}Cr_{29}B_{14}Fe_8Ni_8Si_7$ laser clads by LC [128]. They discovered a decrease in amorphous content and increase in dilution degree with the increase of laser power. Ni et al. studied the effect of scanning speed on $Al_{0.5}Cu_{0.7}$FeNiCoCr-based high-entropy alloy coatings (HEACs) fabricated through LC and found a strong relationship between the melting depth and the scanning speed [129]. They developed defect-free clad with an ultrafine microstructure at the highest scanning speed (630 mm/min) due to the better mixing of the clad and the substrate. Thus, proper tuning of the laser parameters is essential for high-quality LC HEA coatings. Up to now, LC has not been used in the fabrication of HEC coatings.

### 4.3.2 Spray Techniques

Spraying techniques are commonly adopted coating-fabrication processes. Spraying processes such as plasma spraying and cold spraying (CS) are reported to fabricate HEM-based coating materials with enhanced mechanical and functional performances. Huang et al. prepared the multi-principal-element alloys, AlSiTiCrFe CoNi$Mo_{0.5}$ and AlSiTiCrFeNi$Mo_{0.5}$, through an atmosphere plasma spray system [130]. They used Ar as the primary gas and He as the secondary gas to prevent the oxidation of alloys. In the as-sprayed coatings, FCC phases were suppressed since the BCC phase was the primary phase during solidification. The microstructure of as-sprayed coatings revealed the typical lamellar structure built up by plasma spraying, as shown in Figure 4.12. This structure feature exhibits almost in every plasma-sprayed HEA coatings [131–133]. Although plasma spraying has been proven to be efficient in fabricating thick HEA coatings, one drawback of this technique is the inevitable pores and oxides. Ang et al. prepared thick coatings of high entropy AlCoCrFeNi and MnCoCrFeNi alloys by plasma spraying [134]. They found that the lamellar structure of these two alloys was littered with numerous voids that

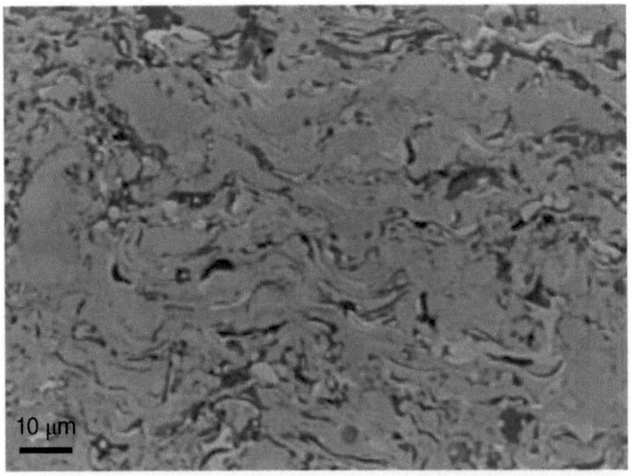


**Figure 4.12** A typical lamellar structure of AlSiTiCrFeCoNi$Mo_{0.5}$ HEA built up by plasma spraying. Source: Huang et al. [130], Reproduced with permission from John Wiley & Sons.

could be classified as pores and interlamellar cracks. In the high magnification microstructure images, two aluminum-rich oxides were identified, i.e. $Al_2O_3$ and $Al_2MO_4$ ($M = Cr_{0.5}+Fe_{0.5}$). The existence of aluminum-rich oxides resulted in the depletion of aluminum in the AlCoCrFeNi HEA matrix and phase transformation from the major BCC phase before spraying to the major FCC phase after the spraying. The lamellar structure also resulted in the anisotropic mechanical response of the HEA coatings.

To lower the porosity and oxygen levels, high velocity oxy-fuel (HVOF) spraying is a good alternative due to its high particle-jet velocity and relatively low processing temperatures [135]. Different from plasma spraying, the heat source of HVOF comes from the combustion heat of the fuel and oxygen. The most used fuels in HVOF are kerosene [135] and LPG [136]. Li et al. compared the microstructure and wear behavior of $FeCoCrNiMo_{0.2}$ high-entropy coatings prepared by air plasma spraying and HVOF processes [137]. The cross-sectional morphologies of the thermal sprayed HEA coatings are displayed in Figure 4.13. Typical lamellar structures were observed in both coatings. Energy dispersive spectrometer results indicated that dark gray region was

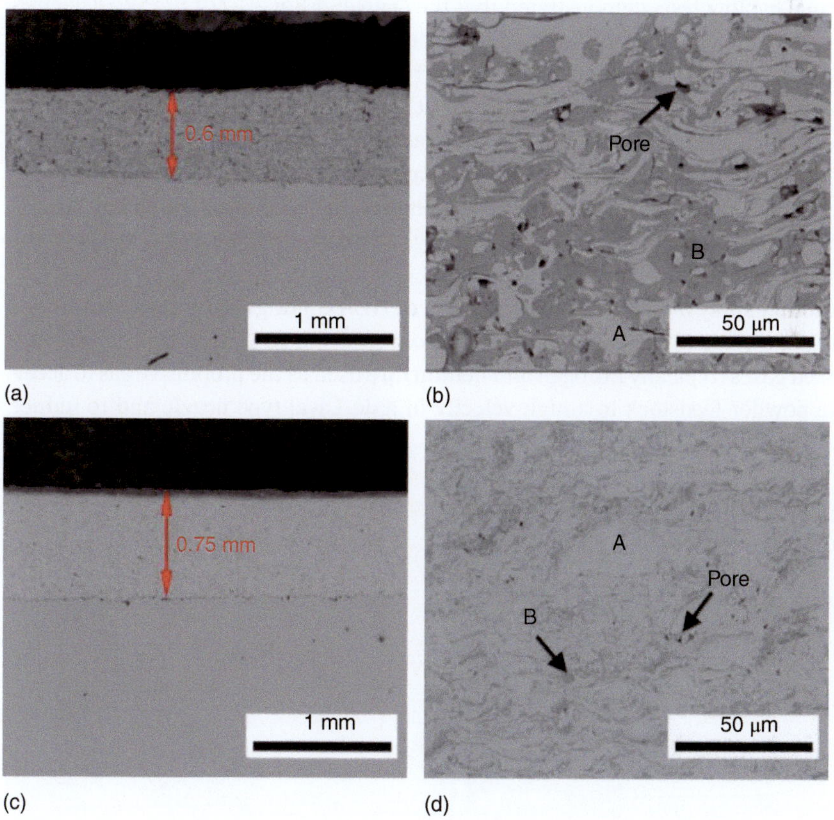


**Figure 4.13** Cross-sectional scanning electron microscope images of (a, b) atmospheric plasma spraying coating and (c, d) high-velocity oxy-fuel spraying (HVOF) coating at low and high magnifications. Source: Li et al. [137], MDPI, CC BY 4.0.

the dispersed oxide phase (labeled as B). Clearly, the air plasma spraying coating had a high concentration of oxides with more microstructural defects, such as pores and inter-splat interfaces, than that in the HVOF coating (the overall porosity in plasma sprayed coating was 3.1%, while that in HVOF coating was 0.6%). Moreover, the hardness and Young's modulus of HVOF coating were higher. Similar results are also obtained by Hsu et al. in their research on the thermal sprayed $Ni_{0.2}Co_{0.6}Fe_{0.2}CrSi_{0.2}AlTi_{0.2}$ HEA coatings [138]. The as-sprayed HVOF coating exhibited denser structure and less oxidation stringers than the plasma-sprayed coatings.

Plasma spraying can also be employed in the preparation of HEC coatings. Zhou et al. deposited a $(LaNdSmEuGd)_2Zr_2O_7$/yttria stabilized zirconia (YSZ) double-ceramic-layer (DCL) coating on an Ni-based superalloy by atmospheric plasma spraying [139]. Different from thermal sprayed HEA coatings, no lamellar structure was observed in the dense HEC coating with a porosity of 9.17%, as shown in Figure 4.14. A phase transformation from pyrochlore to defective fluorite structure was observed after the spraying, which could attribute to the high processing temperature and ultrafast cooling rate. A slight composition deviation from the nominal near-equimolar ratio was detected due to their different volatilities. Nevertheless, thermal cycling tests demonstrated that the $(LaNdSmEuGd)_2Zr_2O_7$/YSZ DCL coating exhibited improved thermal stability due to its good sintering resistance and close thermal expansion coefficient of HEC to YSZ. The phase transformation during the spray was also observed by Zhang et al. in their work concerning the preparation of HE $La_2(ZrCeHfSnTi)_2O_7$ ceramic as top coating by plasma spraying [140]. Park et al. successfully used the plasma spray technique to prepare rare-earth oxide coating as thermal barrier coatings [141]. The coating was stable without any phase transformation after spraying. Likewise, slight composition deviation existed in the prepared coating.

Another spray method in the preparation of HEA coatings is the newly developed solid-state coating deposition technique, cold spraying (CS). In this process, compressed gases (typically nitrogen and helium) are used as the propulsive gas to accelerate powder feedstock to a high velocity in a de Laval type nozzle and to induce deposition when the powders impact onto a substrate at a velocity over a critical

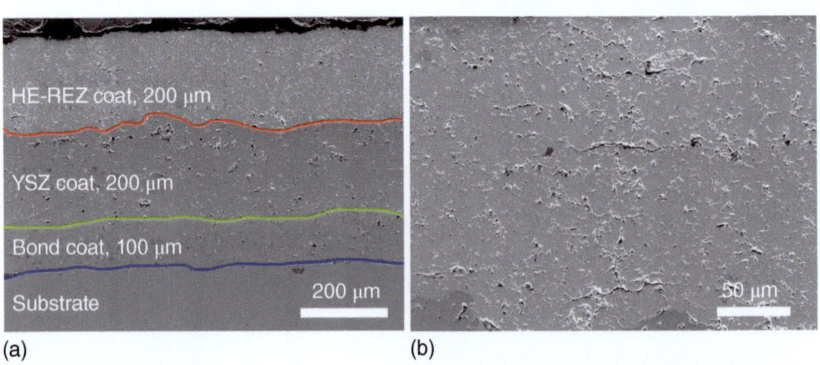


**Figure 4.14** SEM images of the cross-section of (a) the coating system and (b) the high-entropy rare-earth zirconate ceramic layer. Source: Zhou et al. [139], Reproduced with permission from Elsevier.

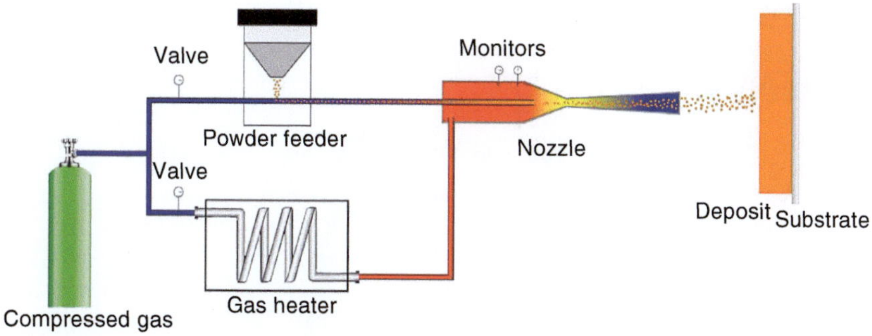


**Figure 4.15** Schematic of a cold spraying system and its working principle. Source: Reproduced with permission from Yin et al. [142]/Elsevier.

value [142]. The working principle of CS is shown in Figure 4.15. The advantage of this technique is the low processing temperature, which allows for the avoidance of defects such as oxidation, residual thermal stress and phase transformation in the final product. Using this method, Yin et al. deposited FCC-type FeCoNiCrMn HEA coatings on an Al substrate [142]. A dense coating (with a porosity of 0.47%) was produced without any phase transformation. Significantly refined HEA grains were obtained due to the dynamic recrystallization under the combined action of adiabatic heating and plastic deformation. Their work has proved that solid-state CS technology is promising for the fabrication of thick and dense HEA coatings. Due to the lack of deformability of ceramic particles, they can only be embedded in the substrates to form a thin layer in CS process. There are limited research studies on the CS of ceramic particles, and the ceramic bonding mechanism is not fully understood yet [143].

### 4.3.3 Vapor Deposition

#### 4.3.3.1 Magnetron Sputtering

In the pursue of HEM thin films, the most used technique is the sputtering deposition, including magnetron sputtering (MS), reactive magnetron-sputtering, and vacuum arc deposition (VAD). A scheme of MS is shown in Figure 4.16 [144]. At the very beginning, Yeh et al. had deposited a $Cu_{0.5}$CoNiCrAl high-entropy film on silicon substrates from casted alloy ingots by radio frequency (RF) magnetron sputtering [67]. The film consisted of a simple BCC structure with an average grain size of 7 nm. Afterward, films in other systems have also been deposited by RF or direct current (DC) magnetron sputtering, e.g. AlMoNbSiTaTiVZr, TiTaHfNbZr, $AlCoCrCu_{0.5}FeNi$, and $CoCrFeNiAl_{0.3}$ [145–148].

In a typical MS process, the microstructure and phase composition of the deposited films are affected by the deposition parameters such as working pressure, substrate bias, target-substrate distance, substrate temperature, and sputtering power. Braeckman et al. investigated the effects of working pressure and target–substrate distance on the formation of AlCoCrCuFeNi high-entropy alloy thin films (HEATFs) [149]. They found that the target–substrate distance and sputtering pressure barely influenced the film composition, which was understandable

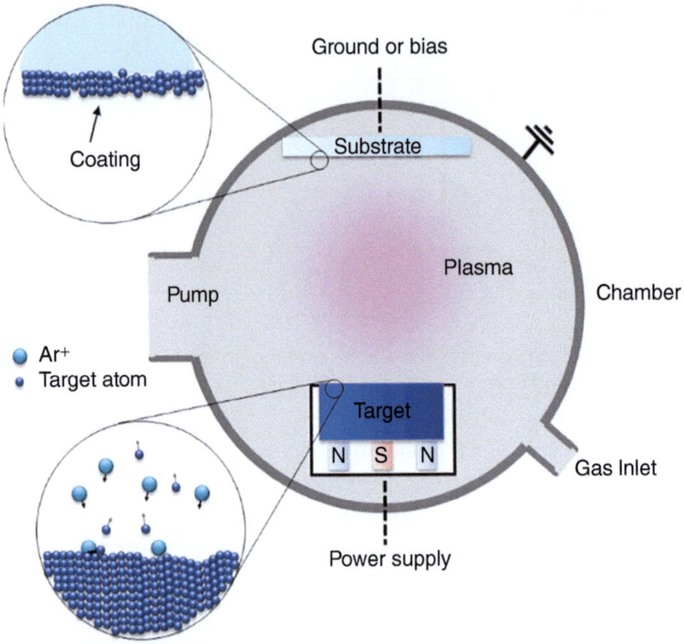


**Figure 4.16** Schematic of a magnetron sputtering system. Source: Reproduced with permission from Velasco et al. [144]/Elsevier.

considering the similarity in the effects of these two parameters on the transport of elements in the first-order approximation. However, the microstructure of the films was closely related to the working pressure, where high working pressure led to rough surface and high porosity.

Khan et al. deposited the $AlCoCrCu_{0.5}FeNi$ HEATFs with different working pressure and sputtering power using RF-MS from a single stoichiometric target with BCC structure [147, 150]. They found that the deposition rate of HEA films did not vary linearly with working pressure. The rate was the highest when the pressure was controlled to a moderate value (10 mTorr), which could not only increase the number of gas ions striking the target but also avoid the deflections of sputtered atoms caused by the collisions between these atoms and gas molecules at high pressure. Due to the high deposition rate, the grain size and surface roughness of the film deposited with 10 mTorr working pressure were the largest. Besides, there was a preferential sputtering of Al from the composite target at low working pressure due to significantly lower atomic weight of Al compared to other metal elements. Further, relatively high concentration of Al in the film deposited at low work pressure (5 mTorr) caused amorphousization of the film due to the size-induced excessive lattice distortion. With regard to the sputtering power, high deposition power favored high crystallinity structure containing more BCC component and a rough porous surface with a large grain size. Moreover, increasing the sputtering power resulted in a significant decrease in Al concentration in the films, which was possibly induced by the re-sputtering of deposited films from energetic depositing atoms and gas ions.

The substrate temperature in the MS process is also vital in determining the microstructure and phase selection of the deposited HEA films. As demonstrated in CrMnFeCoNi and HfNbTiVZr systems, the phase formation and growth mechanisms in magnetron sputtering were believed to be different from those of bulk HEAs [151, 152]. In bulks, a high enough cooling rate kinetically hindered the precipitation of other competing phases, and the homogenous single FCC structure could reserve to room temperature despite of its thermodynamical metastability. While in MS process, the atoms are immediately quenched to the substrate temperature during sputtering. Thus, if the substrate temperature was high, the high energy input allowed for rapid diffusion of atoms on the surface of the growing film and in grain boundaries, resulting in the elemental segregation and formation of equilibrium phases (C15 Laves phase [152] or $\sigma$-phase, $\chi$-phase [151]). At low substrate temperatures, the diffusion rate at the surface and in grain boundaries was reduced significantly both due to the lower energy input, indicating that the phase selection would be a kinetic process. The deposited films were more homogenous, and in most cases, an amorphous structure was obtained [146, 147, 153, 154].

The substrate bias also has notable effects on the phase and microstructure of films. Sha et al. deposited FeMnNiCoCr HEA films using direct current magnetron sputtering under the different substrate bias [155]. Besides the dominant FCC phase, a secondary HCP phase was also observed in the films, and the content of the HCP phase increased with the bias. The high levels of compressive stress induced by energized ion bombardment on the film surface attributed to this stress-induced FCC-to-HCP martensitic transformation. The compressive stress increased with the bias voltage, resulting in a higher volume fraction of the HCP phase. Another effect caused by the ion bombardment was the back-scattering re-sputtering, which led to the decreased deposition rate at high substrate bias voltage, and microstructural evolution from zone I (elongated columnar grains) to zone T (V-shaped columnar grains). The re-sputtering also induced the variations in the chemical composition due to the sputtering yields of different elements, which were calculated as 1.704, 2.859, 1.682, 1.584, and 2.009 for Fe, Mn, Ni, Co and Cr, respectively. The highest sputtering yield for Mn suggests the largest re-sputtering yield, through which a significant amount of Mn was ejected from the deposited film and a Mn-depleted steady-state composition was generated. Contrariwise, the composition for Co increased significantly as the bias voltage increased. They also observed the formation of nano-twin boundaries in the deposited film with high substrate bias due to the enhanced level of compressive stress, surface mobilities, implanting of impinging particles, and displacement of lattice atoms.

Normally, the HEA films are deposited by MS from a single stoichiometric alloy target. The targets are usually prepared by arc melting, and re-melting is needed to improve homogeneity of the target, which is time- and energy-consuming. Two alternatives have been reported, i.e. multiple targets and pressed powder targets. In the preparation of an AlCoCrCuFeNi thin film, Dolique et al. used three mosaic targets, where Fe, Co, and Ni were on target 1, Cu and Cr on target 2 and Al on target 3 according to their sputtering yield and magnetic nature [156]. The advantages of this deposition technique include variation of the stoichiometry by adjusting the target powers and introduction of a concentration gradient in the films. Zou et al. employed co-sputtering technique to deposit NbMoTaW HEA films from elemental targets [157]. The powers of the magnetrons were adjusted to obtain the equal arriving ratio of Nb, Mo, Ta, and W, and the substrate was rotated to homogenize alloy composition and film thickness. An ion gun was employed in the deposition to reduce deposition rate and decrease grain size. To ensure the composition, delicate controlling of the deposition parameters is needed when multiple targets are used. Braeckman et al. introduced a pressed powder target prepared by uniaxial cold pressing the elemental powders of Al, Co, Cr, Cu, Fe, and Ni [149, 158]. Although the compositions of the films did not perfectly match those of the targets, the difference were limited to 5 at.% at the most.

Most of the reported multicomponent ceramics have been synthesized by reactive magnetron sputtering of HEA targets. With different reactive gas introduced, such as $N_2$, $CH_4$, and $O_2$, high-entropy nitrides [159–163], carbides [164, 165], and oxides [166] films can be synthesized. Chen et al. deposited high-entropy nitride films from two HEA systems, Fe–Co–Ni–Cr–Cu–Al–Mn and Fe–Co–Ni–Cr–Cu–$Al_{0.5}$, with the nitrogen flow changed from 0 to 20 sccm and substrate bias changed from −50 to −150 V [159]. Following the increase in the nitrogen flow for both targets, the crystallinity of the nitride films decreased and approached amorphous, and/or their grain sizes became smaller, which also verified by the smooth and dense morphology from their cross-section images. The effect of bias on the film structure was not evident due to the amorphous structure in all films. Nevertheless, a slight decrease in the film growth rate with increasing bias voltage was observed, which was similar to that in the preparation of HEA films [151]. With regard to the effect of substrate temperature, Lai et al. deposited (AlCrTaTiZr)N films by reactive MS at different temperatures [160]. The films possessed the same FCC structure with the lattice parameter constrained at higher temperatures, which was contributed to the reduction of nitrogen content with the increase in substrate temperature. Moreover, a decreased crystallite size and increased orientation preference were observed. Besides, non-reactive magnetron sputtering is also employed in the preparation of HEC films. Mayrhofer et al. prepared $(ZrTiHfVTa)B_2$ film from five corresponding diboride composite targets [167]. A single-phased solid solution was formed with slight composition deviation from stoichiometry. After annealing at 1500 °C, the formation of lower boron containing relatives was observed due to the significant B-loss and O impinging in the high-entropy borides. Nevertheless, this work has demonstrated the possibility of preparation of HEC films from their composite targets.

#### 4.3.3.2 Pulsed Laser Deposition

Pulsed laser deposition (PLD) is a physical vapor deposition (PVD) technique where a high-power pulsed laser beam is focused to strike a target. The target material is evaporated to create a plasma plume and deposits as a thin film on a substrate. This process can occur in ultra-high vacuum or in the presence of a background gas. A scheme of PLD is shown in Figure 4.17. It is a quite new technique in the preparation of HEM thin films; thus, no comprehensive research has been conducted.

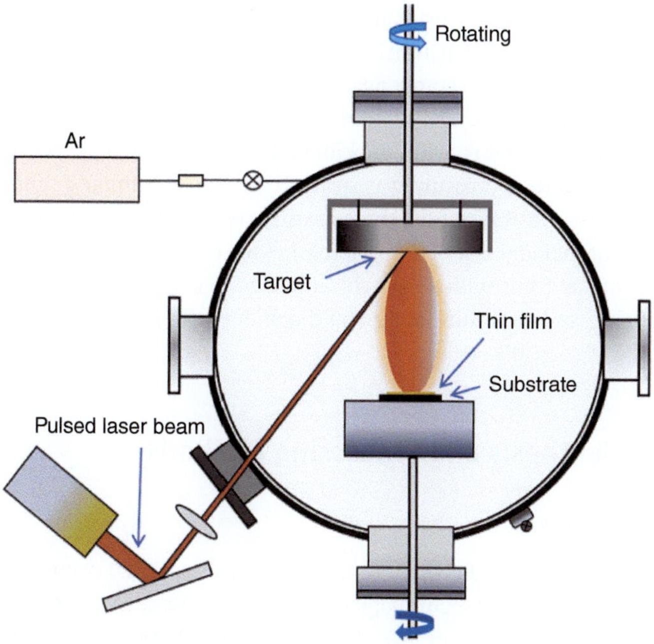


**Figure 4.17** Schematic of a PLD system. Source: Reproduced with permission from Liu et al. [168]/Elsevier.

Cropper prepared a thin film of AlCrFeCoNiCu on borosilicate glass by PLD at different temperatures [169]. A pulsed and frequency-doubled Quanta Ray Nd:YAG laser operating at 10 Hz and wavelength 532 nm was used as the laser source. Results showed that large crystal sizes of both FCC and BCC structures with texture microstructure were obtained at room temperature. With increasing deposition temperature, the uniaxial texture fell to being effectively zero at 300 °C. The depth profiles showed the existence of surface segregation of Al and Cr in alloys. Lu et al. studied the mechanical properties of the $CoCrFeNiAl_{0.3}$ HEATFs on a silicon substrate under vacuum atmosphere, room temperature, and different deposition times [168]. Smooth surface with some granules was observed for all samples. An enrichment of Al and depletion of Cr was found. The deposited films crystallized in FCC structure with an amorphous structure co-existed.

With regard to the application of PLD in the preparation of HEC thin films, Meisenheimer et al. deposited $(Mg_{0.25(1-x)}Co_xNi_{0.25(1-x)}Cu_{0.25(1-x)}Zn_{0.25(1-x)})O$ thin film on single crystal MgO (001) substrates from their sintered targets [170]. Using a similar procedure, Sharma et al. successfully prepared high-quality single-crystal epitaxial thin films of $Ba(ZrSnTiHfNb)O_3$ on $SrTiO_3$ and MgO substrates [171]. To identify the effect of kinetic bombardment, partial $O_2$ pressure, and epitaxy on the phase and structure of deposited (MgNiCoCuZnSc)O films, several sets of thin films were prepared by Kotsonis et al. [172]. They observed a phase transition with the variation in gas pressure and gas composition. Generally, low total pressure led to single phase rock salt structure, while high pressure induced phase separation and surface roughness. Jacobson et al. investigated the pressure and temperature dependence of film structure in (MgCoNiCuZn)O system [173]. In their work, separate CuO and ZnO targets were used to compensate the loss of these two elements in PLD and grow stoichiometric films. They found that the structure of deposited films was kinetic-energy-related, that is, with low kinetic energy, the film deposited at low temperature and low oxygen pressure was strained, and no columnar grains were observed. Moreover, it is feasible to tune the microstructure of deposited films by controlling the energy of incident laser [174]. Increasing the energy fluence leads to better crystallization of the films with larger columnar grains, islands and pore size. The droplets reducing and film thickening could achieve with the increasing energy fluence of laser.

## 4.4 Other Novel Synthesis and Processing Methods

### 4.4.1 Additive Manufacturing

Additive manufacturing (AM), also referred to as three-dimensional (3D) printing, is a series of advanced manufacturing technologies used to fabricate physical parts in a discrete point-by-point, line-by-line, or layer-by-layer additive manner from 3D CAD models that are digitally sliced into 2D cross sections [175]. This layer-by-layer nature of AM processes implies some intrinsic advantages such as complex geometry fabrication, design freedom, and higher heating and cooling rates [176]. Two types of AM technique, i.e. direct laser deposition (DLD) and powder bed fusion (PBF), are used in the fabrication of HEA parts, as schematically shown in Figure 4.18 [177, 178]. DLD may also be known as laser additive manufacturing (LAM), laser metal deposition (LMD), direct metal deposition (DMD), LENS, direct laser fabrication (DLF), directed energy deposition (DED), and etc. In a successful AM process, high quality powders with spherical morphology, smooth surface, and narrow size distribution are required. The most used processes to develop HEA powders have been introduced in Section 4.1.1.

DLD is the most popular printing process to develop HEA products, where a laser or electron beam was used to melt alloys in the form of powders upon deposition [176]. LC, which is employed in the synthesis of thick HEM coatings, can be regarded as a variation of DLD. This technique is well reviewed in Section 4.3.1. Joseph et al. manufactured bulk $Al_xCoCrFeNi$ HEAs by DLD from elemental blend powders [179]. The chemical composition of deposited products showed good agreement with the elemental composition of the starting powder mixture, and the phase transformed from FCC to BCC with the increase in the content of Al. The FCC solid solution alloy produced by DLD exhibited columnar elongated grains longitudinal to the build direction and equi-axed grains transverse to the build direction, which was also observed in $Al_{0.85}CoCrFeNi$ BCC solid solution. EBSD orientation data confirmed a strong <001> fiber texture along the deposition direction for FCC and BCC samples. A small amount of very fine Ni- and Al-rich particles could be detected at grain boundaries. The microstructure, segregation behavior, chemical composition,

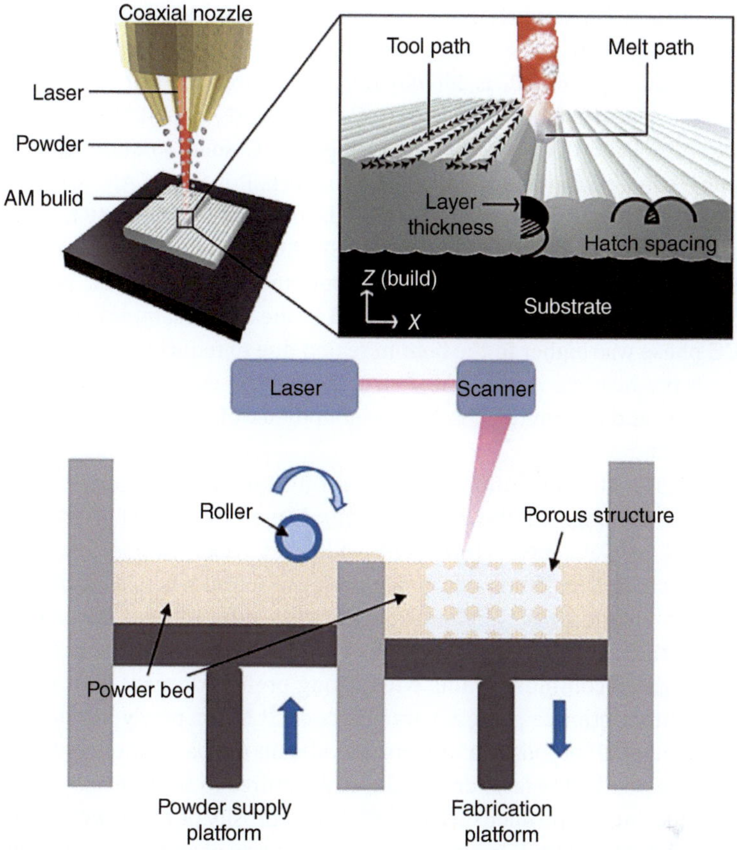


**Figure 4.18** Schematic of a (a) direct laser deposition system and (b) powder bed fusion system. Source: (a) Reproduced with permission from Wolff et al. [177]/Elsevier; (b) Zhao et al. [178]/Frontiers Media S.A./CC BY-4.0.

and homogeneity of the printed HEAs are greatly controlled by the processing parameters of DLD, such as laser power, scanning speed, hatching distance, and layer thickness. As concluded by Moghaddam et al. the laser power and scanning speed determined the cooling rate of the melts and subsequently the phases and microstructural characteristics of printed products [180]. Decreasing laser power and increasing scanning speed resulted in a high cooling rate. Phase segregation and elemental partitioning could be greatly suppressed at high cooling rates. Besides, the high cooling rate increased the nucleation rate due to the boosted solidification driving force (supercooling) and hence provided a refined microstructure for printed HEAs [180]. Thus, finding the optimal process parameters for a given alloy through an iterative parametric study is needed for a desired microstructure and properties. Due to its large layer thickness and laser diameter, DLD offers a high productivity with the cost of low resolution and is useful in fabricating large components.

In PBF, the bulk component is fabricated in a layer-by-layer mode through selectively melting and fusing the metal powders on a thin powder bed by a high energy laser or electron beam, i.e. selective laser melting (SLM) [181–183] and selective electron beam melting (SEBM) [184, 185]. The main difference between DLD and PBF is the powder feeding method, as demonstrated in Figure 4.18. Brif et al. demonstrated the first successful fabrication of FeCoCrNi bulk HEA via SLM from pre-alloyed, gas-atomized alloy powders [181]. A single BCC solid solution with a uniform chemical composition was obtained in their work. Printing of AlCoCrFeNi HEA products by SEBM was first reported by Fujieda and his coworkers [184, 185]. Due to the pre-heating of powder bed, the HEA by SEBM exhibited different microstructure and phase composition in the top and bottom regions. An FCC phase, which disappeared in the casting ingot, was formed in the SEBM samples, and the content of FCC phase was higher in the bottom region due to reduced temperature gradient through pre-heating. A modulated structure within equiaxed grains was observed in both top and bottom regions, but the abundance of equiaxed grains was greater at the bottom.

The laser parameters strongly influence the heat transfer and grain growth during PBF. Li et al. investigated the microstructural evolution of SLM-printed CoCrFeMnNi and AlCoCrFeNi HEA with different laser energy density [182, 183]. In both cases, densification improvement was found when laser energy density increasing. Owing to the highest volatilization ratio of Mn, the lattice parameter of CoCrFeMnNi decreased with laser energy density. In the printed products, the elongated columnar grains with strong preferred orientation were formed. To efficiently optimize process parameters of SLM for newly developed HEA powders, Zhou et al. proposed a universal model on the basis of energy balance and transfer to predict the laser energy density required to achieve near-fully dense parts, considering the powder surface structure, porosity of the powder bed, vaporization, and heat loss [186]. In their model, the energy absorption $Q_a$ by the local powder bed should be approximately 3–8 times greater than energy consumption $Q_c$ to achieve near-full density parts using SLM for the various metallic materials studied. The value of $Q_a/Q_c$ highly related to laser absorptivity, latent heat of melting and specific heat capacity of designed HEAs. The validation and accuracy of this model were determined through the printing of a CoCrFeMnNi powder, where a near-full density sample was achieved using the calculated line energy density.

Although AM has been widely used in the fabrication of metal, ceramic, and polymer components, no investigation on the additive manufactory of HECs has yet been reported.

### 4.4.2 Carbothermal Shock Synthesis

The carbothermal shock (CTS) synthesis, which was first proposed by Hu and coworkers, has brought about the preparation of a new paradigm of metal alloys and nanomaterials with excellent properties [187]. This method employs flash heating and cooling (temperature of ~2000 K, shock duration of ~55 ms, and ramp rates on the order of $10^5$ K/s) of metal precursors on oxygenated carbon support to produce high-entropy-alloy nano-particles (HEA-NPs) with up to eight dissimilar metallic elements (Pt, Pd, Ni, Co, Fe, Au, Cu, and Sn). The HEA-NPs are of nanoscale dimensions in FCC crystal structures with compositional uniformity. The HEA-NP composition has a small deviation from the ideal composition based on the initial precursor salt molar ratios, due to vapor loss at high temperature, which can be compensated by compositional control. In a typical CTS process, two sequential facile steps were required: first, the metal salt precursors were mixed into a solution and loaded onto a conductive carbon support, such as carbon nanofibers (CNFs). Then, the precursor-loaded sample was exposed to a rapid thermal shock (55 ms) in an Ar-filled glovebox under high reaction temperature (~2000 K) and rapid heating and cooling rates (~$10^5$ K/s).

Based on the facts that defects and metal catalysts play a key role during the CTS process, the authors proposed a catalytically driven carbon metabolism reaction to explain the formation of uniformly dispersed alloy nanoparticles from single-element, micron-sized liquid metal droplets:

$$C + O^* \rightarrow CO\uparrow \tag{4.2}$$

where O* denotes surface-bound residual oxygen. The carbon metabolism reaction involved C (fuel), O* (oxidizer), and metal (catalyst), which correlated with the surface defect concentration and the metal's catalytic activity to the final nanoparticle size and level of dispersity. During the process, the liquid metal droplets actively traveled around and split ("fission") to harvest the dispersed O* on the carbon surface based on the catalytically driven reaction. The liquid metal movement allowed different droplet compositions to continually meet and fuse into single-phase alloys. With the depleted O* concentrations, the mobility of liquid metal droplets decreased and slow coarsening occurred to reduce the total surface energy. A dynamic steady state of particle fusion and fission achieved in the entire process. The scheme of this carbon metabolism reaction is shown in Figure 4.19. Using this technique, highly efficient catalysts for ammonia decomposition and oxygen electrocatalysis have been successfully synthesized [188–191].

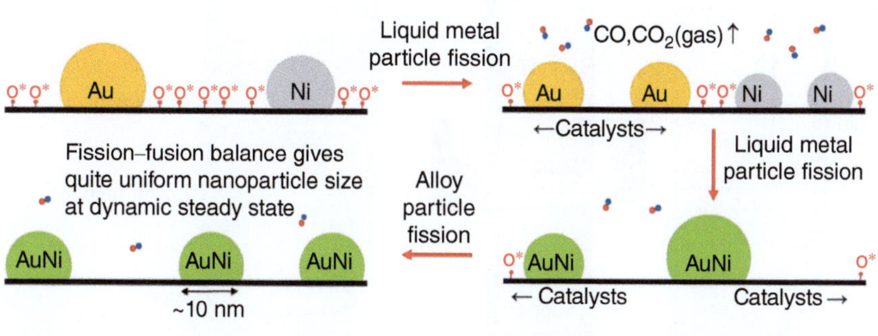


**Figure 4.19** An illustration of the catalysis-driven particle fission/fusion mechanism to synthesize uniformly dispersed HEA nanoparticles. Source: Reproduced with permission from Yao et al. [187]/American Association for the Advancement of Science.

### 4.4.3 Severe Plastic Deformation Process

Processing by severe plastic deformation (SPD) refers to various experimental procedures of metal forming that may be used to impose very high strains on materials leading to exceptional grain refinement [192]. Two unique feature of SPD processing are high strain imposed without any significant change in the overall dimensions and significant hydrostatic pressure produced by special tool geometries. The major methods already established for the fabrication of HEMs are high-pressure torsion (HPT) and equal-channel angular pressing (ECAP).

HPT refers to processing in which a thin disk is subjected to torsional straining under a high hydrostatic pressure, and the principle of HPT is illustrated in Figure 4.20. The plastic torsional straining is achieved by rotation of the anvils [192]. HPT has already been used for the processing of several HEAs, such as CoCrFeMnNi [193, 194], $Al_{0.3}CoCrFeNi$ [195], $Al_{0.3}Cu_{0.5}CoCrFeNi$ [196], TiZrNbHfTa [197], CoCrFeNi [198], and TiAlFeCoNi [199]. In a typical HPT process, due to the formation of twining, especially in FCC HEAs, a dramatic decrease in grain size from hundreds of micrometers to tens of nanometers can be observed, even when the sample is processed only for 1/2 turn [198, 200]. The distribution of the elements in the deformed materials is homogenous with rather blurry grain boundaries (GBs) due to the formation of non-equilibrium GBs. Moreover, significant enhanced hardness is achieved, which can be attributed to the severe strains, and show clear temperature-dependency due to the decomposition of the matrix and formation of precipitates. Besides a process to enhance the mechanical properties of HEAs, HPT can also be employed to synthesize HEA from raw metal powders [201]. The multicomponent mixtures of Co, Cr, Fe, Mn, and Ni powder were subjected to a pre-compaction step followed by HPT deformation at the pressure of 5 GPa. After 100 rotations, an FCC single-phase structure with an average grain size of 50 nm remains. No large fluctuations in concentration are observed. With 5%

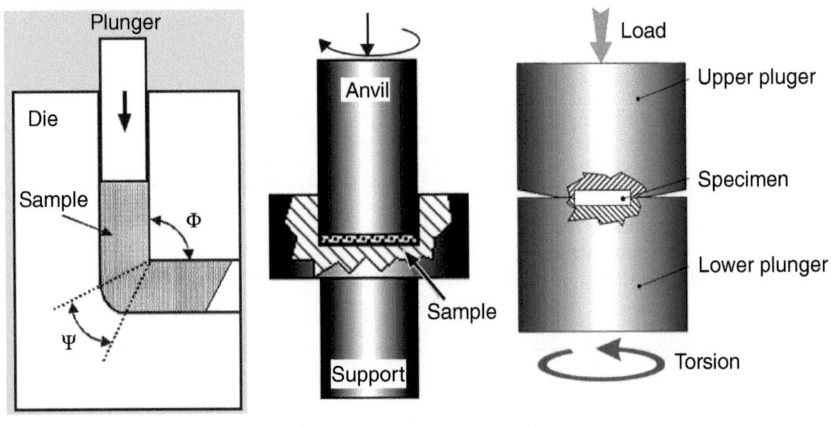


**Figure 4.20** The principle of (a) ECAP and (b, c) HPT. Reproduced with permission from Valiev et al. [192]/Springer Nature.

chromium oxide in the size range of 7–10 nm precipitated, exceptional high hardness of CoCrFeMnNi HEA can be obtained. TiZrHfNbTa HEA powders prepared by HPT have also been reported, and the alloyed powders can be transformed into oxide and oxynitride by oxidizing and nitriding the HEA powders [202, 203].

Equal-channel angular pressing is the most developed SPD processing technique. In a typical ECAP, a rod-like sample is pressed through a die constrained within a channel which is bent at an abrupt angle [192]. Hammond et al. tried to consolidate the mechanical alloyed AlFeMgTiZn HEA powders via ECAP [204]. The microstructures of HEA bulks pressed at different temperatures were quite complex, consisting of a mixture of solid solutions and secondary phases, as well as intermetallics and/or fine precipitates. Nevertheless, reasonably high hardness and adequate mechanical properties with a strain hardening behavior were observed. In the attempts to prepare bulk CoCrFeNiMn HEA by ECAP, Shahmir et al. observed a build-up of dislocation density and shrinkage of grain size with several passes, which resulted from the high internal stresses and elastic distortions within the crystal lattice [205]. Consistent results between calculation and experiments revealed that grain boundary strengthening was the most important strengthening mechanism in ECAP samples.

The use of the SPD technique to manipulate the shape and mechanical properties of HEMs requires large deformation in the crystal grains. Considering the brittleness of HECs, it would be a challenge to apply SPD in the process of HECs. Thus, no efforts regarding processing HECs by SPD have been reported yet.

## References


1 Suryanarayana, C. (2001). Mechanical alloying and milling. *Prog. Mater Sci.* 46: 1–184.

2 Varalakshmi, S., Kamaraj, M., and Murty, B.S. (2008). Synthesis and characterization of nanocrystalline AlFeTiCrZnCu high entropy solid solution by mechanical alloying. *J. Alloys Compd.* 460: 253–257.

3 Varalakshmi, S., Kamaraj, M., and Murty, B.S. (2010). Formation and stability of equiatomic and nonequiatomic nanocrystalline CuNiCoZnAlTi high-entropy alloys by mechanical alloying. *Metall. Mater. Trans. A* 41: 2703–2709.

4 Chen, Y., Hu, Y., Tsai, C. et al. (2009). Alloying behavior of binary to octonary alloys based on Cu–Ni–Al–Co–Cr–Fe–Ti–Mo during mechanical alloying. *J. Alloys Compd.* 477: 696–705.

5 Chen, Y., Hu, Y., and Hsieh, C. (2009). Competition between elements during mechanical alloying in an octonary multi-principal-element alloy system. *J. Alloys Compd.* 481: 768–775.

6 Vaidya, M., Muralikrishna, G., and Murty, B.S. (2019). High-entropy alloys by mechanical alloying: a review. *J. Mater. Res.* 34 (5): 664–686.

7 Salemi, F., Abbasi, M., and Karimzadeh, F. (2016). Synthesis and thermodynamic analysis of nanostructured CuNiCoZnAl high entropy alloy produced by mechanical alloying. *J. Alloys Compd.* 685: 278–286.

8 Tan, X., Zhang, G., Zhi, Q. et al. (2016). Effects of milling on the microstructure and hardness of $Al_2NbTi_3V_2Zr$ high-entropy alloy. *Mater. Des.* 109: 27–36.

9 Wang, H., Gao, T., Niu, J. et al. (2016). Microstructure, thermal properties, and corrosion behaviors of FeSiBAlNi alloy fabricated by mechanical alloying and spark plasma sintering. *Int. J. Miner. Metall. Mater.* 23: 77.

10 Vaidya, M., Prasad, A., Parakh, A. et al. (2017). Influence of sequence of elemental addition on phase evolution in nanocrystalline AlCoCrFeNi: novel approach to alloy synthesis using mechanical alloying. *Mater. Des.* 126: 37–46.

11 Kunce, I., Polanski, M., Karczewski, K. et al. (2015). Microstructural characterisation of high-entropy alloy AlCoCrFeNi fabricated by laser engineered net shaping. *J. Alloys Compd.* 648: 751–758.

12 Liu, Y., Wang, J., Fang, Q. et al. (2016). Preparation of superfine-grained high entropy alloy by spark plasma sintering gas atomized powder. *Intermetallics* 68: 16–22.

13 Ding, P., Mao, A., Zhang, X. et al. (2017). Preparation, characterization and properties of multicomponent $AlCoCrFeNi_{2.1}$ powder by gas atomization method. *J. Alloys Compd.* 721: 609–614.

14 Li, C., Chang, K., and Yeh, A. (2019). On the microstructure and properties of an advanced cemented carbide system processed by selective laser melting. *J. Alloys Compd.* 782: 440–450.

15 Zhou, P., Xiao, D., Wu, Z. et al. (2019). $Al_{0.5}FeCoCrNi$ high entropy alloy prepared by selective laser melting with gas-atomized pre-alloy powders. *Mater. Sci. Eng. A* 739: 86–89.

16 Zhou, S., Zhang, P., Xue, Y. et al. (2018). Microstructure evolution of $Al_{0.6}CoCrFeNi$ high entropy alloy powder prepared by high pressure gas atomization. *Trans. Nonferrous Met. Soc. China* 28: 939.

17 Yang, C., Chau, J., Weng, C. et al. (2017). Preparation of high-entropy AlCoCrCuFeNiSi alloy powders by gas atomization process. *Mater. Chem. Phys.* 202: 151–158.

18 Singh, M. and Srivastava, C. (2015). Synthesis and electron microscopy of high entropy alloy nanoparticles. *Mater. Lett.* 160: 419–422.

19 Niu, B., Zhang, F., Ping, H. et al. (2017). Sol–gel autocombustion synthesis of nanocrystalline high-entropy alloys. *Sci. Rep.* 7: 3421.

20 Bondesgaard, M., Broge, N., Mamakhel, A. et al. (2019). General solvothermal synthesis method for complete solubility range bimetallic and high-entropy alloy nanocatalysts. *Adv. Funct. Mater.* 29: 1905933.

21 Yang, Y., Song, B., Ke, X. et al. (2020). Aerosol synthesis of high entropy alloy nanoparticles. *Langmuir* 36: 1985–1992.

22 Lee, W., Park, K., Yi, K. et al. (2019). Synthesis of spherical V–Nb–Mo–Ta–W high-entropy alloy powder using hydrogen embrittlement and spheroidization by thermal plasma. *Metals* 9: 1296.

23 Robertson, I.M., Sofronis, P., Nagao, A. et al. (2015). Hydrogen embrittlement understood. *Metall. Mater. Trans. A* 46: 2323–2341.

24 Park, K., Park, J., Kim, Y. et al. (2021). Spark plasma sintering behavior of TaNbHfZrTi high-entropy alloy powder synthesized by hydrogenation-dehydrogenation reaction. *Intermetallics* 130: 107077.

**25** Moskovskikh, D., Vorotilo, S., Sedegov, A. et al. (2020). High-entropy (HfTaTiNbZr)C and (HfTaTiNbMo)C carbides fabricated through reactive high-energy ball milling and spark plasma sintering. *Ceram. Int.* 46: 19008–19014.

**26** Chicardi, E., García-Garrido, C., Hernandez-Saz, J. et al. (2020). Synthesis of all equiatomic five-transition metals high entropy carbides of the IVB (Ti, Zr, Hf) and VB (V, Nb, Ta) groups by a low temperature route. *Ceram. Int.* 46: 21421–21430.

**27** Chicardi, E., García-Garrido, C., and Gotor, F. (2019). Low temperature synthesis of an equiatomic $(TiZrHfVNb)C_5$ high entropy carbide by a mechanically-induced carbon diffusion route. *Ceram. Int.* 45: 21858–21863.

**28** Guan, J., Li, D., Yang, Z. et al. (2020). Synthesis and thermal stability of novel high-entropy metal boron carbonitride ceramic powders. *Ceram. Int.* 46: 26581–26589.

**29** James, S.L., Adams, C.J., Bolm, C. et al. (2012). Mechanochemistry: opportunities for new and cleaner synthesis. *Chem. Soc. Rev.* 41: 413–447.

**30** Jin, T., Sang, X., Unocic, R. et al. (2018). Mechanochemical-assisted synthesis of high-entropy metal nitride via a soft urea strategy. *Adv. Mater.* 30: 1707512.

**31** Chen, H., Lin, W., Zhang, Z. et al. (2019). Mechanochemical synthesis of high entropy oxide materials under ambient conditions: dispersion of catalysts via entropy maximization. *ACS Mater. Lett.* 1: 83–88.

**32** Sarkar, A., Djenadic, R., Usharani, N. et al. (2016). Nanocrystalline multicomponent entropy stabilised transition metal oxides. *J. Eur. Ceram. Soc.* 37: 747–754.

**33** Biesuz, M., Spiridigliozzi, L., Dell'Agli, G. et al. (2018). Synthesis and sintering of (Mg, Co, Ni, Cu, Zn)O entropy-stabilized oxides obtained by wet chemical methods. *J. Mater. Sci.* 53: 8074–8085.

**34** Zhao, Z., Xiang, H., Dai, F. et al. (2019). $(La_{0.2}Ce_{0.2}Nd_{0.2}Sm_{0.2}Eu_{0.2})_2Zr_2O_7$: a novel high-entropy ceramic with low thermal conductivity and sluggish grain growth rate. *J. Mater. Sci. Technol.* 35: 2647–2651.

**35** Zhao, Z., Xiang, H., Chen, H. et al. (2020). High-entropy $(Nd_{0.2}Sm_{0.2}Eu_{0.2}Y_{0.2}Yb_{0.2})_4Al_2O_9$ with good high temperature stability, low thermal conductivity, and anisotropic thermal expansivity. *J. Adv. Ceram.* 9: 595–605.

**36** Zhao, Z., Chen, H., Xiang, H. et al. (2019). $(La_{0.2}Ce_{0.2}Nd_{0.2}Sm_{0.2}Eu_{0.2})PO_4$: a high-entropy rare-earth phosphate monazite ceramic with low thermal conductivity and good compatibility with $Al_2O_3$. *J. Mater. Sci. Technol.* 35: 2892–2896.

**37** Nair, G. and Dhoble, S. (2020). *The Fundamentals and Applications of Light-Emitting Diodes*. Duxford: Elsevier.

**38** Karhunen, T., Lähde, A., Leskinen, J. et al. (2011). Transition metal-doped lithium titanium oxide nanoparticles made using flame spray pyrolysis. *Int. Sch. Res. Notices* 2011: 180821.

**39** Sarkar, A., Djenadic, R., Wang, D. et al. (2018). Rare earth and transition metal based entropy stabilised perovskite type oxides. *J. Eur. Ceram. Soc.* 38: 2318–2327.

**40** Wang, J., Stenzel, D., Stenzel, D. et al. (2020). Spinel to rock-salt transformation in high entropy oxides with Li incorporation. *Electrochemistry* 1: 60–74.

**41** Djenadic, R., Sarkar, A., Clemens, O. et al. (2017). Multicomponent equiatomic rare earth oxides. *Mater. Res. Lett.* 5: 102–109.

**42** Tseng, K., Yang, Q., McCormack, S. et al. (2020). High-entropy, phase-constrained, lanthanide sesquioxide. *J. Am. Ceram. Soc.* 103: 569–576.

**43** Yoon, B., Avila, V., Raj, R. et al. (2020). Reactive flash sintering of the entropy-stabilized oxide $Mg_{0.2}Ni_{0.2}Co_{0.2}Cu_{0.2}Zn_{0.2}O$. *Scr. Mater.* 181: 48–52.

**44** Du, B., Liu, H., and Chu, Y. (2020). Fabrication and characterization of polymer-derived high-entropy carbide ceramic powders. *J. Am. Ceram. Soc.* 103: 4063–4068.

**45** Sun, Y., Chen, F., Qiu, W. et al. (2020). Synthesis of rare earth containing single-phase multicomponent metal carbides via liquid polymer precursor route. *J. Am. Ceram. Soc.* 103: 6081–6087.

**46** Mao, A., Xiang, H., Zhang, Z. et al. (2019). Solution combustion synthesis and magnetic property of rock-salt $(Co_{0.2}Cu_{0.2}Mg_{0.2}Ni_{0.2}Zn_{0.2})O$ high-entropy oxide nanocrystalline powder. *J. Magn. Magn. Mater.* 484: 245–252.

**47** Okejiri, F., Zhang, Z., Liu, J. et al. (2020). Room-temperature synthesis of high-entropy perovskite oxide nanoparticle catalysts via ultrasonication-based method. *ChemSusChem* 13: 111–115.

**48** Dong, Y., Ren, K., Lu, Y. et al. (2019). High-entropy environmental barrier coating for the ceramic matrix composites. *J. Eur. Ceram. Soc.* 39: 2574–2579.

**49** Wang, D., Liu, Z., Du, S. et al. (2019). Low-temperature synthesis of small-sized high-entropy oxides for water oxidation. *J. Mater. Chem. A* 7: 24211.

**50** Rost, C.M., Sachet, E., Borman, T. et al. (2015). Entropy-stabilized oxides. *Nat. Commun.* 6: 8485.

**51** Liu, D., Liu, H., Ning, S. et al. (2019). Synthesis of high-purity high-entropy metal diboride powders by boro/carbothermal reduction. *J. Am. Ceram. Soc.* 102: 7071–7076.

**52** Chen, H., Zhao, B., Zhao, Z. et al. (2020). Achieving strong microwave absorption capability and wide absorption bandwidth through a combination of high entropy rare earth silicide carbides/rare earth oxides. *J. Mater. Sci. Technol.* 2020 (47): 216–222.

**53** Gild, J., Zhang, Y., Harrington, T. et al. (2016). High-entropy metal diborides: a new class of high-entropy materials and a new type of ultrahigh temperature ceramics. *Sci. Rep.* 6: 37946.

**54** Liu, D., Wen, T., Ye, B. et al. (2019). Synthesis of superfine high-entropy metal diboride powders. *Scr. Mater.* 167: 110–114.

**55** Zhang, Y., Guo, W., Jiang, Z. et al. (2019). Dense high-entropy boride ceramics with ultra-high hardness. *Scr. Mater.* 164: 135–139.

**56** Chen, H., Xiang, H., Dai, F. et al. (2019). Porous high entropy $(Zr_{0.2}Hf_{0.2}Ti_{0.2}Nb_{0.2}Ta_{0.2})B_2$: a novel strategy towards making ultrahigh temperature ceramics thermal insulating. *J. Mater. Sci. Technol.* 35: 2404–2408.

**57** Tallarita, G., Licheri, R., Garroni, S. et al. (2019). Novel processing route for the fabrication of bulk high-entropy metal diborides. *Scr. Mater.* 158: 100–104.

**58** Feng, L., Fahrenholtz, W.G., Hilmas, G.E. et al. (2019). Synthesis of single-phase high-entropy carbide powders. *Scr. Mater.* 162: 90–93.

**59** Ye, B., Wen, T., Huang, K. et al. (2019). First-principles study, fabrication, and characterization of $(Hf_{0.2}Zr_{0.2}Ta_{0.2}Nb_{0.2}Ti_{0.2})C$ high-entropy ceramic. *J. Am. Ceram. Soc.* 102: 4344–4352.

**60** Chen, H., Xiang, H., Dai, F. et al. (2019). High porosity and low thermal conductivity high entropy ($Zr_{0.2}Hf_{0.2}Ti_{0.2}Nb_{0.2}Ta_{0.2}$)C. *J. Mater. Sci. Technol.* 35: 1700–1705.

**61** Peng, C., Gao, X., Wang, M. et al. (2019). Diffusion-controlled alloying of single-phase multi-principal transition metal carbides with high toughness and low thermal diffusivity. *Appl. Phys. Lett.* 114: 011905.

**62** He, Y., Peng, C., Xin, S. et al. (2020). Vacancy effect on the preparation of high-entropy carbides. *J. Mater. Sci.* 55: 6754–6760.

**63** Wright, A.J., Wang, Q., Huang, C. et al. (2020). From high-entropy ceramics to compositionally-complex ceramics: a case study of fluorite oxides. *J. Eur. Ceram. Soc.* 40: 2120–2129.

**64** Gu, J., Zou, J., Sun, S. et al. (2019). Dense and pure high-entropy metal diboride ceramics sintered from self-synthesized powders via boro/carbothermal reduction approach. *Sci. China Mater.* 62: 1898–1909.

**65** Song, J., Chen, G., Xiang, H. et al. (2022). Regulating the formation ability and mechanical properties of high-entropy transition metal carbides by carbon stoichiometry. *J. Mater. Sci. Technol.* 121: 181–189.

**66** Cantor, B., Chang, I.T.H., Knight, P. et al. (2004). Microstructural development in equiatomic multicomponent alloys. *Mater. Sci. Eng., A* 375–377: 213–218.

**67** Yeh, J., Chen, S., Lin, S.J. et al. (2004). Nanostructured high-entropy alloys with multiple principal elements: novel alloy design concepts and outcomes. *Adv. Eng. Mater.* 6: 299–303.

**68** Zhang, Y., Ma, S., and Qiao, J. (2012). Morphology transition from dendrites to equiaxed grains for AlCoCrFeNi high-entropy alloys by copper mold casting and Bridgman solidification. *Metall. Mater. Trans. A* 43: 2625–2630.

**69** Gao, M.C. and Alman, D.E. (2013). Searching for next single-phase high-entropy alloy compositions. *Entropy* 15: 4504–4519.

**70** Santodonato, L.J., Zhang, Y., Feygenson, M. et al. (2015). Deviation from high-entropy configurations in the atomic distributions of a multi-principal-element alloy. *Nat. Commun.* 6: 5964.

**71** Baldenebro-Lopez, F.J., Herrera-Ramírez, J.M., Arredondo-Rea, S.P. et al. (2015). Simultaneous effect of mechanical alloying and arc-melting processes in the microstructure and hardness of an AlCoFeMoNiTi high-entropy alloy. *J. Alloys Compd.* 643: S250–S255.

**72** Hu, L., Liu, B., Ye, F. et al. (2011). Ca–Mg–Zn–(Ag) bulk metallic glasses prepared by unidirectional quenching. *Intermetallics* 19: 662–665.

**73** Ma, S., Zhang, S., Gao, M. et al. (2013). A successful synthesis of the $CoCrFeNiAl_{0.3}$ single-crystal, high-entropy alloy by Bridgman solidification. *JOM* 65: 1751–1758.

**74** Ma, S., Zhang, S., Qiao, J. et al. (2014). Superior high tensile elongation of a single-crystal $CoCrFeNiAl_{0.3}$ high-entropy alloy by Bridgman solidification. *Intermetallics* 54: 104–109.

**75** Zhang, K., Fu, Z., Zhang, J. et al. (2010). Characterization of nanocrystalline CoCrFeNiTiAl high-entropy solid solution processed by mechanical alloying. *J. Alloys Compd.* 495: 33–38.

**76** Praveen, S., Murty, B.S., and Kottada, R.S. (2012). Alloying behavior in multi-component AlCoCrCuFe and NiCoCrCuFe high entropy alloys. *Mater. Sci. Eng., A* 534: 83–89.

**77** Fu, Z., Chen, W., Xiao, H. et al. (2013). Fabrication and properties of nanocrystalline $Co_{0.5}FeNiCrTi_{0.5}$ high entropy alloy by MA-SPS technique. *Mater. Des.* 44: 535–539.

**78** Ji, W., Wang, W., Wang, H. et al. (2015). Alloying behavior and novel properties of CoCrFeNiMn high-entropy alloy fabricated by mechanical alloying and spark plasma sintering. *Intermetallics* 56: 24–27.

**79** Fu, Z., Chen, W., Wen, H. et al. (2016). Microstructure and strengthening mechanisms in an FCC structured single-phase nanocrystalline $Co_{25}Ni_{25}Fe_{25}Al_{7.5}Cu_{17.5}$ high-entropy alloy. *Acta Mater.* 107: 59–71.

**80** Zhang, A., Han, J., Meng, J. et al. (2016). Rapid preparation of AlCoCrFeNi high entropy alloy by spark plasma sintering from elemental powder mixture. *Mater. Lett.* 181: 82–85.

**81** Qiu, H., Fang, G., Wen, Y. et al. (2019). Nanoporous high-entropy alloys for highly stable and efficient catalysis. *J. Mater. Chem. A* 7: 6499–6506.

**82** Li, S., Tang, X., Jia, H. et al. (2020). Nanoporous high-entropy alloys with low Pt loadings for high-performance electrochemical oxygen reduction. *J. Catal.* 383: 164–171.

**83** Jin, Z., Lv, J., Jia, H. et al. (2019). Nanoporous Al–Ni–Co–Ir–Mo high-entropy alloy for record-high water splitting activity in acidic environments. *Small* 15: 1904180.

**84** Cai, Z., Goou, H., Ito, Y. et al. (2021). Nanoporous ultra-high-entropy alloys containing fourteen elements for water splitting electrocatalysis. *Chem. Sci.* 12: 11306.

**85** Yang, L., Li, Y., Wang, Z. et al. (2019). Nanoporous quasi-high-entropy alloy microspheres. *Metals* 9: 345.

**86** Joo, S., Bae, J., Park, W. et al. (2019). Beating thermal coarsening in nanoporous materials via high-entropy design. *Adv. Mater.* 32: 1906160.

**87** Tang, J., Xu, J., Ye, Z. et al. (2021). Microwave sintered porous CoCrFeNiMo high entropy alloy as an efficient electrocatalyst for alkaline oxygen evolution reaction. *J. Mater. Sci. Technol.* 79: 171–177.

**88** Xu, Z., Zhu, Z., Wang, P. et al. (2020). Fabrication of porous CoCrFeMnNi high entropy alloy using binder jetting additive manufacturing. *Addit. Manuf.* 35: 101441.

**89** Xiang, H., Xing, Y., Dai, F. et al. (2021). High-entropy ceramics: present status, challenges, and a look forward. *J. Adv. Ceram.* 10: 385–441.

**90** Qiu, S., Xiang, H., Dai, F. et al. (2021). Preparation and molten salts corrosion behavior of high-entropy $(Ca,Sr,Ba)TMO_3$ (TM = Zr, Hf and Ta): promising protecting materials of reference electrode for metal smelting through fused salts electrolysis. *Corr. Comm.* 3: 10–22.

**91** Zhu, H., Liu, L., Xiang, H. et al. (2022). Improved thermal stability and infrared emissivity of high-entropy $REMgAl_{11}O_{19}$ and $LaMAl_{11}O_{19}$ (RE = La, Nd, Gd, Sm, Pr, Dy; M = Mg, Fe, Co, Ni, Zn). *J. Mater. Sci. Technol.* 104: 131–144.

**92** Sun, Y., Xiang, H., Dai, F. et al. (2021). Preparation and properties of CMAS resistant bixbyite structured high-entropy oxides $RE_2O_3$ (RE = Sm, Eu, Er, Lu, Y, and Yb): promising environmental barrier coating materials for $Al_2O_{3f}/Al_2O_3$ composites. *J. Adv. Ceram.* 10: 596–613.

**93** Zhao, Z., Chen, H., Xiang, H. et al. (2020). High entropy defective fluorite structured rare-earth niobates and tantalates for thermal barrier applications. *J. Adv. Ceram.* 9: 303–311.

**94** Chen, H., Zhao, Z., Xiang, H. et al. (2020). High entropy $(Y_{0.2}Yb_{0.2}Lu_{0.2}Eu_{0.2}Er_{0.2})Al_5O_{12}$: a novel high temperature stable thermal barrier material. *J. Mater. Sci. Technol.* 48: 57–62.

**95** Zhao, Z., Chen, H., Xiang, H. et al. (2020). High-entropy $(Y_{0.2}Nd_{0.2}Sm_{0.2}Eu_{0.2}Er_{0.2})AlO_3$: a promising thermal/environmental barrier material for oxide/oxide composites. *J. Mater. Sci. Technol.* 47: 45–51.

**96** Chen, H., Xiang, H., Dai, F. et al. (2020). High entropy $(Yb_{0.25}Y_{0.25}Lu_{0.25}Er_{0.25})_2SiO_5$ with strong anisotropy in thermal expansion. *J. Mater. Sci. Technol.* 36: 134–139.

**97** Zhao, Z., Chen, H., Xiang, H. et al. (2020). $(Y_{0.25}Yb_{0.25}Er_{0.25}Lu_{0.25})_2(Zr_{0.5}Hf_{0.5})_2O_7$: a defective fluorite structured high entropy ceramic with low thermal conductivity and close thermal expansion coefficient to $Al_2O_3$. *J. Mater. Sci. Technol.* 39: 167–172.

**98** Bérardan, D., Franger, S., Meena, A.K. et al. (2016). Room temperature lithium superionic conductivity in high entropy oxides. *J. Mater. Chem. A* 4: 9536.

**99** Yu, D., Yin, J., Zhang, B. et al. (2021). Pressureless sintering and properties of $(Hf_{0.2}Zr_{0.2}Ta_{0.2}Nb_{0.2}Ti_{0.2})C$ high-entropy ceramics: the effect of pyrolytic carbon. *J. Eur. Ceram. Soc.* 41: 3823–3831.

**100** Rezaei, F., Kakroudi, M.G., Shahedifar, V. et al. (2017). Consolidation and mechanical properties of hot pressed TaC–HfC-VC composites. *Ceram. Int.* 43: 15537–15543.

**101** Chen, L., Zhang, W., Tan, Y. et al. (2021). Influence of vanadium content on the microstructural evolution and mechanical properties of (TiZrHfVNbTa)C high-entropy carbides processed by pressureless sintering. *J. Eur. Ceram. Soc.* 41: 60–67.

**102** Chen, X. and Wu, Y. (2020). High-entropy transparent fluoride laser ceramics. *J. Am. Ceram. Soc.* 103: 750–756.

**103** Wang, K., Chen, L., Xu, C. et al. (2020). Microstructure and mechanical properties of (TiZrNbTaMo)C high-entropy ceramic. *J. Mater. Sci. Technol.* 39: 99–105.

**104** Wen, T., Ye, B., Nguyen, M.C. et al. (2020). Thermophysical and mechanical properties of novel high-entropy metal nitride-carbides. *J. Am. Ceram. Soc.* 103: 6475–6489.

**105** Liu, J., Shen, X., Wu, Y. et al. (2020). Mechanical properties of hot-pressed high-entropy diboride-based ceramics. *J. Adv. Ceram.* 9: 503–510.

**106** Feng, L., Fahrenholtz, W.G., and Hilmas, G.E. (2019). Low-temperature sintering of single-phase, high-entropy carbide ceramics. *J. Am. Ceram. Soc.* 102: 7217–7224.

**107** Li, M., Zhao, X., Shao, G. et al. (2021). Oscillatory pressure sintering of high entropy $(Zr_{0.2}Ta_{0.2}Nb_{0.2}Hf_{0.2}Mo_{0.2})B_2$ ceramic. *Ceram. Int.* 47: 8707–8710.

**108** Tallarita, G., Licheri, R., Garroni, S. et al. (2020). High-entropy transition metal diborides by reactive and non-reactive spark plasma sintering: a comparative investigation. *J. Eur. Ceram. Soc.* 40: 942–952.

**109** Zhang, Y., Jiang, Z., Sun, S. et al. (2019). Microstructure and mechanical properties of high-entropy borides derived from boro/carbothermal reduction. *J. Eur. Ceram. Soc.* 39: 3920–3924.

**110** Gild, J., Kaufmann, K., Vecchio, K. et al. (2019). Reactive flash spark plasma sintering of high-entropy ultrahigh temperature ceramics. *Scr. Mater.* 170: 106–110.

**111** Liu, D., Peng, X., Liu, J. et al. (2020). Ultrafast synthesis of entropy-stabilized oxide at room temperature. *J. Eur. Ceram. Soc.* 40: 2504–2508.

**112** Kumar, A., Sharma, G., Aftab, A. et al. (2020). Flash assisted synthesis and densification of five component high entropy oxide (Mg, Co, Cu, Ni, Zn)O at 350 °C in 3 min. *J. Eur. Ceram. Soc.* 40: 3358–3362.

**113** Liu, J., Ren, K., Ma, C. et al. (2020). Dielectric and energy storage properties of flash-sintered high-entropy $(Bi_{0.2}Na_{0.2}K_{0.2}Ba_{0.2}Ca_{0.2})TiO_3$ ceramic. *Ceram. Int.* 46: 20576–20581.

**114** Mao, H., Guo, R., Cao, Y. et al. (2021). Ultrafast densification of high-entropy oxide $(La_{0.2}Nd_{0.2}Sm_{0.2}Eu_{0.2}Gd_{0.2})_2Zr_2O_7$ by reactive flash sintering. *J. Eur. Ceram. Soc.* 41: 2855–2860.

**115** Failla, S., Galizia, P., Fu, S. et al. (2020). Formation of high entropy metal diborides using arc-melting and combinatorial approach to study quinary and quaternary solid solutions. *J. Eur. Ceram. Soc.* 40: 588–593.

**116** Chen, H., Xiang, H., Dai, F. et al. (2019). Low thermal conductivity and high porosity ZrC and HfC ceramics prepared by in situ reduction reaction/partial sintering method for ultrahigh temperature applications. *J. Mater. Sci. Technol.* 35: 2778–2784.

**117** Chen, H., Xiang, H., Dai, F. et al. (2019). High strength and high porosity $YB_2C_2$ ceramics prepared by a new high temperature reaction/partial sintering process. *J. Mater. Sci. Technol.* 35: 2883–2891.

**118** Guo, Q., Xiang, H., Sun, X. et al. (2015). Preparation of porous $YB_4$ ceramics using a combination of in situ borothermal reaction and high temperature partial sintering. *J. Eur. Ceram. Soc.* 35: 3411–3418.

**119** Chen, H., Zhao, Z., Xiang, H. et al. (2020). Effect of reaction routes on the porosity and permeability of porous high entropy $(Y_{0.2}Yb_{0.2}Sm_{0.2}Nd_{0.2}Eu_{0.2})B_6$ for transpiration cooling. *J. Mater. Sci. Technol.* 38: 80–85.

**120** Wang, G., Qin, J., Feng, Y. et al. (2020). Sol–gel synthesis of spherical mesoporous high-entropy oxides. *ACS Appl. Mater. Interfaces* 12: 45155–45164.

**121** Zhang, X., Xue, L., Yang, F. et al. (2021). $(La_{0.2}Y_{0.2}Nd_{0.2}Gd_{0.2}Sr_{0.2})CrO_3$: a novel conductive porous high-entropy ceramic synthesized by the sol-gel method. *J. Alloys Compd.* 863: 158763.

**122** Zhang, H., Pan, Y., and He, Y. (2011). Synthesis and characterization of FeCoNiCrCu high-entropy alloy coating by laser cladding. *Mater. Des.* 32: 1910–1915.

**123** Zhang, H., Pan, Y., He, Y. et al. (2011). Microstructure and properties of 6FeNiCoSiCrAlTi high-entropy alloy coating prepared by laser cladding. *Appl. Surf. Sci.* 257: 2259–2263.

**124** Shu, F., Yang, B., Dong, S. et al. (2018). Effects of Fe-to-Co ratio on microstructure and mechanical properties of laser cladded FeCoCrBNiSi high-entropy alloy coatings. *Appl. Surf. Sci.* 450: 538–544.

**125** Arif, Z.U., Khalid, M.Y., Rehman, E. et al. (2021). A review on laser cladding of high-entropy alloys, their recent trends and potential applications. *J. Manuf. Processes* 68: 225–273.

**126** Chao, Q., Guo, T., Jarvis, T. et al. (2017). Direct laser deposition cladding of $Al_xCoCrFeNi$ high entropy alloys on a high-temperature stainless steel. *Surf. Coat. Technol.* 332: 440–451.

**127** Guo, Y. and Liu, Q. (2018). MoFeCrTiWAlNb refractory high-entropy alloy coating fabricated by rectangular-spot laser cladding. *Intermetallics* 102: 78–87.

**128** Shu, F., Zhang, B., Liu, T. et al. (2019). Effects of laser power on microstructure and properties of laser cladded CoCrBFeNiSi high-entropy alloy amorphous coatings. *Surf. Coat. Technol.* 358: 667–675.

**129** Ni, C., Shi, Y., Liu, J. et al. (2018). Characterization of $Al_{0.5}FeCu_{0.7}NiCoCr$ high-entropy alloy coating on aluminum alloy by laser cladding. *Opt. Laser Technol.* 105: 257–263.

**130** Huang, P., Yeh, J., Shun, T. et al. (2004). Multi-principal-element alloys with improved oxidation and wear resistance for thermal spray coating. *Adv. Eng. Mater.* 6: 74–78.

**131** Wang, L., Chen, C., Yeh, J. et al. (2011). The microstructure and strengthening mechanism of thermal spray coating $Ni_xCo_{0.6}Fe_{0.2}Cr_ySi_zAlTi_{0.2}$ high-entropy alloys. *Mater. Chem. Phys.* 126: 880–885.

**132** Yue, T., Xie, H., Lin, X. et al. (2013). Microstructure of laser re-melted AlCoCrCuFeNi high entropy alloy coatings produced by plasma spraying. *Entropy* 15: 2833–2845.

**133** Wang, W., Qi, W., Xie, L. et al. (2019). Microstructure and corrosion behavior of $(CoCrFeNi)_{95}Nb_5$ high-entropy alloy coating fabricated by plasma spraying. *Materials* 12: 694.

**134** Ang, A.S.M., Berndt, C.C., Sesso, M.L. et al. (2014). Plasma-sprayed high entropy alloys: microstructure and properties of AlCoCrFeNi and MnCoCrFeNi. *Metall. Mater. Trans. A* 46A: 791–800.

**135** Löbel, M., Lindner, T., Mehner, T. et al. (2017). Microstructure and wear resistance of AlCoCrFeNiTi high-entropy alloy coatings produced by HVOF. *Coat* 7: 144.

**136** Srivastava, M., Jadhav, M., Chethan et al. (2019). Synthesis and properties of high velocity oxy-fuel sprayed $FeCoCrNi_2Al$ high entropy alloy coating. *Surf. Coat. Technol.* 378: 124950.

**137** Li, T., Liu, Y., Liu, B. et al. (2017). Microstructure and wear behavior of $FeCoCrNiMo_{0.2}$ high entropy coatings prepared by air plasma spray and the high velocity oxy-fuel spray processes. *Coatings* 7: 151.

**138** Hus, W., Murakami, H., Yeh, J. et al. (2017). On the study of thermal-sprayed $Ni_{0.2}Co_{0.6}Fe_{0.2}CrSi_{0.2}AlTi_{0.2}$ HEA overlay coating. *Surf. Coat. Technol.* 316: 71–74.

**139** Zhou, L., Li, F., Liu, J. et al. (2020). High-entropy thermal barrier coating of rare-earth zirconate: a case study on $(La_{0.2}Nd_{0.2}Sm_{0.2}Eu_{0.2}Gd_{0.2})_2Zr_2O_7$ prepared by atmospheric plasma spraying. *J. Eur. Ceram. Soc.* 40: 5731–5739.

**140** Zhang, D., Ye, Y., Feng, X. et al. (2022). Thermal barrier coatings with high-entropy oxide as a top coat. *Ceram. Int.* 40: 1349–1359.

**141** Park, T., Adomako, N.K., Ashong, A. et al. (2021). Interfacial structure and physical properties of high-entropy oxide coatings prepared via atmospheric plasma spraying. *Coatings* 11: 755.

**142** Yin, S., Li, W., Song, B. et al. (2019). Deposition of FeCoNiCrMn high entropy alloy (HEA) coating via cold spraying. *J. Mater. Sci. Technol.* 35: 1003–1007.

**143** Moridi, A., Hassani-Gangaraj, S.M., Guagliano, M. et al. (2014). Cold spray coating: review of material systems and future perspectives. *Surf. Eng.* 36: 369–395.

**144** Velasco, S.C., Cavaleiro, A., and Carvalho, S. (2016). Functional properties of ceramic-Ag nanocomposite coatings produced by magnetron sputtering. *Prog. Mater Sci.* 84: 158–191.

**145** Tsai, M., Yeh, J., and Gan, J. (2008). Diffusion barrier properties of AlMoNbSiTaTiVZr high-entropy alloy layer between copper and silicon. *Thin Solid Films* 516: 5527–5530.

**146** Tüten, N., Canadinc, D., Motallebzadeh, A. et al. (2019). Microstructure and tribological properties of TiTaHfNbZr high entropy alloy coatings deposited on Ti–6Al–4V substrates. *Intermetallics* 105: 99–106.

**147** Khan, N.A., Akhavan, B., Zhou, H. et al. (2019). High entropy alloy thin films of $AlCoCrCu_{0.5}FeNi$ with controlled microstructure. *Appl. Surf. Sci.* 495: 143560.

**148** Liao, W., Lan, S., Gao, L. et al. (2017). Nanocrystalline high-entropy alloy ($CoCrFeNiAl_{0.3}$) thin-film coating by magnetron sputtering. *Thin Solid Films* 638: 383–388.

**149** Braeckman, B.R., Boydens, F., Hidalgo, H. et al. (2015). High entropy alloy thin films deposited by magnetron sputtering of powder targets. *Thin Solid Films* 580: 71–76.

**150** Khan, N.A., Akhavan, B., Zhou, C. et al. (2020). RF magnetron sputtered $AlCoCrCu_{0.5}FeNi$ high entropy alloy (HEA) thin films with tuned microstructure and chemical composition. *J. Alloys Compd.* 836: 155348.

**151** Medina, L.Z., Riekehr, L., and Jansson, U. (2020). Phase formation in magnetron sputtered CrMnFeCoNi high entropy alloy. *Surf. Coat. Technol.* 403: 126323.

**152** Fritze, S., Koller, C.M., Fieandt, L. et al. (2019). Influence of deposition temperature on the phase evolution of HfNbTiVZr high-entropy thin films. *Materials* 12: 587.

**153** Tsai, C., Lai, S., Cheng, K. et al. (2012). Strong amorphization of high-entropy AlBCrSiTi nitride film. *Thin Solid Films* 520: 2613–2618.

**154** Hruška, P., Lukác, F., Cichoň, S. et al. (2021). Oxidation of amorphous HfNbTaTiZr high entropy alloy thin films prepared by DC magnetron sputtering. *J. Alloys Compd.* 869: 157978.

**155** Sha, C., Zhou, Z., Xie, Z. et al. (2020). High entropy alloy FeMnNiCoCr coatings: enhanced hardness and damage-tolerance through a dual-phase structure and nanotwins. *Surf. Coat. Technol.* 385: 125435.

**156** Dolique, V., Thomann, A.L., Brault, P. et al. (2009). Complex structure/composition relationship in thin films of AlCoCrCuFeNi high entropy alloy. *Mater. Chem. Phys.* 117: 142–147.

**157** Zou, Y., Ma, H., and Spolenak, R. (2015). Ultrastrong ductile and stable high-entropy alloys at small scales. *Nat. Commun.* 6: 7748.

**158** Braeckman, B.R. and Depla, D. (2015). Structure formation and properties of sputter deposited $Nb_x$-CoCrCuFeNi high entropy alloy thin films. *J. Alloys Compd.* 646: 810–815.

**159** Chen, T.K., Shun, T.T., Yeh, J. et al. (2004). Nanostructured nitride films of multi-element high-entropy alloys by reactive DC sputtering. *Surf. Coat. Technol.* 188, 189: 193–200.

**160** Lai, C., Tsai, M., Lin, S. et al. (2007). Influence of substrate temperature on structure and mechanical, properties of multi-element (AlCrTaTiZr)N coatings. *Surf. Coat. Technol.* 201: 6993–6998.

**161** Chang, H., Huang, P., Davison, A. et al. (2008). Nitride films deposited from an equimolar Al–Cr–Mo–Si–Ti alloy target by reactive direct current magnetron sputtering. *Thin Solid Films* 516: 6402–6408.

**162** Liu, L., Zhu, J., Hou, C. et al. (2013). Dense and smooth amorphous films of multicomponent FeCoNiCuVZrAl high-entropy alloy deposited by direct current magnetron sputtering. *Mater. Des.* 46: 675–679.

**163** Hahn, R., Kirnbauer, A., Bartosik, M. et al. (2019). Toughness of Si alloyed high-entropy nitride coatings. *Mater. Lett.* 251: 238–240.

**164** Braic, M., Balaceanu, M., Vladescu, A. et al. (2013). Deposition and characterization of multi-principal-element (CuSiTiYZr)C coatings. *Appl. Surf. Sci.* 284: 671–678.

**165** Malinovskis, P., Fritze, S., Riekehr, L. et al. (2018). Synthesis and characterization of multicomponent (CrNbTaTiW)C films for increased hardness and corrosion resistance. *Mater. Des.* 149: 51–62.

**166** Kirnbauer, A., Spadt, C., Koller, C.M. et al. (2019). High-entropy oxide thin films based on Al–Cr–Nb–Ta–Ti. *Vacuum* 168: 108850.

**167** Mayrhofer, P.H., Kirnbauer, A., Ertelthaler, P. et al. (2018). High-entropy ceramic thin films; a case study on transition metal diborides. *Scr. Mater.* 149: 93–97.

**168** Lu, T., Feng, C., Wang, Z. et al. (2019). Microstructures and mechanical properties of $CoCrFeNiAl_{0.3}$ high-entropy alloy thin films by pulsed laser deposition. *Appl. Surf. Sci.* 494: 72–79.

**169** Cropper, M.D. (2018). Thin films of AlCrFeCoNiCu high-entropy alloy by pulsed laser deposition. *Appl. Surf. Sci.* 455: 153–159.

**170** Meisenheimer, P.B., Kratofl, T.J., and Heron, J.T. (2017). Giant enhancement of exchange coupling in entropy-stabilized oxide heterostructures. *Sci. Rep.* 7: 13344.

**171** Sharma, Y., Musico, B., Gao, X. et al. (2018). Single-crystal high entropy perovskite oxide epitaxial films. *Phys. Rev. Mater.* 2: 060404.

**172** Kotsonis, G.N., Rost, C.M., Harris, D.T. et al. (2018). Epitaxial entropy-stabilized oxides: growth of chemically diverse phases via kinetic bombardment. *MRS Comm.* 8: 1371–1377.

**173** Jacobson, V., Diercks, D., To, B. et al. (2021). Thin film growth effects on electrical conductivity in entropy stabilized oxides. *J. Eur. Ceram. Soc.* 41: 2617–2624.

**174** Guo, H., Wang, X., Dupuy, A.D. et al. (2022). Growth of nanoporous high-entropy oxide thin films by pulsed laser deposition. *J. Mater. Res.* 37: 124–135.

**175** He, R., Zhou, N., Zhang, K. et al. (2021). Progress and challenges towards additive manufacturing of SiC ceramic. *J. Adv. Ceram.* 10: 637–674.

**176** Han, C., Fang, Q., Shi, Y. et al. (2020). Recent advances on high-entropy alloys for 3D printing. *Adv. Mater.* 32: e1903855.

**177** Wolff, S.J., Lin, S., Faierson, E. et al. (2017). A framework to link localized cooling and properties of directed energy deposition (DED)-processed Ti–6Al–4V. *Acta Mater.* 132: 106–117.

**178** Zhao, Y., Hou, Y., Li, Z. et al. (2020). Powder-based 3D printed porous structure and its application as bone scaffold. *Front. Mater.* 7: 150.

**179** Joseph, J., Jarvis, T., Wu, X. et al. (2015). Comparative study of the microstructures and mechanical properties of direct laser fabricated and arc-melted $Al_xCoCrFeNi$ high entropy alloys. *Mater. Sci. Eng., A* 633: 184–193.

**180** Moghaddam, A.O., Shaburova, N.A., Samodurova, M.N. et al. (2021). Additive manufacturing of high entropy alloys: a practical review. *J. Mater. Sci. Technol.* 77: 131–162.

**181** Brif, Y., Thomas, M., and Todd, I. (2015). The use of high-entropy alloys in additive manufacturing. *Scr. Mater.* 99: 93–96.

**182** Li, R., Niu, P., Yuan, T. et al. (2018). Selective laser melting of an equiatomic CoCrFeMnNi high-entropy alloy: processability, non-equilibrium microstructure and mechanical property. *J. Alloys Compd.* 746: 125–134.

**183** Niu, P., Li, R., Yuan, T. et al. (2019). Microstructures and properties of an equimolar AlCoCrFeNi high entropy alloy printed by selective laser melting. *Intermetallics* 104: 24–32.

**184** Fujieda, T., Shiratori, H., Kuwabara, K. et al. (2015). First demonstration of promising selective electron beam melting method for utilizing high-entropy alloys as engineering materials. *Mater. Lett.* 159: 12–15.

**185** Shiratori, H., Fujieda, T., Yamanaka, K. et al. (2016). Relationship between the microstructure and mechanical properties of an equiatomic AlCoCrFeNi high-entropy alloy fabricated by selective electron beam melting. *Mater. Sci. Eng., A* 656: 39–46.

**186** Zhou, Y., Zhang, Z., Wang, Y. et al. (2018). Selective laser melting of typical metallic materials: an effective process prediction model developed by energy absorption and consumption analysis. *Addit. Manuf.* 25: 204–217.

**187** Yao, Y., Huang, Z., Xie, P. et al. (2018). Carbothermal shock synthesis of high-entropy-alloy nanoparticles. *Science* 359: 1489–1494.

**188** Lacey, S., Dong, Q., Huang, Z. et al. (2019). Stable multimetallic nanoparticles for oxygen electrocatalysis. *Nano Lett.* 19: 5149–5158.

**189** Xie, P., Yao, Y., Huang, Z. et al. (2019). Highly efficient decomposition of ammonia using high-entropy alloy catalysts. *Nat. Commun.* 10: 4011.

**190** Yao, Y., Huang, Z., Li, T. et al. (2020). High-throughput, combinatorial synthesis of multimetallic nanoclusters. *Proc. Natl. Acad. Sci. U.S.A.* 117: 6316–6322.

**191** Yao, Y., Liu, Z., Xie, P. et al. (2020). Computationally aided, entropy-driven synthesis of highly efficient and durable multi-elemental alloy catalysts. *Sci. Adv.* 6: eaaz0510.

**192** Valiev, R.Z., Estrin, Y., Horita, Z. et al. (2006). Producing bulk ultrafine-grained materials by severe plastic deformation. *JOM* 58: 33–39.

**193** Schuh, B., Mendez-Martin, F., Völker, B. et al. (2015). Mechanical properties, microstructure and thermal stability of a nanocrystalline CoCrFeMnNi high-entropy alloy after severe plastic deformation. *Acta Mater.* 96: 258–268.

**194** Shahmir, H., He, J., Lu, Z. et al. (2016). Effect of annealing on mechanical properties of a nanocrystalline CoCrFeNiMn high-entropy alloy processed by high-pressure torsion. *Mater. Sci. Eng., A* 676: 294–303.

**195** Tang, Q., Huang, Y., Huang, Y. et al. (2015). Hardening of an $Al_{0.3}$CoCrFeNi high entropy alloy via high-pressure torsion and thermal annealing. *Mater. Lett.* 151: 126–129.

**196** Yuan, H., Tsai, M., Sha, G. et al. (2016). Atomic-scale homogenization in an FCC-based high-entropy alloy via severe plastic deformation. *J. Alloys Compd.* 686: 15–23.

**197** Schuh, B., Völker, B., Todt, J. et al. (2018). Thermodynamic instability of a nanocrystalline, single-phase TiZrNbHfTa alloy and its impact on the mechanical properties. *Acta Mater.* 142: 201–212.

**198** Gubicza, J., Hung, P.T., Kawasaki, M. et al. (2019). Influence of severe plastic deformation on the microstructure and hardness of a CoCrFeNi high-entropy alloy: a comparison with CoCrFeNiMn. *Mater. Charact.* 154: 304–314.

**199** Edalati, P., Floriano, R., Tang, Y. et al. (2020). Ultrahigh hardness and biocompatibility of high-entropy alloy TiAlFeCoNi processed by high-pressure torsion. *Mater. Sci. Eng., C* 112: 110908.

**200** Lee, D., Choi, I., Seok, M. et al. (2015). Nanomechanical behavior and structural stability of a nanocrystalline CoCrFeNiMn high-entropy alloy processed by high-pressure torsion. *J. Mater. Res.* 30: 2804–2815.

**201** Kilmametov, A., Kulagin, R., Mazilkin, A. et al. (2019). High-pressure torsion driven mechanical alloying of CoCrFeMnNi high entropy alloy. *Scr. Mater.* 158: 29–33.

**202** Edalati, P., Wang, Q., Razavi-Khosroshahi, H. et al. (2020). Photocatalytic hydrogen evolution on a high-entropy oxide. *J. Mater. Chem. A* 8: 3814–3821.

**203** Edalati, P., Shen, X., Watanabe, M. et al. (2021). High-entropy oxynitride as a low-bandgap and stable photocatalyst for hydrogen production. *J. Mater. Chem. A* 9: 15076.

**204** Hammond, V.H., Atwater, M.A., Darling, K.A. et al. (2014). Equal-channel angular extrusion of a low-density high-entropy alloy produced by high-energy cryogenic mechanical alloying. *JOM* 66: 2021–2029.

**205** Shahmir, H., Mousavi, T., He, J. et al. (2017). Microstructure and properties of a CoCrFeNiMn high-entropy alloy processed by equal-channel angular pressing. *Mater. Sci. Eng., A* 705: 411–419.

# 5

# Characterization of High-Entropy Materials

The fascinating properties of high-entropy materials (HEMs) are shaped by their microstructures and compositional elements. The large structural and compositional complexity of HEMs provides great space for tuning their properties; however, the design and preparation of HEMs with desired properties still hinge on the clear understanding of the composition–structure–property relationships at microscale and nanoscale. In this chapter, we focus on the advanced methods for the characterization of the phase composition, elemental distribution, lattice distortion, and valence state of the elements in HEMs.

## 5.1 Phase Identification

The phase composition is closely related to the microstructures and properties of materials. For HEMs, to maximize the mixing entropy, all elements distribute homogeneously on their own sites to form a single solid solution, ideally. Identification of phase composition of these solid solutions is commonly and conveniently conducted by X-ray diffraction (XRD) because XRD pattern is a fingerprint of a specific phase. Cantor et al. verified the face-centered cubic (FCC) structure of FeCrMnNiCo alloy by XRD and determined the lattice parameter for this FCC high-entropy alloy (HEA) to be 0.359 nm from its XRD trace [1]. Yeh et al. identified the simple solid-solution structures of $CuCoNiCrAl_xFe$ alloys, essentially FCC and body-centered cubic (BCC), from their XRD patterns [2]. These alloys with aluminum contents from $x = 0$ to 0.5 showed a simple FCC structure. As $x$ exceeded 0.8, a BCC structure appeared in addition to the FCC. A single BCC structure was subsequently obtained for $x > 2.8$. The same technique has also been used to observe the phase evolution of entropy-stabilized oxide (MgNiCoCuZn)O with temperatures, as shown in Figure 5.1, and the threshold temperature for complete solubility of this high-entropy ceramic (HEC) was 875 °C [3].

XRD is proved to be a powerful method in phase identification; however, due to the detection limitation, it is problematic to tell phases with the same Bravais lattice apart if the lattice parameter was close and to identify minor and/or unknown

*High-Entropy Materials: From Basics to Applications,* First Edition.
Huimin Xiang, Fu-Zhi Dai, and Yanchun Zhou.

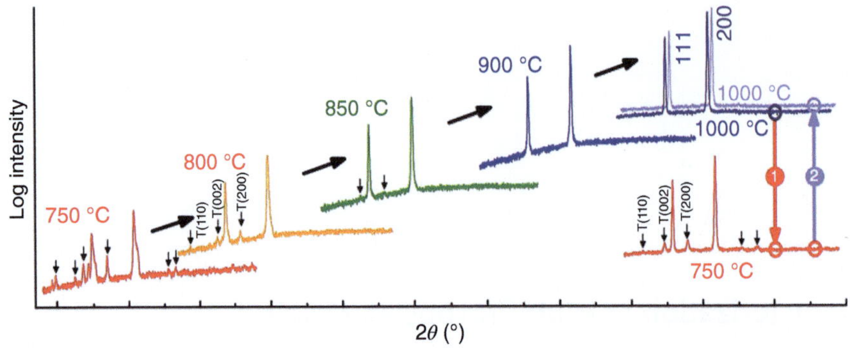


**Figure 5.1** X-ray diffraction patterns for entropy-stabilized oxide (MgNiCoCuZn)O at different temperatures. Peaks indexed with (T) and with (rocksalt [RS]) correspond to tenorite and rocksalt phases, respectively. Source: Rost et al. [3]/Springer Nature/CC BY-4.0.

phases. Other techniques are needed to identify the phase composition and microstructure accurately. As demonstrated by Yeh et al. from the XRD pattern of CuCoNiCrAlFe alloy, two simple structures could be identified, namely an FCC and an order BCC structure [2]. With the help of scanning electron microscopy (SEM), a dendritic microstructure that was commonly observed in the casting alloys was observed. Using transmission electron microscopy (TEM) and selected area electron diffraction (SAED), more details were revealed. A spinodal decomposition occurred in BCC phase, leading to a modulated structure composed of ordered (B2) and disorder (A2) BCC phases with very close lattice parameters. Also, nanoprecipitations in the spinodal plate and inter-spinodal plate were observed, with diameters of 7–50 and 3 nm, respectively. Gild et al. utilized a reactive flash spark plasma sintering (ReaFSPS) to synthesize, densify, and homogenize a high-entropy metal diboride, $(HfZrTaNbTi)B_2$ and a high-entropy metal carbide, (HfZrTaNbTi)C from the binary carbides and borides [4]. They added a small amount of graphitic carbon to eliminate the oxide layer of the raw materials and increase the density of sintered samples. In the XRD patterns of high-entropy boride with 3 wt% C, only an additional cubic carbide phase was observed besides the main boride phase. While for high-entropy carbide, minor $(Hf/Zr)O_2$ was observed. In the microstructure of these two samples characterized by SEM, residual graphite which was not detected by XRD was clearly displayed. Nevertheless, XRD is an efficient method to identify the phase composition inspite of these drawbacks.

The SAED pattern obtained in TEM is a powerful tool to identify the phase composition of a material at the micro scale. Pickering et al. characterized the phase evolution of coarse-grained CrMnFeCoNi following prolonged exposure at 700 °C [5]. They observed two Cr-rich precipitates formed on the grain boundaries (GBs) of this material through heat treatment alone for the first time. Using SAED, they revealed that the crystal structure of one precipitate was FCC, with a lattice parameter of approximately 10.6 Å, consistent with $M_{23}C_6$. The other precipitate was determined to be tetragonal, with lattice parameters of $a \approx 8.8$ Å and $c \approx 4.5$ Å, consistent with the σ phase, as shown in Figure 5.2. Their work not only demonstrated the capability of SAED in phase identification of precipitates in HEAs, but

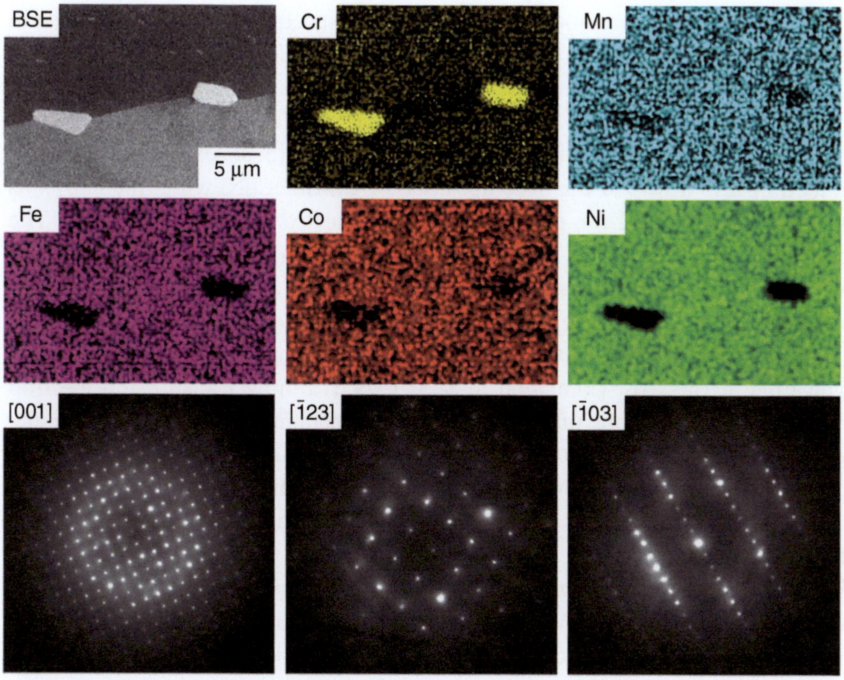


**Figure 5.2** Back-scattered electron (BSE) image, EDS maps and SAEDs from the grain boundary precipitate that appeared bright in BSE imaging in CrMnFeCoNi homogenized and then heat treated for 1000 hours at 700 °C, demostrating the exsitence of a tetragonal σ phase. Source: Pickering et al. [5]/ELSEVIER/CC BY 4.0.

also invalidated the general impression that CrMnFeCoNi was thermodynamically stable as a single solid solution at all temperatures below its solidus. These results have also been verified by Otto et al. who annealed CrMnFeCoNi at temperatures of 500, 700, and 900 °C for 500 days [6]. Besides the σ phase observed at 700 °C, they also discovered a more complex decomposition of CrMnFeCoNi at 500 °C. The precipitates were determined as NiMn phase with a tetragonal $L1_0$ crystal structure, a Cr-rich phase with BCC structure, and an FeCo phase with B2 structure by SAED and JEMS simulations. This technique has also been used to identify the phase composition in HECs. Zhou et al. synthesized (TiZrHfNbTa)C high-entropy carbide using spark plasma sintering (SPS) [7]. They proved the existence of a pure FCC phase and the homogeneity of the solid-solution phase by SAED. Dong et al. also employed SAED in the identification of the single monoclinic crystalline phase of $(5RE_{0.2})_2Si_2O_7$ [8].

## 5.2 Elemental Distribution

As we have discussed in Chapter 2.1.3, the elemental distribution in HEMs is characterized by chemical short-range order (CSRO) and segregation. Revealing these structure feature relies on the accurate probe of elemental distribution. Several techniques have been employed in the characterization of atomic distribution.

The frequently used method is energy dispersive spectroscopy (EDS) equipped in SEM. In the investigations of phase stability of CrMnFeCoNi alloy at high temperatures, Pickering et al. and Otto et al. observed homogenous distribution of the constituent elements with no evidence of segregation or clustering in the EDS maps when the annealing temperature was higher than 800 °C [5, 6]. Using the same technique, they also observed the decomposition of CrMnFeCoNi when the alloy was annealed at 700 °C and the presence of blocky faceted precipitates on GBs. The structure of the precipitates was identified by SAED. Employing this method to demonstrate the homogenous distribution of elements is also applicable to bulk and porous HECs [9, 10]. Although it is feasible to probe the elemental distribution and large precipitates by SEM-EDS, it is still quite a challenge to characterize element segregation at the GBs. Compared to SEM-EDS, scanning transmission electron microscopy (STEM) is one of the state-of-the-art techniques to probe into the elemental distribution of HEMs at the atomic scale. Sarkar et al. showed the homogeneity of the elemental distribution at nanoscale in a high-entropy perovskite oxide by analyzing the contrast in the atomic-resolution high angle annular dark field (HAADF) image obtained in STEM [11]. Regarding the segregation of elements at GBs, Wang et al. had done a good work on the synergic grain boundary segregation and precipitation in W- and W-Mo-containing high-entropy borides using HAADF-STEM [12]. In the baseline $(Ti_{0.2}Zr_{0.2}Hf_{0.2}Nb_{0.2}Ta_{0.2})B_2$ specimen, the atomic-resolution HAADF-STEM image suggested no short-range ordering and distributions of all the constituent elements were homogeneous across the GB. By replacing Ti in $(Ti_{0.2}Zr_{0.2}Hf_{0.2}Nb_{0.2}Ta_{0.2})B_2$ with W, they found substantial precipitates formed along the GB while no obvious short-range order (SRO) was observed. The precipitates at GB were chemically determined to be W-rich precipitates rather than single element diborides. High-resolution EDS maps across the GB showed evident segregation of W to the GB, as shown in Figure 5.3. The coherency of the interface between precipitate and the matrix was confirmed by electron diffraction patterns and geometrical phase analysis. They concluded that W had a high tendency to segregate to GBs, and once the segregation exceeds the solubility, W-rich phases start to precipitate along the GB. The same phenomena were also observed in W-Mo-containing high-entropy borides.

EDS obtained in the HAADF-STEM mode can also be used to demonstrate the CSRO in HEAs. Ding et al. investigated the atomic-scale element distributions in the CrMnFeCoNi Cantor alloy and a new CrFeCoNiPd HEA by using EDS [13]. In the atomic-resolution HAADF images from TEM and corresponding EDS maps of the CrMnFeCoNi alloy with [110] zone axis, little obvious evidence of assemblies of a particular element was observed. This random element distribution with a small variation was confirmed by the line profiles of atomic fraction and calculated pair correlation functions of the atomic fraction for each element. In contrast, in CrFeCoNiPd HEA, as shown in Figure 5.4, they found strong inhomogeneous fluctuations with local aggregations in all five EDS maps, and the corresponding line profiles show strong fluctuations in the atomic fraction of all elements in regions nearby. The calculated pair correlation functions revealed a strong correlation peak

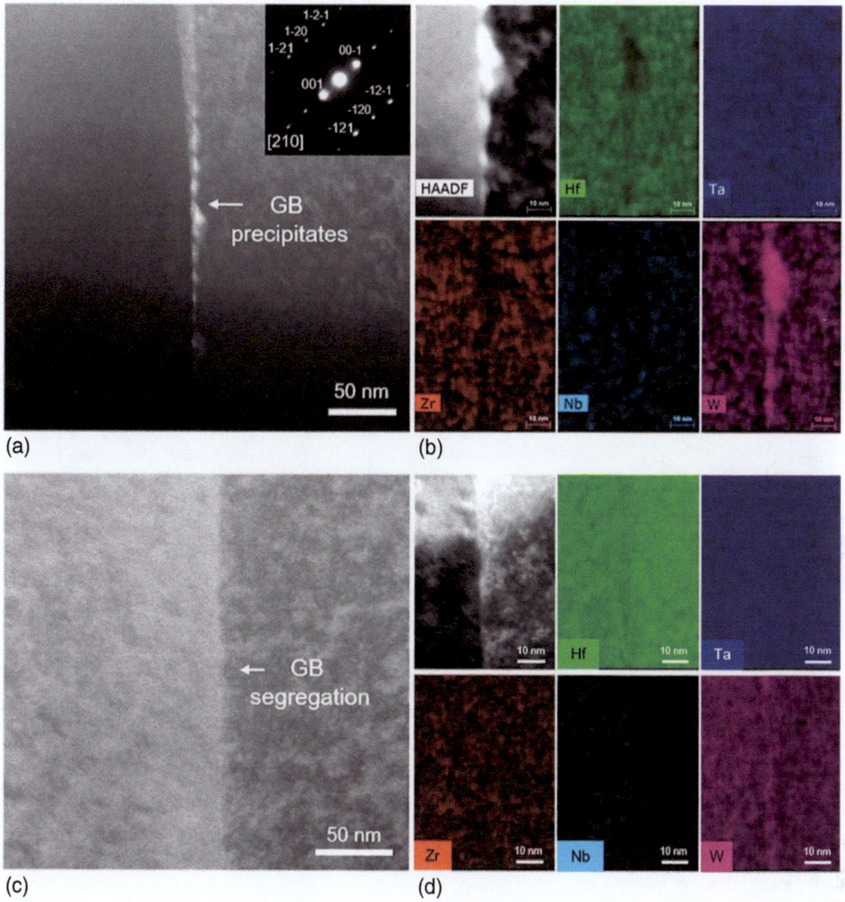


**Figure 5.3** Grain boundary segregation and precipitation in $(Zr_{0.2}Hf_{0.2}Nb_{0.2}Ta_{0.2}W_{0.2})B_2$. (a and c) HAADF-STEM images of GB with and without precipitates, respectively, and (b and d) EDS maps of all elements across the GB. Source: Wang et al. [12], Reproduced with permission from ELSEVIER.

for Pd, Co, and Fe at the concentration wavelength $r$ around 1 nm, a similarly strong correlation peak for Ni at $r$ around 2.5 nm, and a few weak peaks for Cr. The existence of CSRO is also directly observed by Chen et al. [14]. They discovered extra disks in the electron diffraction patterns of VCoNi alloy with [112] zone axis. These disks were highly diffuse but visible halfway in between the transmission spot (000) and the {311} spots. Existence of these extra reflections indicated the presence of CSRO. In the dark-field TEM image taken using the extra reflections, the area of CSRO revealed themselves, and the vast majority (90%) of the lighting up regions were less than 1 nm in size, with an average size of $d \approx 0.6$ nm. To unveil the detailed arrangements of the three elemental species constituting the CSROs, they carried out EDS mapping and found that the CSRO could be best described as two V-enriched $(\bar{3}11)$ planes sandwiching one V-depleted $(\bar{3}11)$ plane, as displayed in Figure 5.5.

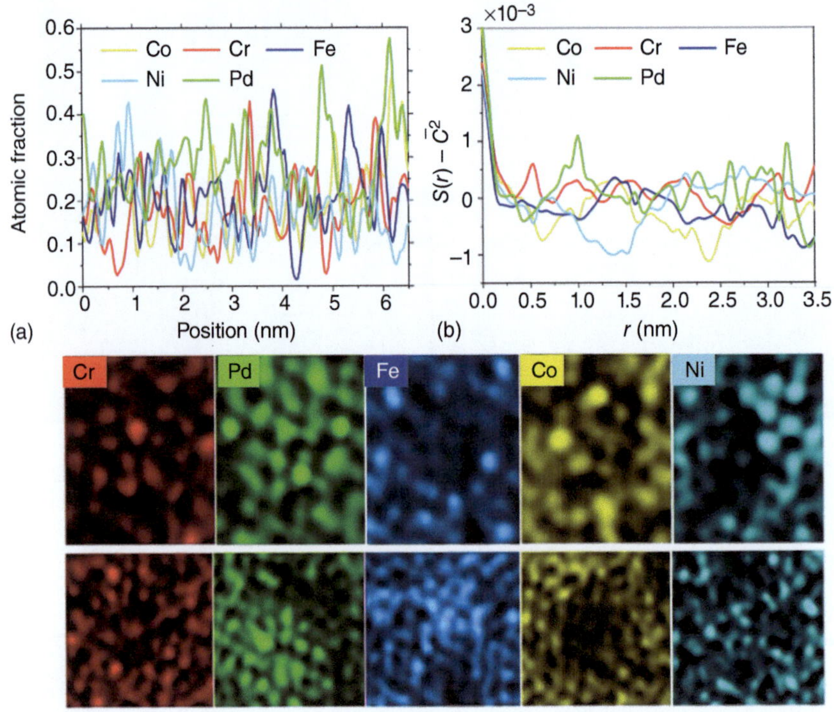


**Figure 5.4** (a) Line profiles of atomic fraction of individual elements taken from respective EDS maps for CoCrFeNiPd alloy, (b) Plots of pair correlation functions $S(r)$ of individual elements against concentration wavelength $r$, (c) Comparison of the local concentration distribution of individual elements for the same region. Source: Ding et al. [13], Reproduced with permission from Springer Nature.

This spatial configuration could explain the presence of extra reflections at the $1/2\{\bar{3}11\}$ positions.

Besides EDS, other strategies are also applicable in the elemental distribution probe. Zhang et al. examined the local structure of NiCoCr alloy with neutron and X-ray pair distribution function (PDF) analysis and X-ray absorption fine structure (EXAFS) measurements [15]. The EXAFS analysis revealed that Cr in the solid solution alloys has a tendency to bond with Ni and Co and form short-range ordering. Atom probe tomography (APT) is another powerful and direct method to characterize the elemental distribution of materials at the atomic scale in three-dimension (3D), and has been widely used in probing the precipitates, segregation and interface in HEMs. Using this technique, Santodonato et al. observed a complex microstructure of $Al_{1.3}CoCrCuFeNi$ alloy at room temperature (RT): the matrix consisted of BCC/B2 phase mixture and two types of submicron Cu-rich precipitates [16]. The BCC phase was Cr–Fe-enriched alloy and the B2 phase was Al–Ni-enriched one. Ming et al. gathered the APT data of GB in CrMnFeCoNi HEA after the tensile tests

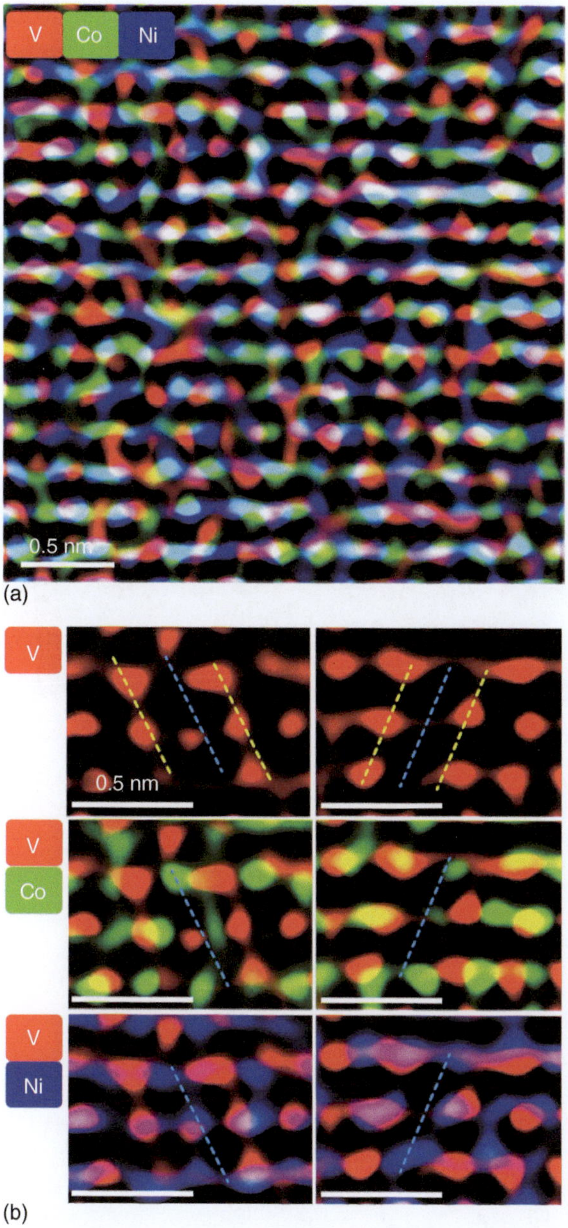


**Figure 5.5** Chemical mapping of VCoNi alloy indicating element-specific enrichment on alternating atomic planes. (a) EDS maps showing element distribution, (b) close-up maps of V, V–Co, and V–Ni, respectively. Source: Chen et al. [14], Reproduced with permission from Springer Nature.

at 700 °C [17]. They found substantial enrichment of Cr, Ni, and Mn in nanosized regions at the GB. Ni and Mn likely co-segregated at GBs, while Cr and Ni segregated separately, as shown in Figure 5.6. They believed that the segregation of elements at GBs in CrMnFeCoNi system HEAs, leading to the embrittlement of GBs, was responsible for the ductility loss with increasing deformation temperature. Chellali et al. confirmed the homogeneity of element distribution in three different HECs, including $(Co_{0.2}Cu_{0.2}Mg_{0.2}Ni_{0.2}Zn_{0.2})O$ with rock-salt structure, $(Ce_{0.2}La_{0.2}Pr_{0.2}Sm_{0.2}Y_{0.2})O_{2-\delta}$ with fluorite structure, and $(Gd_{0.2}La_{0.2}Nd_{0.2}Sm_{0.2}Y_{0.2})(Co_{0.2}Cr_{0.2}Fe_{0.2}Mn_{0.2}Ni_{0.2})O_3$ with perovskite structure by using APT [18].

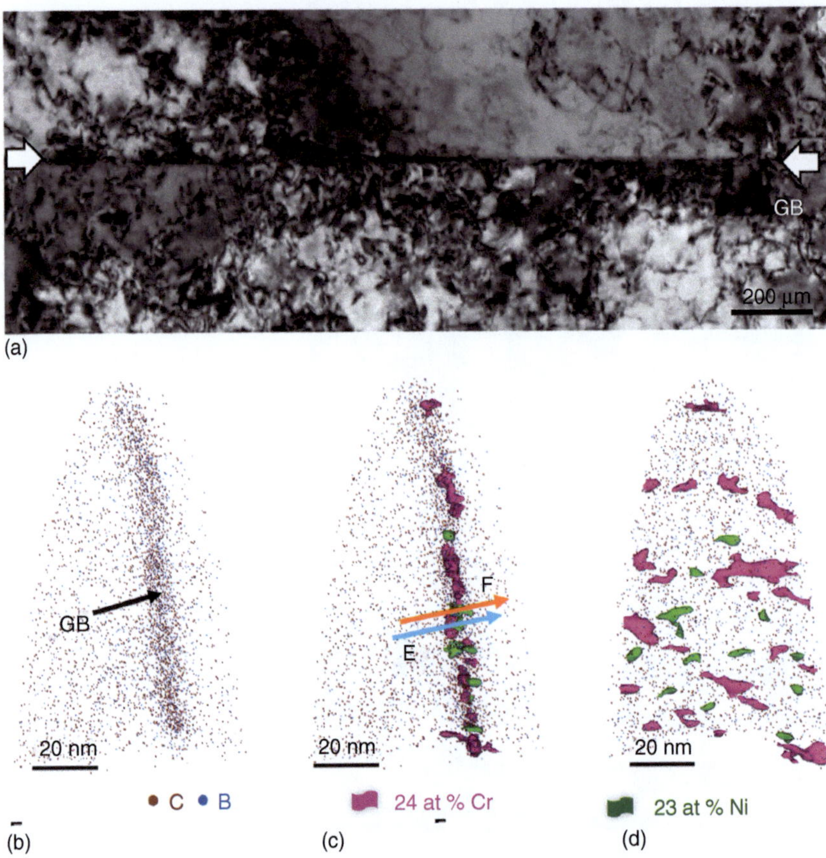


**Figure 5.6** Nanosegregation at GBs in CrMnFeCoNi HEA. (a) TEM image of the single-phase coarse-grained $Cr_{20}Mn_{20}Fe_{20}Co_{20}Ni_{20}$ sample after tensile test at 700 °C. (b) Atom map of the tip showing GB segregation of impurities C and B. Cr (24 at.%) and Ni (23 at.%) isocomposition surfaces viewed with the GB edge-on (c) and viewed from the GB normal (d). Source: Ming et al. [17], Reproduced with permission from American Association for the Advancement of Science - AAAS.

## 5.3 Lattice Distortion

As discussed in Chapter 2.1.4, the lattice distortion of HEMs has been examined by experiments and simulations. The first attempt to provide experimental data to support the concept of the highly distorted lattice in HEAs was made by Yeh et al. using XRD [19]. They related the significant deterioration of XRD peak intensities to the lattice distortion by mathematical treatment [20]. As pointed by Owen and Jones [21], when performing XRD on laboratory equipment, there could be large sources of error in the accurate measurement of the intensity of the Bragg peaks, which was highly problematic due to the relatively poor instrumental resolution and high background signal of such instruments. It could also be affected by the chemical homogeneity or the presence of any preferred crystallographic orientation, which could be controlled to a certain extent through careful sample preparation. Further, since X-rays scatter from the electron cloud of a material, there was a known angular dependency on the magnitude of scattering from an atom, which also led to an angular decay in the intensity of the Bragg peaks, thereby making the determination of the decay that results from local lattice strain more challenging.

To reduce the influence of these complicating factors, the use of high-energy synchrotron radiation is a feasible way, and the radiation source is either X-ray or neutrons. Scattering from materials is typically separated into two components: the diffuse and the Bragg scattering. The Bragg data provides the average crystallographic description of the unit cell. The diffuse scattering provides information of the local environment of a material, which is not immediately accessible without extensive modeling [20]. More useful in the analysis of total scattering is the use of the PDF, which mathematically given by the Fourier transform of the structure function $S(Q)$:

$$G(r) = \frac{2}{\pi}\int_0^\infty Q\left[S(Q) - 1\right]\sin(Qr)dQ \tag{5.1}$$

where $Q$ is the magnitude of the scattering vector in a reciprocal space and $r$ is the interatomic distance [22]. Peaks in the PDF are indicative of interatomic distances within a material, with the area being dependent on the number and type of coordination atom at that distance. The peak shape will be a function of the distribution of atoms at that distance, and so, the width of the peak will provide a measure of the displacement of the atoms away from one another [20]. Guo et al. extracted the information on both the total and local structures of ZrNbHf medium entropy alloy from high-energy synchrotron X-ray and neutron diffraction by using the PDF analysis [23]. As shown in Figure 5.7a, PDFs obtained from XRD and neutron diffraction agreed well with each other, and the neutron data did not exhibit large noise and preferred grain orientation. The model PDF, calculated based upon the BCC structure with all equivalent atomic sizes, generally agreed with the neutron data except for some details, which were clearly resolved by merging the first two peaks in the

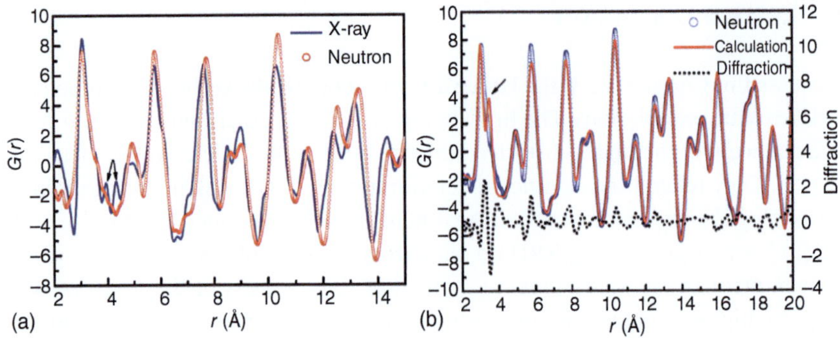


**Figure 5.7** (a) The pair distribution function of the ZrNbHf thin film obtained from X-ray diffraction and the bulk sample from neutron scattering. (b) The PDFs obtained from the neutron and the calculation. Source: Reproduced with permission from Guo et al. [23]/ Springer Nature.

calculated PDF into one peak in the neutron PDF, as shown in Figure 5.7b. The difference originated from the atomic size differences, which blurred the distinction between the first and second nearest neighbors. Owen et al. compared the neutron PDFs of CrMnFeCoNi HEA, Ni powder, Ni–Cr binary alloys, and Ni–Co–Cr ternary alloy to provide a clear picture of the local environment in HEAs [24]. They found no obvious difference between the different materials studied and good peak definition over the complete $r$ range, which suggested that all the materials have well-defined lattices, with no significant distortion. Elaborative analysis on the full width at half maximum (FWHM) of the first six coordination shells for all materials exhibited no clear trend with respect to the relationship between PDF peak width and composition. Their results suggested that it was clear that the HEA lattice contained some level of local strain, but it did not appear to be disproportionately larger than those found in the lattices of other, compositionally simpler alloys. Such technique has also been used in the lattice distortion of pyrochlore high-entropy oxides, where local $TiO_6$ octahedral distortion and some degree of short-range Ti clustering were identified in the $Nd_2(Ti_{0.2}Nb_{0.2}Sn_{0.2}Hf_{0.2}Zr_{0.2})_2O_{7+x}$ [25].

Other probes of the local environment, most notably extended X-ray absorption fine structure (EXAFS), can also be used to provide information about the local distortions in a structure. EXAFS considers the oscillations that occur after an X-ray absorption edge, and a mathematical analysis of this fine structure can yield local structural information [20]. It is possible to extract element-specific local environment information corresponding to an element-specific-edge transition. EXAFS measurements were acquired from a homogenized sample of FCC FeCoNiCrMn, and the results were compared with ab initio models [26]. Good agreement between the experimental and theoretical data was found for the element-resolved mean lattice distortions. Nevertheless, the standard deviations of the local bond lengths were an order of magnitude greater than the mean difference between the local bond length and the global average bond length. This phenomenon was found to be particularly pronounced for the bonds surrounding Cr and Mn, and the anti-ferromagnetic nature of these two elements was believed to be responsible for this phenomenon. Their results suggested that alloying multi-component alloys with Cr and Mn can induce serious local bond fluctuations, although the overall mean alloy distortions appear to be rather small, which might need more experimental evaluations. Rost et al. illustrated the existence of severe distortion around the Cu–O polyhedra in $(Mg_{0.2}Ni_{0.2}Co_{0.2}Cu_{0.2}Zn_{0.2})O$ using EXAFS [27]. Combining with theoretical calculations, they concluded that the lattice distortion in this high-entropy oxide was induced by the disorder in the oxygen sublattice.

High-resolution TEM has also been employed in the investigation of lattice distortion of refractory metal HEA NbMoTaW [28]. A slight amount of kinks and bends were found in the Fourier-filtered image of the lattice fringes, which was attributed to the local lattice distortions. However, these structure features did not exclusively resulted from the lattice distortions, since the thickness variation of the foil [20] and strain field around defects [29] may also give rise to similar effects in the Fourier-filtered images.

## 5.4 Microstructure Evolutions

Microstructure, closely related to the composition and processing method, determines the properties of a material. Accurate characterization of microstructure with proper techniques plays an important role in the comprehension of performance of materials. Here, taking $Al_xCoCrCuFeNi$ alloy as an example, the use of different combinations of microstructure characterization methods is illustrated.

As discussed in Section 5.1, varying the content of Al affects not only the phase composition of $Al_xCoCrCuFeNi$ HEA but also the microstructure. To demonstrate the progress of microstructure with Al content, Tong et al. carefully examined the microstructure of $Al_xCoCrCuFeNi$ HEA with different Al contents [30]. Using SEM, typical cast dendrite and interdendrite structures were observed in these alloys. The chemical composition probed by EDS revealed that the dendrites of large-volume fractions were basically composed of multiprincipal elements, despite copper segregation at the interdendrite regions, particularly for low Al contents ($x < 0.8$). As the aluminum content increased, a spinodal decomposition resulted in netlike structures developed both in the dendrites and interdendrites, which led to the drastic change in the micro-morphologies of these alloys. When the aluminum content exceeded 2.3, vein structures in polycrystalline structure were clearly observed in the SEM images. The microstructure evolution correlated with the XRD results, that is, FCC structure for the alloy with low aluminum content, mixed-phase FCC/BCC/B2 at medium aluminum content, and BCC structure at high Al content. More specific characterizations of the microstructures were conducted by TEM.

Both the dendrites and interdendrites of cast $Al_xCoCrCuFeNi$ alloys with aluminum contents from $x = 0$ to 0.5 were composed of FCC structure, as confirmed by TEM bright-field image and the corresponding SAED pattern. In addition, superlattice spot existed in the SAED pattern, indicating possible ordering in the FCC phase. With the help of TEM dark-field imaging, ordered FCC nanoprecipitates in the disorder FCC matrix with the size of 5–10 nm were revealed (Figure 5.8a,d). The extended interdendrites decomposed into mixed FCC and BCC phases in a eutectic reaction when the aluminum content reached $x = 0.8$, and the BCC phase in the eutectic interdendrites further decomposed into a spinodally decomposed netlike structure. As the Al content increased to 1.0, the formation of modulated plate structures was observed, and again, the superlattice spot of order BCC structure appeared in the SAED pattern. Combining the results of TEM dark-field observation and SAED analyses, the modulated plate structures due to the spinodal decomposition in the BCC phase were characterized as being composed of ordered BCC plates (B2, 100 nm thick) and disordered BCC interpolates (A2, 70 nm thick) (Figure 5.8b,e). With the Al content higher than 1.0, the BCC structures dominated, and order BCC matrix with disorder BCC formed long Widmanstätten precipitates (Figure 5.8c,f). These disordered long Widmanstätten precipitates transformed into spherical shapes with further increase in the Al content. Based on the microstructure

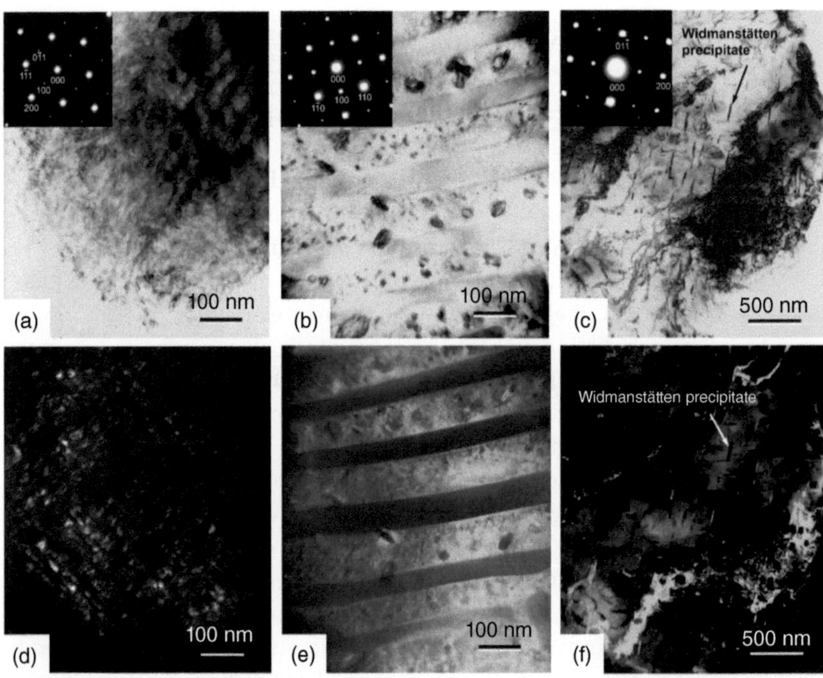


**Figure 5.8** TEM microstructures of the dendrite of as-cast $Al_{0.5}$CoCrCuFeNi alloy: (a) bright-field image with the SAED pattern of FCC [011] zone axis and (d) dark-field image corresponding to the (100) superlattice spot of the SAED pattern shown in (a). TEM microstructures of as-cast $Al_{1.0}$CoCrCuFeNi alloy: (b) bright-field image of modulated plates and interplates with the SAED pattern of the BCC [001] zone axis and (e) dark-field image corresponding to the (010) superlattice spot of the SAED pattern shown in (b). TEM microstructures of as-cast $Al_{2.0}$CoCrCuFeNi alloy: (c) bright-field image with the SAED pattern of the BCC [011] zone axis and (f) dark-field image corresponding to the (100) superlattice spot of the SAED pattern shown in (c). Source: Tong et al. [30], Reproduced with permission from Springer Nature.

evolution and data obtained by differential thermal analyses (DTA), they constructed the phase diagram of the $Al_xCoCrCuFeNi$ alloy system, as shown in Figure 5.9. Their research has demonstrated the power of microstructure characterization methods that integrated XRD, SEM, and TEM in correlating the crystal structures with the nanoscale features of the microstructure.

The microstructure of HEAs is also temperature-dependent. Park et al. examined the microstructure and phase evolution behaviors of annealed $Al_{0.5}CoCrCuFeNi$ alloy at 1000, 1100, and 1200 °C [31]. Additional BCC peaks were found in the samples annealed at 1000 and 1100 °C. Using the SEM and electron backscatter diffraction (EBSD), the grain sizes of FCC and BCC phases, high-angle GBs, and annealing twin boundaries were illustrated, as shown in Figure 5.10. All the alloys were fully recrystallized to form equiaxed grains, and abundant annealing twins could easily be found, which indicated the low stacking-fault energy of this alloy. Due to the existence of the second phase with BCC structure, the grain sizes of samples annealed at 1000 and 1100 °C (2 and 8.5 μm, respectively) were much smaller than the one annealed at 1200 °C (100 μm). EDS analysis demonstrated that Al and Ni concentrated in the BCC phase, and the concentrations of Cr and Fe were relatively low. Considering the strong tendency to crystallize in BCC structure of Al-containing HEAs, the existence of the BCC phase was comprehensible. Detailed microstructure analysis was conducted using TEM. SAED of the BCC phase exhibited clear superlattice reflections, indicating the ordering nature of this BCC phase (B2 structure). The structure of post-tensile samples was recorded by a digital image correlation (DIC) technique using an optical 3-D deformation analysis system equipped on

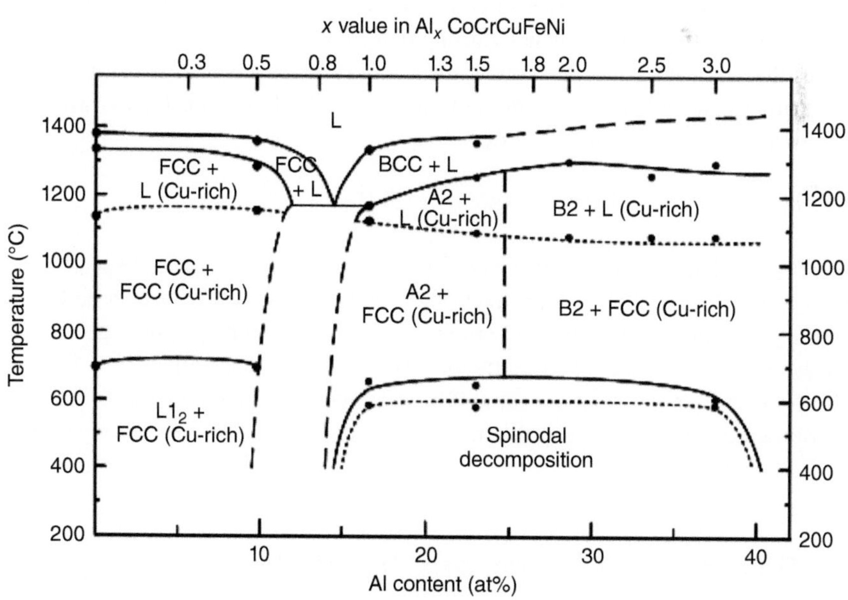


**Figure 5.9** Predicted phase diagram of the $Al_xCoCrCuFeNi$ alloy system with different aluminum contents ($x$ values). L: liquid phase. Source: Reproduced with permission from Tong et al. [30]/Springer Nature.

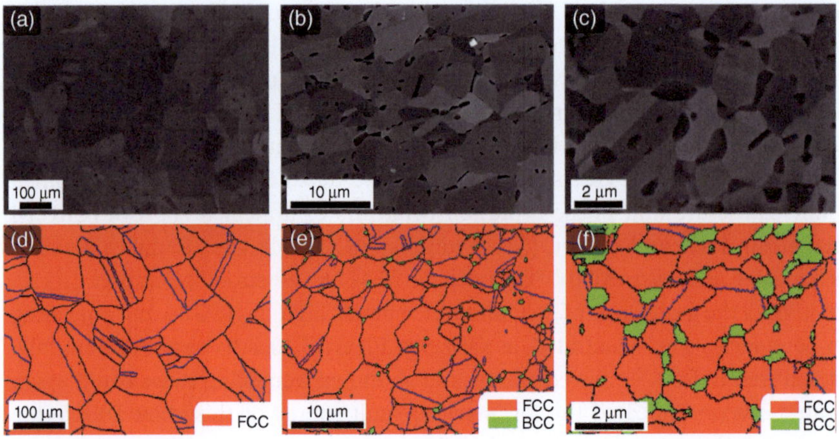


**Figure 5.10** Microstructure of the $Al_{0.5}$CoCrFeMnNi alloy annealed at 1200 °C (a and d), 1100 °C (b and e), and 1000 °C (c and f): (a–c) SEM-BSE images, (d–f) EBSD phase maps. The FCC and BCC phases are indicated by red and green colors, respectively. Source: Park et al. [31], Reproduced with permission from ELSEVIER.

a universal testing machine. The samples annealed at a higher temperature exhibited longer necking regions and more pronounced cup-and-cone-shaped fracture surfaces. The EBSD inverse pole figure (IPF) around the location with the largest local strain of the samples illustrated that most of the grains with FCC structure were elongated along the tensile directions, while the B2 phase remained its original ones. The IPF also indicated that large engineering strain of sample annealed at 1200 °C resulted from the profuse twinning deformation, which was scarce in the samples that annealed at lower temperatures. To correlate the yield strengths with microstructure, they deduced an equation from Hall–Petch relationship to estimate the strength of $Al_{0.5}$CoCrCuFeNi alloy:

$$\sigma_{YS,overall} = (215 + 490 \times d_{FCC}^{-1/2}) \times V_{FCC} + (2141 + 61 \times d_{B2}^{-1/2}) \times V_{B2} \tag{5.2}$$

where $d_{FCC}$ and $d_{B2}$ are the diameter of grain for FCC and B2 phases, respectively, and $V_{FCC}$ and $V_{B2}$ are the volume fractions of the two phases, respectively. They found a good agreement between estimation and experiments, implying the B2 phase was the main source of strengthening of $Al_{0.5}$CoCrCuFeNi alloy.

These works demonstrated the capability and necessity of combinatorial characterization methods in examining the microstructure of HEAs and correlating them with their properties.

## 5.5 Other Advanced Characterization Methods

High-throughput manufacturing of HEMs has already been realized by diffusion-multiple approach, additive manufacturing, etc. Accordingly, high-throughout characterizations have also been developed to accelerate the screening of novel materials with desired properties. Coury et al. produced a compositionally graded CrMnFeCoNi sample by a diffusion-multiple approach [32]. To extract the interested property, yield strength, from high-throughput screening technique, nanoindentation, they provided a pathway that combines the high-throughput methodologies for predicting and measuring properties. First, the Toda-Caraballo (TC) model with the atomic radii set determined by an Effective Atomic Radii for Strength (EARS) approach was used to predict the yield strengths of the compositionally graded sample with no other inputs besides the chemical composition measured by EDS. This calculation provided a qualitative trend on the yield strength with a reasonable match to the experimental hardness data. Then, using this prediction as guide, high-throughput nanoindentation was used to extract quantitative value of yield strength. During the process, the conversion of hardness to yield strength was treated using the Clausner–Richter–Hollomon model with the strain hardening exponent $n$ varying with the hardness by $n_{var} = 0.37 - \mathrm{H}\cdot 0.02$. When this integrated method was applied to the prediction of yield strength of alloys in the CrFeMnCoNi system, the agreement between the strength from their methodology and the experimental tensile ones was quite satisfactory. Hardness testing on the deposited samples by high-throughput additive manufacturing [33] and laser deposition-based fabrication technique [34] are used for high-throughput rapid screening of complex concentrated alloys, and optimization for structural applications has also been conducted.

Micro-compression technique, which is a well-established method in the mechanical testing of other materials [35, 36], has also been employed in the characterization of HEMs. Liu et al. synthesized $Al_{0.8}$CoCrCuFeNi HEA by arc-melting a mixture of the constituent elements [37]. They found an additional FCC phase that had not been reported in this HEA with the lattice parameter of 3.631 Å by neutron diffraction. Combining the results of EDS and microstructure observation, this new FCC phase was assumed to solidify in the Cu-rich interdendrite region via spinodal decomposition. Further, they characterized the mechanical properties of three distinctive regions using the micro-compression technique. The micropillars were machined by the focused ion beam (FIB) technique and then compressed using a flat-end diamond punch under the load-controlled condition. The micropillars before and after the tests are shown in Figure 5.11. The extracted yield strengths for the three regions were largely different, with the highest yield strength for the region containing BCC and FCC phases and lowest for the region containing one FCC phase. Nevertheless, an apparent size effect in the yield strengths was observed due to much higher values than that from the macro-compression tests. Moreover, plastic deformation of the pillars was observed. Using the same strategy, Csanádi et al. fabricated micropillars of (HfTaZrNb)C high-entropy carbide in visibly pore free regions of large grains of a specific orientation ($\Phi \sim 14°$ and $\varphi_2 \sim 45°$) to make sure the Schmid factor was equal for both the $\{110\}\langle 1\bar{1}0\rangle$ and $\{111\}\langle 1\bar{1}0\rangle$ slip systems [38], which would be easy to determine the slip system family that had lower critical resolved shear stress under the uniaxial stress conditions from one set of measurements, as shown in Figure 5.12. Compression results demonstrated a considerable yield strength enhancement in HEC with a value of 6.20 GPa. Microstructure

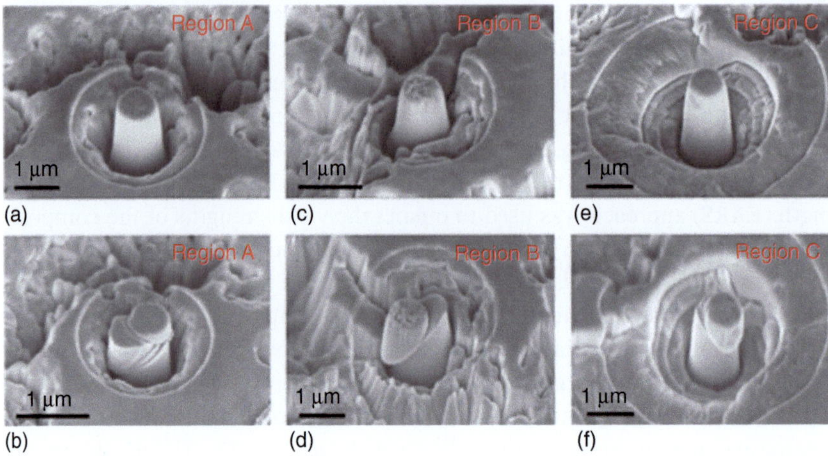


**Figure 5.11** SEM images of the $Al_{0.8}CoCrCuFeNi$ alloy micropillars before (a, c, and e) and after (b, d, and f) the micro-compression tests. Regions A, B, and C correspond to dendritic, Cu-depleted interdendritic and Cu-rich interdendritic regions, respectively. Source: Liu et al.[37], Reproduced with permission from ELSEVIER.

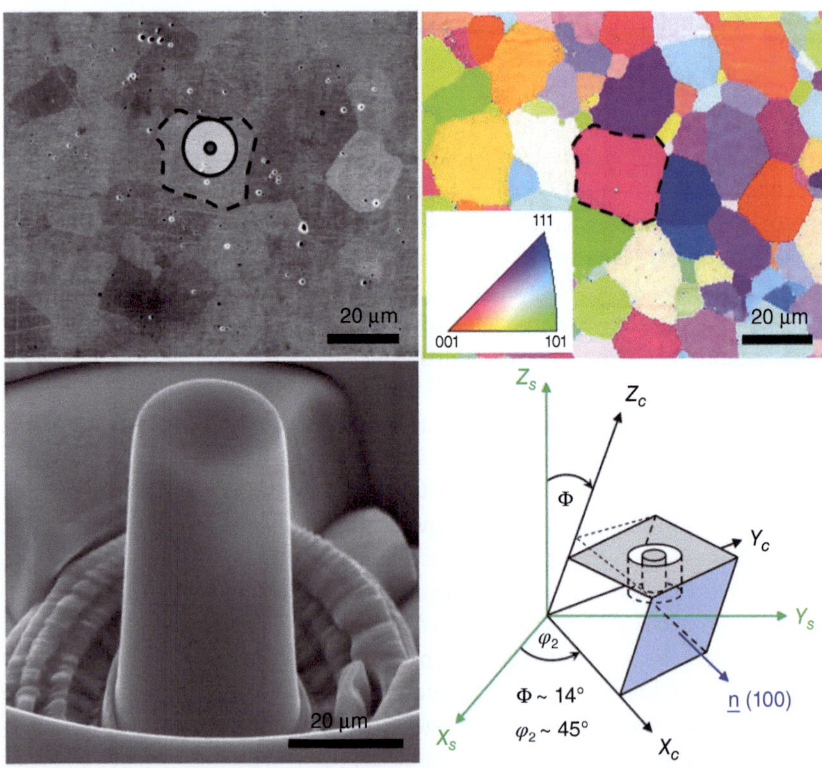


**Figure 5.12** SEM of microstructure of (Hf–Ta–Zr–Nb)C, corresponding EBSD map, and a FIB-milled micropillar located in a selected grain marked on both the SEM and EBSD images. The spatial orientation of the micropillar is illustrated in the schematic. Source: Csanádi et al. [38], Springer Nature, CC BY 4.0.

observation illustrated that HEC exhibited larger work hardening than HfC and showed similar ductility to TaC with more pronounced slip patterns, thus suggesting that the nature of the bonding in the high-entropy carbide was similar to TaC and showing a more metallic character than that of HfC. Analysis of the active slip systems unambiguously confirmed the operation of the $\{110\}\langle 1\bar{1}0\rangle$ slip system during the compression, which was different from the dominant $\{111\}\langle 1\bar{1}0\rangle$ slip system in nanoindentation. The difference was attributed to the different Schmid factors due to the different stress fields between nanoindentation and micropillar compression.

Besides, with the development of characterization method, direct observation of more detailed structural characteristics of materials is becoming reality. For instance, Gao et al. directly observed the local charge density distribution at the interface of $SrTiO_3/BiFeO_3$ using aberration-corrected STEM (AC-STEM) with sub-angström resolution [39]. Yan et al. experimentally observed defect-induced local phonon resonances and an associated red energy shift by using a combination of atomic resolution and angle-resolved electron energy-loss spectroscopy (EELS) [40]. Yet, such attractive techniques have not been applied to the characterization of HEMs; however, the development of new characterization techniques has provided an exciting possibility in further understanding the composition–structure–property relationships of HEMs.

## References


1 Cantor, B., Chang, I.T.H., Knight, P. et al. (2004). Microstructural development in equiatomic multicomponent alloys. *Mater. Sci. Eng., A* 375–377: 213–218.

2 Yeh, J., Chen, S.K., Lin, S.J. et al. (2004). Nanostructured high-entropy alloys with multiple principal elements: novel alloy design concepts and outcomes. *Adv. Eng. Mater.* 6: 299–303.

3 Rost, C.M., Sachet, E., Borman, T. et al. (2015). Entropy-stabilized oxides. *Nat. Commun.* 6: 8485.

4 Gild, J., Kaufmann, K., Vecchio, K. et al. (2019). Reactive flash spark plasma sintering of high-entropy ultrahigh temperature ceramics. *Scr. Mater.* 170: 106–110.

5 Pickering, E.J., Muñoz-Moreno, R., Stone, H.J. et al. (2016). Precipitation in the equiatomic high-entropy alloy CrMnFeCoNi. *Scr. Mater.* 113: 106–109.

6 Otto, F., Dlouhý, A., Pradeep, K.G. et al. (2016). Decomposition of the single-phase high-entropy alloy CrMnFeCoNi after prolonged anneals at intermediate temperatures. *Acta Mater.* 112: 40–52.

7 Zhou, J., Zhang, J., Zhang, F. et al. (2018). High-entropy carbide: A novel class of multicomponent ceramics. *Ceram. Int.* 44: 22014–22018.

8 Dong, Y., Ren, K., Lu, Y. et al. (2019). High-entropy environmental barrier coating for the ceramic matrix composites. *J. Eur. Ceram. Soc.* 39: 2574–2579.

9 Chen, H., Xiang, H., Dai, F. et al. (2019). Porous high entropy $(Zr_{0.2}Hf_{0.2}Ti_{0.2}Nb_{0.2}Ta_{0.2})B_2$: a novel strategy towards making ultrahigh temperature ceramics thermal insulating. *J. Mater. Sci. Technol.* 35: 2404–2408.

**10** Zhao, Z., Chen, H., Xiang, H. et al. (2020). High-entropy ($Y_{0.2}Nd_{0.2}Sm_{0.2}Eu_{0.2}Er_{0.2}$) $AlO_3$: a promising thermal/environmental barrier material for oxide/oxide composites. *J. Mater. Sci. Technol.* 47: 45–51.

**11** Sarkar, A., Djenadic, R., Wang, D. et al. (2018). Rare earth and transition metal based entropy stabilised perovskite type oxides. *J. Eur. Ceram. Soc.* 38: 2318–2327.

**12** Wang, C., Qin, M., Lei, T. et al. (2021). Synergic grain boundary segregation and precipitation in W- and W-Mo-containing high-entropy borides. *J. Eur. Ceram. Soc.* 41: 5380–5387.

**13** Ding, Q., Zhang, Y., Chen, X. et al. (2019). Tuning element distribution, structure and properties by composition in high-entropy alloys. *Nature* 574: 223–229.

**14** Chen, X., Wang, Q., Cheng, Z. et al. (2021). Direct observation of chemical short-range order in a medium-entropy alloy. *Nature* 592: 712–716.

**15** Zhang, F., Zhao, S., Jin, K. et al. (2017). Local structure and short-range order in a NiCoCr solid solution alloy. *Phys. Rev. Lett.* 118: 205501.

**16** Santodonato, L.J., Zhang, Y., Feygenson, M. et al. (2015). Deviation from high-entropy configurations in the atomic distributions of a multi-principal-element alloy. *Nat. Commun.* 6: 5964.

**17** Ming, K., Li, L., Li, Z. et al. (2019). Grain boundary decohesion by nanoclustering Ni and Cr separately in CrMnFeCoNi high-entropy alloys. *Sci. Adv.* 5: 0639.

**18** Chellali, M.R., Sarkar, A., Nandam, S.H. et al. (2019). On the homogeneity of high entropy oxides: an investigation at the atomic scale. *Scr. Mater.* 166: 58–63.

**19** Yeh, J., Chen, S., Gan, J. et al. (2004). Formation of simple crystal structures in Cu–Co–Ni–Cr–Al–Fe–Ti–V alloys with multiprincipal metallic elements. *Metall. Mater. Trans. A* 35A: 2533–2536.

**20** Yeh, J., Chang, S., Hong, Y. et al. (2007). Anomalous decrease in X-ray diffraction intensities of Cu–Ni–Al–Co–Cr–Fe–Si alloy systems with multi-principal elements. *Mater. Chem. Phys.* 103: 41–46.

**21** Owen, L.R. and Jones, N.G. (2018). Lattice distortions in high-entropy alloys. *J. Mater. Rev.* 33: 2954–2969.

**22** Gao, M.C., Yeh, J., Liaw, P.K. et al. (2016). *High-Entropy Alloys: Fundamentals and Applications*. Springer International Publishing.

**23** Guo, W., Dmowski, W., Noh, J. et al. (2013). Local atomic structure of a high-entropy alloy: an X-ray and neutron scattering study. *Metall. Mater. Trans. A* 44: 1994–1997.

**24** Owen, L.R., Pickering, E.J., Playford, H.Y. et al. (2017). An assessment of the lattice strain in the CrMnFeCoNi high-entropy alloy. *Acta Mater.* 122: 11–18.

**25** Jiang, B., Bridges, C.A., Unocic, R.R. et al. (2021). Probing the local site disorder and distortion in pyrochlore high-entropy oxides. *J. Am. Chem. Soc.* 143: 4193–4204.

**26** Oh, H., Ma, D., Leyson, G. et al. (2016). Lattice distortions in the FeCoNiCrMn high entropy alloy studied by theory and experiment. *Entropy* 18: 321.

**27** Rost, C.M., Rák, Z., Brenner, D.W. et al. (2017). Local structure of the $Mg_xNi_xCo_xCu_xZn_xO$ ($x = 0.2$) entropy-stabilized oxide: an EXAFS study. *J. Am. Ceram. Soc.* 100: 2732–2738.

**28** Zou, Y., Maiti, S., Steurer, W. et al. (2014). Size-dependent plasticity in an $Nb_{25}Mo_{25}Ta_{25}W_{25}$ refractory high-entropy alloy. *Acta Mater.* 65: 85–97.

**29** He, Q. and Yang, Y. (2018). On lattice distortion in high entropy alloys. *Front. Mater.* 5: 42.

**30** Tong, C.J., Chen, Y.L., Chen, S.K. et al. (2005). Microstructure characterization of AlxCoCrCuFeNi high-entropy alloy system with multiprincipal elements. *Metall. Mater. Trans. A Phys. Metall. Mater. Sci.* 36A (4): 881–893.

**31** Park, J.M., Moon, J., Bae, J.W. et al. (2018). Effect of annealing heat treatment on microstructural evolution and tensile behavior of $Al_{0.5}$CoCrFeMnNi high-entropy alloy. *Mater. Sci. Eng., A* 728: 251–258.

**32** Coury, F.G., Wilson, P., Clarke, K.D. et al. (2019). High-throughput solid solution strengthening characterization in high entropy alloys. *Acta Mater.* 167: 1–11.

**33** Melia, M.A., Whetten, S.R., Puckett, R. et al. (2020). High-throughput additive manufacturing and characterization of refractory high entropy alloys. *Appl. Mater. Today* 19: 100560.

**34** Li, M., Gazquez, J., Borisevich, A. et al. (2018). Evaluation of microstructure and mechanical property variations in $Al_x$CoCrFeNi high entropy alloys produced by a high-throughput laser deposition method. *Intermetallics* 95: 110–118.

**35** Uchic, M.D. and Dimiduk, D.M. (2005). A methodology to investigate size scale effects in crystalline plasticity using uniaxial compression testing. *Mater. Sci. Eng., A* 400, 401: 268–278.

**36** Uchic, M.D., Shade, P.A., and Dimiduk, D.M. (2009). Plasticity of micrometer-scale single crystals in compression. *Annu. Rev. Mater. Res.* 39: 361–386.

**37** Liu, Z., Guo, S., Liu, X. et al. (2011). Micromechanical characterization of casting-induced inhomogeneity in an $Al_{0.8}$CoCrCuFeNi high-entropy alloy. *Sci. Mater.* 64: 868–871.

**38** Csanádi, T., Castle, E., Reece, M.J. et al. (2019). Strength enhancement and slip behaviour of high-entropy carbide grains during micro-compression. *Sci. Rep.* 9: 10200.

**39** Gao, W., Addiego, C., Wang, H. et al. (2019). Real-space charge-density imaging with sub-ångström resolution by four-dimensional electron microscopy. *Nature* 575: 480–484.

**40** Yan, X., Liu, C., Gadre, C.A. et al. (2021). Single-defect phonons imaged by electron microscopy. *Nature* 589: 65–69.

# 6

# Mechanical Properties

## 6.1 Introduction

High-entropy materials (HEMs), including high-entropy alloys (HEAs), high-entropy ceramics (HECs) and other materials with high chemical configurational entropy, are a new set of materials. The distinct multiprincipal element nature of HEMs makes people believe that materials with new properties and physics may be discovered. Mechanical properties are a very board class of properties, including elastic properties, strength, toughness, creep, fracture, and fatigue, which belong to the most fundamental properties of materials. In this chapter, we are not going to give a board comprehensive review on the mechanical properties of all the reported HEMs. Instead, we will focus on some typical examples where HEMs reveal intriguing mechanical properties and try to uncover the microstructures or mechanisms underlying the novel properties. There have been many excellent review papers on HEMs, especially HEAs, some of which focus on the mechanical properties of HEAs. For example, George et al. [1] briefly introduced mechanical properties based on a specific set of HEAs and emphasized on the underlying mechanism. Then, George et al. [2] gave another more comprehensive review on the strengths of HEAs, which focused on ductile face-centered cubic (FCC) and body-centered cubic (BCC) HEAs and their deformation mechanisms. Li et al. [3] gave a comprehensive review on mechanical properties (including strengths, creep, fracture, and fatigue) of HEAs and the related mechanisms and theories. Readers interested in in-depth understanding of mechanical properties of HEAs can refer to these review papers.

## 6.2 Exceptional Toughness at Cryogenic Temperatures

One of the most well-known outstanding mechanical properties of HEMs is the exceptional damage tolerance at cryogenic temperatures of CoCrFeMnNi HEA [4], which breaks the conventional knowledge that toughness usually decreases with the decrease of temperature. The quinary CoCrFeMnNi HEA is one of the first reported HEAs crystallized as a single-phase FCC solid solution by Cantor et al. [5],

*High-Entropy Materials: From Basics to Applications*, First Edition.
Huimin Xiang, Fu-Zhi Dai, and Yanchun Zhou.

which is named as Cantor alloy for simplicity. Gludovatz et al. [4] revealed that the strength of the Cantor alloy exhibited strong temperature dependence, where both its strength and ductility increased with the decrease of temperature, as shown in Figure 6.1. As stated in their paper [4], with the decrease of temperature from 293 to 77 K, the yield strength $\sigma_y$ and ultimate tensile strength $\sigma_{uts}$ increased by ~85% and ~70% to 759 and 1280 MPa, respectively. Similarly, the tensile ductility (ultimate failure strain, $\varepsilon_{ufs}$) increased by ~25% to >0.7; the strain hardening exponent $n$ remained high at ~0.4, such that there was an enhancement in the fracture energy by more than a factor of 2. Microstructure evolution with the increase in strain was systematically characterized to explore the underlying mechanism, which revealed that the TWIP (twinning-induced plasticity) effect was mainly responsible for the exceptional properties. At room temperature, plastic deformation is dominated by the dislocation mechanism, which is mainly carried out by planner slip on {111} planes of <110>/2 dislocations that dissociate into Shockley partials. At cryogenic temperatures, the microstructure evolution is distinctly different. Initially, plastic deformation is dominated by dislocations, where dislocation distributions and density evolution with strain (<~8%) is similar to those at room temperature. When the strain is >~8%, twinning is activated as an additional mechanism, which increases both strength and plasticity. On the one hand, twinning serves as an alternative mechanism to release deformation strains, which increases the plasticity. On the other hand, twinning reduces the scale of microstructure, and twin boundaries can act as barriers to dislocation motion, which provides an additional hardening

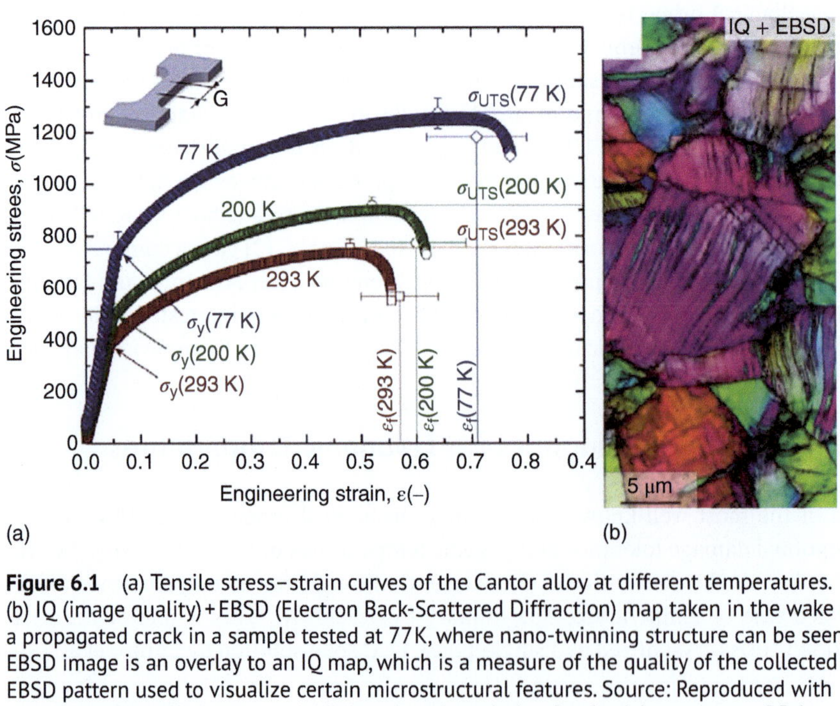


**Figure 6.1** (a) Tensile stress–strain curves of the Cantor alloy at different temperatures. (b) IQ (image quality) + EBSD (Electron Back-Scattered Diffraction) map taken in the wake of a propagated crack in a sample tested at 77 K, where nano-twinning structure can be seen. EBSD image is an overlay to an IQ map, which is a measure of the quality of the collected EBSD pattern used to visualize certain microstructural features. Source: Reproduced with permission from Gludovatz et al. [4]/American Association for the Advancement of Science.

mechanism similar to the Hall–Petch effect. The multiple deformation mechanisms guarantee the exceptional damage tolerance at cryogenic temperatures.

After the discovery of the exceptional toughness of the Cantor alloy at cryogenic temperatures, efforts were carried out to develop similar medium/high-entropy alloys by varying the compositions. The underlying principle to develop new alloys is to reduce the stacking-fault energy (SFE) to promote TWIP and/or TRIP (transformation-induced plasticity) effects. The most notable examples include medium-entropy alloys CoCrNi and CoNiV and non-equiatomic HEA $Co_{10}Cr_{10}Fe_{80-x}Mn_x$. In comparison to the Cantor alloy, the medium-entropy alloys CoCrNi and CoNiV exhibit enhanced strength, ductility, and toughness. As reported by Gludovatz et al. [6], the CoCrNi alloy shows tensile strengths of almost 1 GPa, failure strains of ~70%, and $K_{IC}$ fracture-toughness values above 200 MPa m$^{1/2}$ at room temperature; furthermore, the strength, ductility, and toughness values improve to >1.3 GPa, >90%, and 275 MPa m$^{1/2}$, respectively, at cryogenic temperatures. Different from the Cantor alloy, the TWIP effect is already activated at room temperature in CoCrNi medium-entropy alloy, which may result from the relatively higher yield strength and lower SFE. Based on the idea to increase lattice distortions, Sohn et al. [7] changed CoCrNi to CoNiV. As shown in Figure 6.2, medium-entropy CoNiV alloy exhibits significantly higher lattice distortions than other medium/high-entropy alloys, which results in a near 1 GPa yield strength at room temperature and good ductility, outperforming most conventional solid-solution alloys. However, no TWIP effect was observed in CoNiV medium-entropy alloy at room temperature.

Another effect that is very similar to TWIP is the TRIP effect. Li et al. [8] introduced the TRIP effect into HEA first and designed non-equiatomic HEA $Co_{10}Cr_{10}Fe_{80-x}Mn_x$ as transformation-induced plasticity dual-phase high-entropy alloy (TRIP-DP-HEA). The results reveal that the dominating deformation mechanism changes from dislocations to TWIP and then to TRIP with the content of Mn

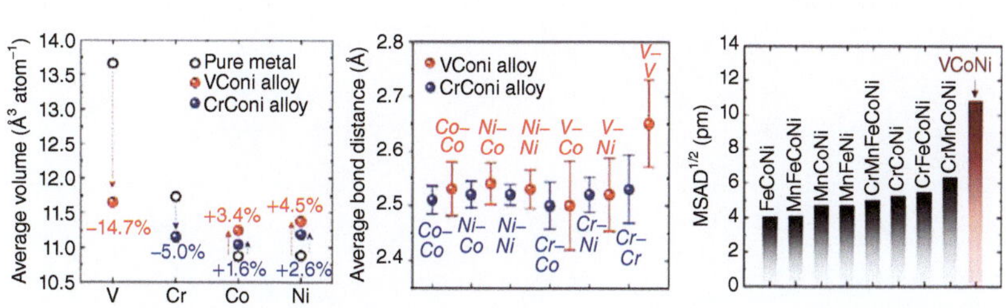


**Figure 6.2** Ab initio calculations for the local lattice distortion effect. (a) Relative changes of an average atomic (Voronoi) volume of the individual elements in CoNiV and CoCrNi alloys in comparison with the volumes of the pure elements in the FCC phases. Relative changes are caused not only by the alloying effect but also by macroscopic lattice constant changes between specific elements in the alloys and the corresponding elemental phases (3.575 Å for CoNiV; 3.544 Å for CoCrNi). (b) First nearest-neighbor distances of each elemental pairs in the CoNiV and CoCrNi alloys. The mean value and their fluctuations were evaluated by fitting the computed bond distances to normal distribution functions. (c) Mean square atomic displacement (MSAD) for various multiprincipal element alloys. Source: Reproduced with permission from Sohn et al. [7]/John Wiley & Sons.

changing from 45 at.% to 40 at.% and to 35 at.%. This means that the SFE reduces with the decrease in Mn content. With the further decrease in Mn content to 30 at.%, they developed a TRIP-DP-HEA, as shown in Figure 6.3. In the TRIP-DP-HEA, the SFE might be close to zero due to the similar stability of the FCC phase and the hexagonal close-packed (HCP) phase. The elimination of SFE guarantees bidirectional TRIP effect, where the FCC phase and the HCP phase can transform forward and backward, and substantially refines the microstructure down to a nanolaminate-type substructure, as illustrated by Lu et al. [9].

It is well established that plastic deformation is dominated by dislocation slip when SFE is high, by twinning when SFE is low, and possibly by transformation to the HCP phase when SFE is further lowered in FCC HEAs (see Figure 6.3). Usually, the TRIP effect in HEA is related to phase transformation between the FCC phase and the HCP phase. The coherent twin boundary in an FCC alloy also changes local FCC stacking sequence into local HCP stacking sequence (see Chapter 3). Thus, the key property in these HEAs with possible TWIP and/or TRIP effect is SFE or the relative stability between the FCC phase and the HCP phase. In addition, the SFE of an FCC phase can be approximated by the energy difference between the FCC phase and its related HCP phase (see Chapter 3). Therefore, evaluating temperature dependence and compositional dependence of SFE is essential for better understanding of the properties of HEAs with TWIP/TRIP effects and for the design of high-performance HEAs. For example, Huang et al. [10] employed density function

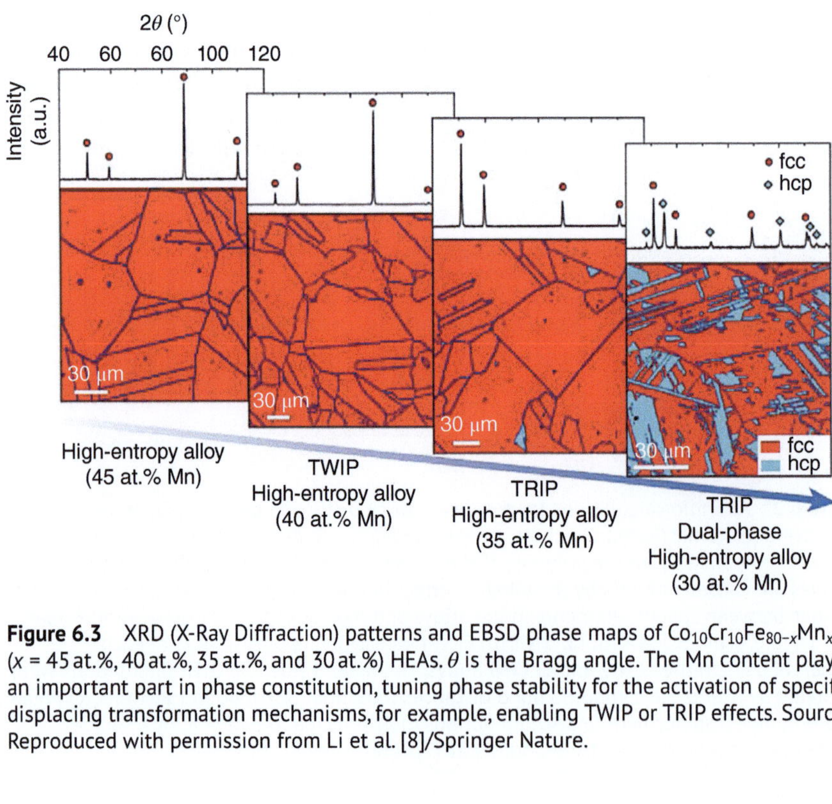


**Figure 6.3** XRD (X-Ray Diffraction) patterns and EBSD phase maps of $Co_{10}Cr_{10}Fe_{80-x}Mn_x$ ($x$ = 45 at.%, 40 at.%, 35 at.%, and 30 at.%) HEAs. $\theta$ is the Bragg angle. The Mn content plays an important part in phase constitution, tuning phase stability for the activation of specific displacing transformation mechanisms, for example, enabling TWIP or TRIP effects. Source: Reproduced with permission from Li et al. [8]/Springer Nature.

theory (DFT)-based methods to calculate the temperature dependence of SFE of the Cantor alloy and separated the total SFE $\gamma_{SF}$ to chemical, magnetic, and strain contributions ($\gamma_{chem}$, $\gamma_{mag}$, and $\gamma_{strain}$), as shown in Figure 6.4a. According to their results, $\gamma_{SF}$ increases as temperature increases with $\gamma_{chem}$ also increasing with temperature; $\gamma_{mag}$ first increases with temperature and then becomes saturated when $T > 300$ K, and $\gamma_{strain}$ shows negligible temperature dependence. The temperature-dependent $\gamma_{chem}$ results from the temperature-dependent formation energy difference between the HCP and FCC phases. The relative stability of the FCC phase increases with temperature, which results in increasing $\gamma_{SF}$ with temperature. The result that SFE decreases with decreasing temperature agrees well with experiments, where the deformation mechanism of the Cantor alloy changes from dislocation dominance at room temperature to twinning dominance at cryogenic temperatures. Zhao et al. [12] employed DFT-based methods to determine both temperature- and composition-dependent SFEs of CrMnFeCoNi and CrFeCoNiPd and subsystems from these two HEAs, as shown in Figure 6.4b. Their results reveal that the SFE of CoCrNi is the lowest, lower than that of CoCrFeMnNi, which agrees well with the experimental result that the TWIP effect can be activated at room temperature in the CoCrNi medium-entropy alloy, while it is not so in the Cantor alloy. After the report of the first TRIP-DP-HEA, Li et al. [13] further employed DFT-based methods to predict relative stability between the FCC phase and the HCP phase versus composition in $Co_{20}Cr_{20}Fe_{40-x}Mn_{20}Ni_x$ ($x$ = 0–20 at.%) HEA. By referring to their TRIP-DP-HEA, they successfully predicted that the stability of the HCP phase decreased with the increase in Ni. Therefore, decreasing the Ni content in CoCrFeMnNi alloy may obtain the HCP+FCC duplex phase TRIP alloy, which was validated by experiments. These results indicate that tailoring the stability between the FCC phase and the HCP phase or the SFE is a valuable method to design high-performance HEAs with TWIP/TRIP effects.

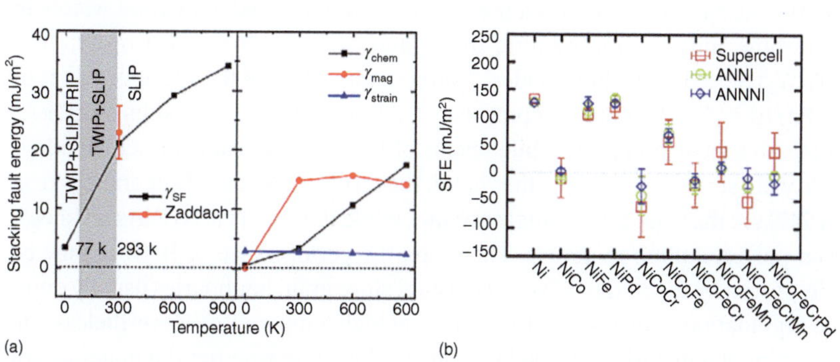


**Figure 6.4** Temperature-dependent SFE of CoCrFeMnNi HEA: (a) left: Total SFE, right: contributions from chemical part $\gamma_{chem}$, magnetic part $\gamma_{mag}$, and strain part $\gamma_{strain}$. In (a), Zaddach means results reported by Zaddach et al. [11], and rough boundaries between different plastic mechanisms (SLIP→SLIP + TWIP→TWIP + SLIP/TRIP) are illustrated. Source: Huang et al. [10]/With permission of Elsevier. (b) SFEs calculated by the supercell method as well as ANNI and ANNNI approximations. Source: Reproduced with permission from Zhao et al. [12]/Elsevier. (see Chapter 3 for more information.)

## 6.3 Superior Performances at Elevated Temperatures

Materials with superior mechanical properties at elevated temperatures are in high demand in many applications. For example, the thrust of an aero-engine depends strongly on its working temperature, where an increase in temperature before the turbine by 55 °C can approximately increase the thrust by 10%. One of the most well-known materials applied in high temperatures is Ni-based superalloy. Currently, the sixth-generation superalloy TMS-238 can keep a unique combination of properties at temperatures up to 1150 °C. However, new technologies require materials that can operate at even higher temperatures. The development of refractory high-entropy alloys (RHEAs) provides a new strategy to design the new generation of high temperature alloys, where intriguing results have been reported.

RHEAs are based on early transition metals in group IVB (Ti, Zr, and Hf), group VB (V, Nb, and Ta), and group VIB (Cr, Mo, and W), sometimes with additions of other elements such as Al, Si, Co, Ni, C, or B to tailor the properties. Among those principal elements, Ti has the lowest melting point (1660 °C), higher than that of Ni (1455 °C). The crystal structure of RHEAs is mainly BCC or ordered B2, and the precipitation of Laves, $Al_xZr_5$, $M_3Si_5$, BCC, and B2 phase is frequently observed. Outstanding high-temperature mechanical properties have been reported for some of the RHEAs. For example, Senkov et al. [14] reported that the yield strength of MoNbTaW and MoNbTaVW RHEAs under compression were 405 and 477 MPa at 1600 °C, respectively. Furthermore, they developed another RHEA $AlMo_{0.5}NbTa_{0.5}TiZr$ with compressive yield strength of 745 MPa at 1000 °C. In a review paper, Miracle and Senkov [15] summarized the reported data on high-temperature mechanical properties of RHEAs and compared the results with those for typical conventional superalloys, as shown in Figure 6.5. In Figure 6.5, Haynes® 230® ($Co_4Cr_{27}Fe_3Mo_1Ni_{60}W_5$) is a solid solution alloy used for static sheet parts, INCONEL® 718 ($(Al,Nb,Ti)_5Co_1Cr_{21}Fe_{19}Mo_2Ni_{52}$) is a precipitation-strengthened alloy used widely in the gas turbine industry for rotating disks, and MAR-M 247® ($Al_{12}Co_{10}Cr_{10}Hf_1$–$Ni_{62}Ta_1Ti_1W_3$) is used for turbine blades. A direct comparison of the strength values or the specific strength values are inappropriate. On the one hand, the results of RHEAs are compressive yield strengths, while results of the three typical commercial alloys are tensile yield strengths. On the other hand, the results of Ni-based superalloys (e.g. MAR-M 247) are measured from single crystals, while the results of RHEAs are measured from polycrystalline materials. Therefore, the performance of RHEAs may be further improved by the design of single crystals, since grain boundaries usually dominate the degradation of mechanical properties at high temperature. Nevertheless, the softening point of a material is a good indicator to illustrate whether the material can be applied in high temperatures and where the MoNbTaW, MoNbTaVW, and $AlMo_{0.5}NbTa_{0.5}TiZr$ RHEAs show promising strength retention at high temperatures (>1000 °C or even 1200 °C). It reveals that RHEAs provide new opportunities for the development of the new generation of high-performance high-temperature alloys.

Senkov et al. [16] summarized the room temperature strength and 1000 °C strength of a plenty of RHEAs (see Figure 6.6) and found no correlation between the two strengths. In addition, they also indicated the number of phases for each reported

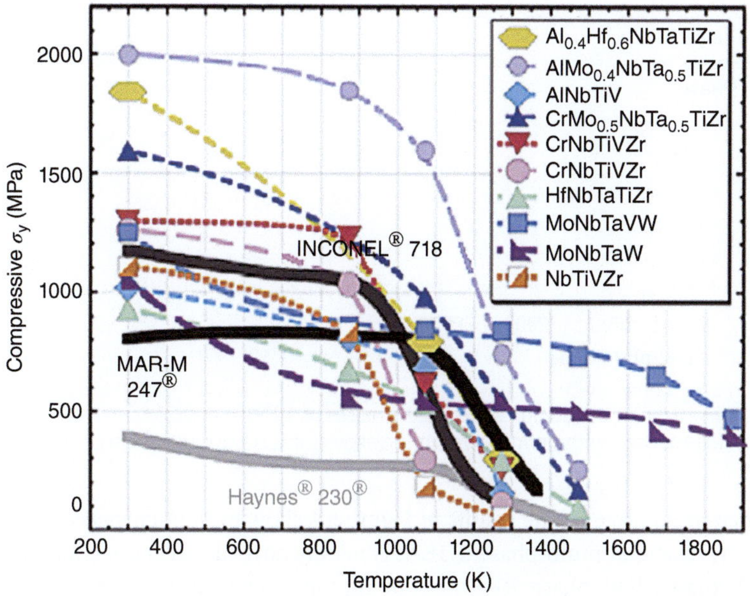


(a)

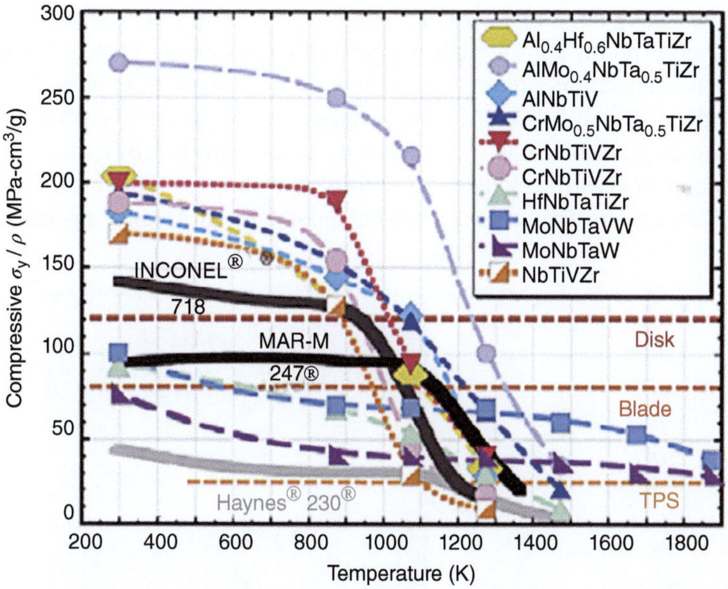


(b)

**Figure 6.5** Temperature dependence of (a) compressive yield strength, $\sigma_y$, and (b) $\sigma_y$ normalized by alloy density, $\rho$, of RHEAs. TPS, thermal protection systems. Source: Reproduced with permission from Miracle and Senkov [15]/Elsevier.

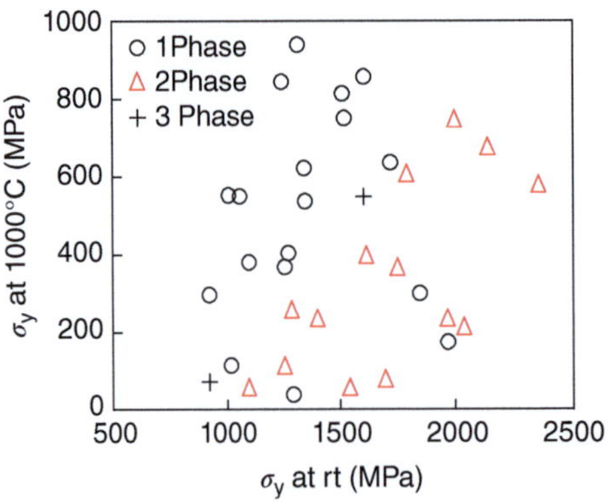


**Figure 6.6** Comparison of the compressive yield stresses at 1000 °C and 25 °C (rt) for RHEAs. Source: Reproduced with permission from Senkov et al. [16]/Springer Nature.

data, as shown in Figure 6.6. The results in Figure 6.6 show evident clustering tendency, which reveals that multi-phase RHEAs generally have higher room temperature strength than single-phase RHEAs with the advantageous flipping over at 1000 °C. This means that the strength of a multi-phase RHEA usually drops more rapidly with the increase in temperature. Therefore, RHEAs with simple phase compositions are more promising for high temperature applications. Those RHEAs with good high-temperature strength retention are either single BCC phase (e.g. MoNbTaW and MoNbTaVW) or B2/BCC duplex phases (e.g. $AlMo_{0.5}NbTa_{0.5}TiZr$). In $AlMo_{0.5}NbTa_{0.5}TiZr$, the matrix is the ordered B2 phase with the BCC phase as coherent, nanometer-sized precipitates, as shown in Figure 6.7. This modulated nanophase B2/BCC microstructure has good thermal stability and high-temperature

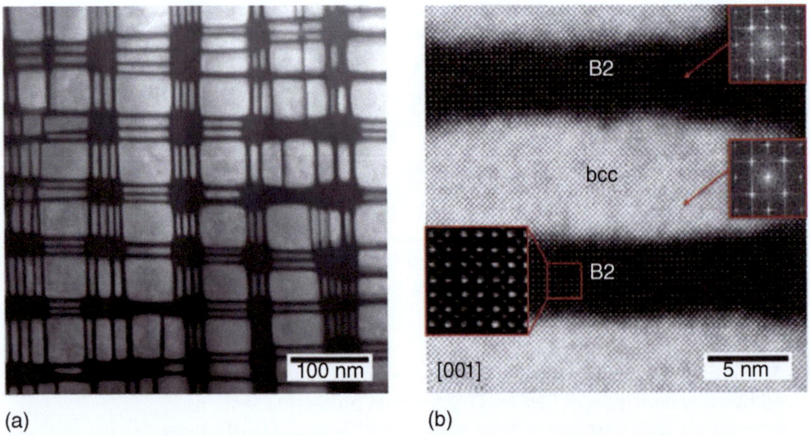


**Figure 6.7** STEM (scanning transmission electron microscope) and HAADF (high-angle annular dark field) images of the nanophase structure present inside the grains of $AlMo_{0.5}NbTa_{0.5}TiZr$. (a) Cuboidal and plate-like precipitates of a disordered BCC phase are separated by continuous channels of an ordered B2 phase. (b) Fast Fourier transformations (shown inside the red squares) reveal an ordered B2 structure for the dark channels and a disordered A2 (BCC) structure for the cuboidal precipitates. Source: Senkov et al. [17], Elsevier.

strength retention, which is quite similar to that of Ni-based superalloys. Due to these similarities, RHEAs with coherent B2/BCC nanostructure are refractory high-entropy superalloys.

Except for RHEAs, it is also reported that superior mechanical performance at elevated temperatures can also be achieved in high-entropy ultra-high temperature ceramics (HE-UHTCs). For example, Demirskyi et al. [18] reported that the flexural strength of medium-entropy $NbTaZrC_3$ was 460±24 MPa at room temperature, 496±44 MPa at 1600 °C, and 366±46 MPa at 1800 °C, as shown in Figure 6.8a.

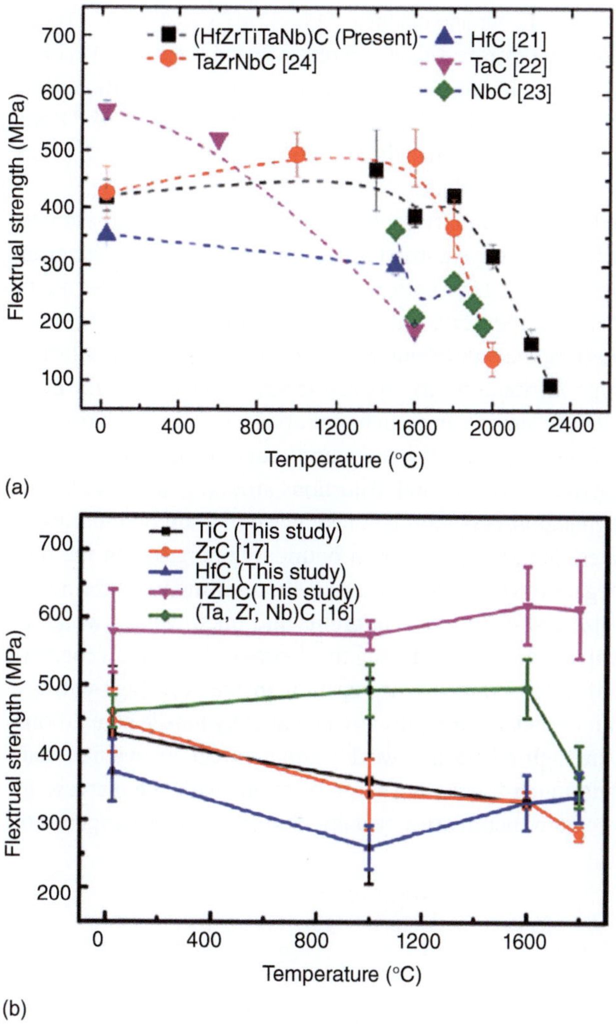


**Figure 6.8** (a) Flexural strength as a function of temperature for $HfNbTaTiZrC_5$ high-entropy ceramics compared to the results of individual carbides and ternary carbides. Source: Reproduced with permission from Feng et al. [19]/John Wiley & Sons. (b) Flexural strength as a function of temperature for $HfTiZrC_3$ medium-entropy ceramics compared to the results of TiC (sintered at 2000 °C), ZrC (sintered at 2000 °C), 17 HfC (sintered at 2200 °C), and NbTaZrC3 (sintered at 1920 °C). Source: Reproduced with permission from Yang et al. [20]/John Wiley & Sons.

Feng et al. [19] reported that the flexural strength of $HfNbTaTiZrC_5$ high-entropy carbide was 421 ± 27 MPa at room temperature, which stayed above ~400 MPa up to 1800 °C and then decreased nearly linearly to 318 ± 21 MPa at 2000 °C and to 93 ± 10 MPa at 2300 °C, as shown in Figure 6.8a. Wang et al. [25] reported that the flexural strength of $HfTiZrC_3$ medium-entropy carbide was 579 ± 62 MPa at room temperature, which increased up to 619 ± 57 MPa at 1600 °C and had no significant degradation even up to 1800 °C, as shown in Figure 6.8b. Further, they [20] reported another medium-entropy carbide $TaTiZrC_3$ with promising high temperature strength, where the flexural strength was 511 ± 52 MPa at room temperature, 493 ± 21 MPa at 1000 °C, and increased significantly from 1600 to 1800 °C, which reached 725 ± 32 MPa at 1800 °C. These results reveal that medium/high-entropy carbides can retain their strength well to high temperatures, which can hardly be obtained in conventional carbides. In addition, Han et al. [26] measured steady-state creep rates of high-entropy $HfNbTaZrC_4$, ranged from $2 \times 10^{-9}$/s to $8 \times 10^{-8}$/s, which was approximately 10 times lower than the published creep rates of the corresponding monocarbides.

Dai et al. [24] simulated grain boundary segregations in medium/high-entropy carbides, $HfTiZrC_3$, $HfNbTaZrC_4$, and $HfNbTaTiZrC_5$, and attributed the superior mechanical performance of medium/high-entropy UHTCs at elevated temperatures to intrinsic grain boundary segregations in medium/high-entropy UHTCs. In medium/high-entropy materials, the multielements can form a random solid solution in grains, where the entropy contribution can suppress the enthalpy effect. However, random distribution of the multielements in grain boundaries is harder to achieve due to the more rugged potential energy profile of grain boundaries (see Figure 6.9), which results from bond breaking/reconstruction and distortions around grain boundaries. Therefore, spontaneous segregation/depletion will take place in grain boundaries of medium/high-entropy materials to minimize grain boundary energies. On the one hand, grain boundary segregation will hinder grain growth, resulting in uniform and fine microstructures (the size and distribution of grains and micro-voids), which is beneficial to the mechanical properties by reducing the degree of local stress concentration. On the other hand, grain boundary segregation in medium/high-entropy UHTCs also enhances grain boundary strengths as calculated by tensile simulations in [24]. Grain boundaries are high-energy and weakly adhered regions in a material, and their properties degrade much faster than those of grains with the increase in temperature. As a result, grain boundaries usually dominate high-temperature

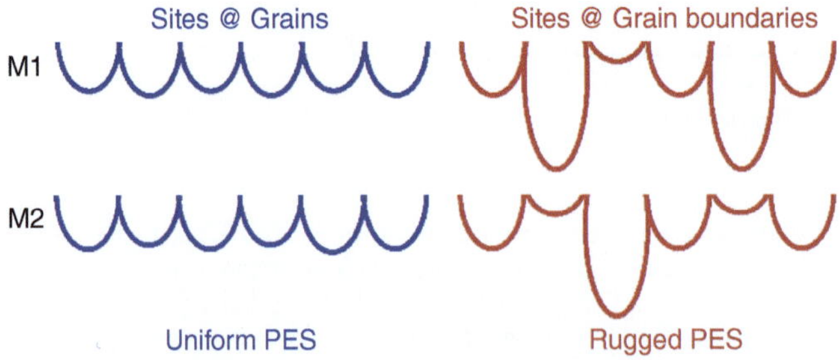


**Figure 6.9** Illustration of local potential energy surface (PES) perceived by different elements (e.g. M1 and M2) in grains or grain boundaries.

mechanical properties of ceramics. Enhancing grain boundaries will improve high temperature strengths. Except for grain boundary segregation, the solid solution state of medium/high-entropy UHTCs may also enhance the strength by pinning dislocation slip.

The principles of designing materials for high-temperature applications are different from those for low-temperature applications. For low-temperature applications, metastable microstructures are fine because of slow kinetics. Sometimes, metastable microstructures are designed to tailor materials' properties, e.g. TRIP alloys. However, for high-temperature applications, microstructure stability is crucial, especially for long-term services at high temperatures. Microstructure transformation will change the materials' properties severely. Decomposition of solid solutions after heat treatments can usually be observed in HEAs. For example, Senkov and Semiatin [27] reported that the single BCC phase HfNbTaTiZr was metastable. Cold rolling followed by annealing at 800 °C resulted in a dual BCC microstructure, which was significantly brittle compared to the single BCC phase obtained by cold rolling and annealing at 1000 °C, as shown in Figure 6.10. Precipitation both within the grains and on grain boundaries was observed when the rolled sheet was annealed for two hours at 1200 °C.

Even though no significant change in microstructure can be observed during processing or service, one cannot say that the microstructure is stable. There may be

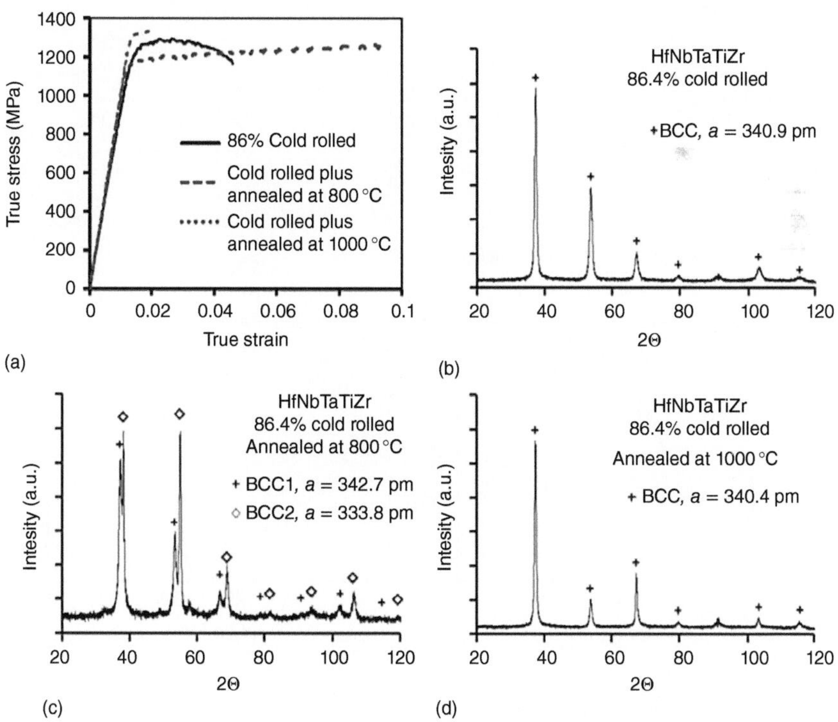


**Figure 6.10** (a) True stress-true strain curves of as-cold-rolled and rolled-and-annealed sheets of HfNbTaTiZr. X-ray diffraction patterns for HfNbTaTiZr: (b) After 86.4% total reduction and following annealing for two hours at: (c) 800 °C or (d) 1000 °C. Source: Reproduced with permission from Senkov et al. [27]/Elsevier.

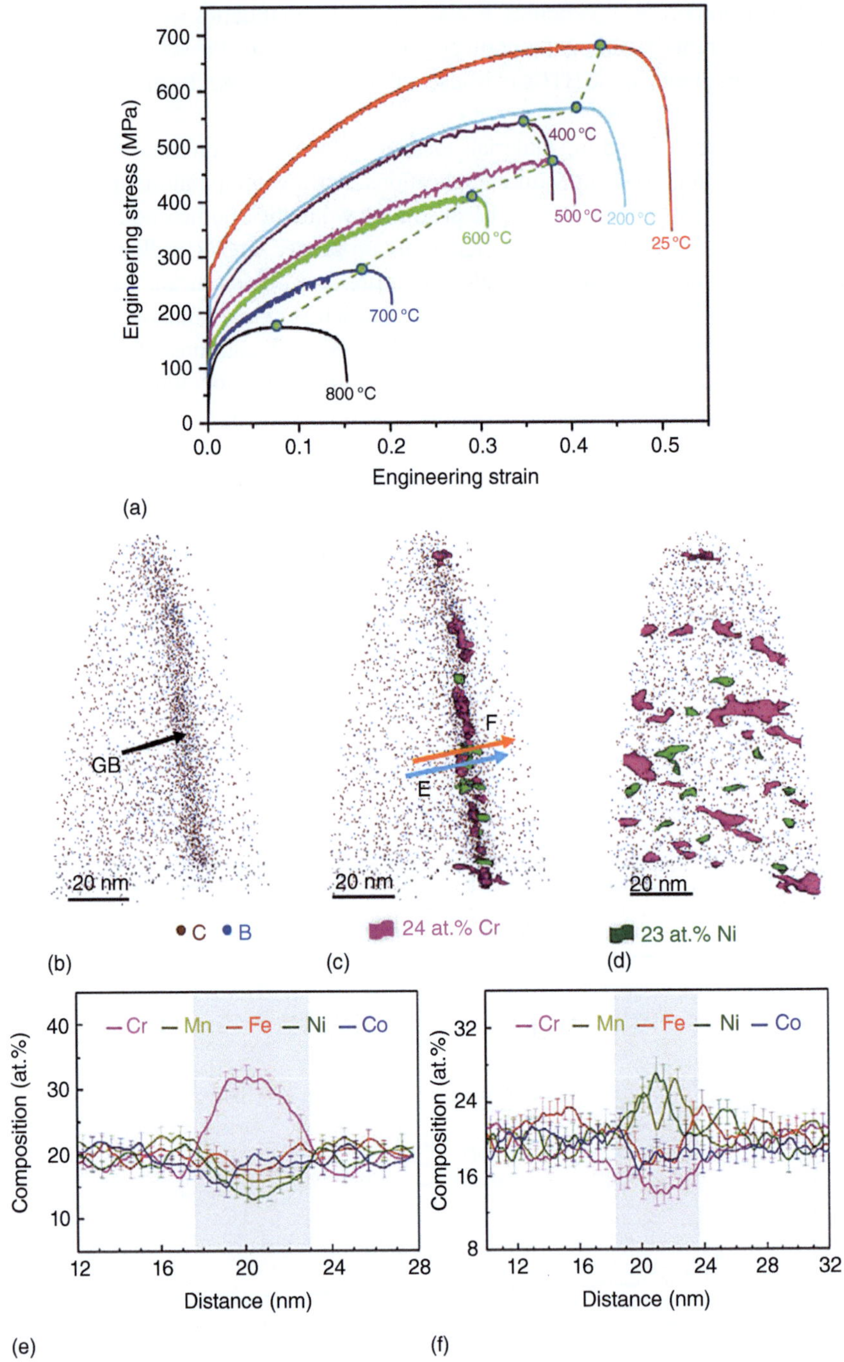


**Figure 6.11** (a) Engineering stress–strain curves of CoCrFeMnNi HEA at different temperatures (b) Atom map of a tip showing grain boundary segregation of impurities C and B. Cr (24 at.%) and Ni (23 at.%) iso-composition surfaces (c) viewed with the grain boundary edge-on and (d) viewed from the grain boundary normal. One dimensional compositional profiles (e) along the cyan arrow and (f) along the orange arrow indicated in (c); the shaded areas represent the approximate grain boundary regions in thickness direction. Source: Reproduced with permission from Ming et al. [29]/American Association for the Advancement of Science.

some compositional or structural changes around defects, e.g. grain boundary segregation, which then alter materials' performances. Even though HEMs have been proposed for nearly 20 years, grain boundaries of HEMs have not been paid due attention until recently. As suggested above, grain boundary segregation/depletion may be intrinsic problems in HEMs due to the rugged potential energy surface (PES) around grain boundaries and the multielement nature of HEMs. Grain boundary segregations have been reported in more and more HEMs, either by experiments or by simulations. For example, grain boundary segregation of Nb in MoNbTaW HEA was reported according to atomistic simulations by Li et al. [28], which was also believed responsible for the outstanding high-temperature strength of the alloy; segregation and spinodal like decomposition of grain boundaries in CoCrFeMnNi HEA were reported by Ming et al. [29], which resulted in the loss of ductility with temperature (see Figure 6.11) and the segregation was confirmed according to atomistic simulations by Hu and Luo [21]; synergic grain boundary segregation and precipitation in W- and W-Mo-containing high-entropy diborides were reported by Wang et al. [30]; and general grain boundary segregations in medium/high-entropy carbides were simulated by Dai et al. [24]. Therefore, special attention should be paid to design grain boundaries or even eliminate grain boundaries (similar to single crystal superalloys) in the design of high-temperature materials, since grain boundaries usually serve as nucleation sites for voids, flaws at high temperatures.

## 6.4 Improved Hardness: Toward Super Hard Materials

Materials with high hardness are in high demand in industries, e.g. cutting and polishing tools, wear-resistant coatings, molds, dies, and abrasives. Compared to conventional materials, HEMs usually exhibit improved resistance to deformation due to severe lattice distortions, which pin dislocations due to both volume distortions and chemical bond mismatch. This means that HEMs are usually harder than conventional materials. Therefore, HEMs are naturally suitable as new hard and wear-resistant materials. In addition, as mentioned in Section 6.3, HEMs may exhibit better property retention at high temperatures, which is also beneficial to their applications. For example, cutting tools serve at high temperatures during high-speed dry cutting.

Gild et al. [22] successfully synthesized six single-phase solid solution diborides with $AlB_2$ structure and measured their hardness; the measured hardness values were generally greater than the rule-of-mixture averages of the hardness values measured from individual metal diborides that were fabricated via the same route, as shown in Figure 6.12a. Similarly, Harrington et al. [31] synthesized a series of high-entropy carbides of group IVB, VB, and VIB refractory transition metals and measured the mechanical properties via nanoindentation. Their results revealed that the high-entropy carbides showed significantly enhanced hardness when compared with the rule-of-mixtures average of the constituent binary carbides and to the highest hardness of the binary constituents, as shown in Figure 6.12b. Except for these two sets of results, enhanced hardness has also been proven by other researchers in similar systems. Especially, the enhancing effect depends on the lattice distortions, e.g. bigger lattice mismatch between the constitute metal elements usually

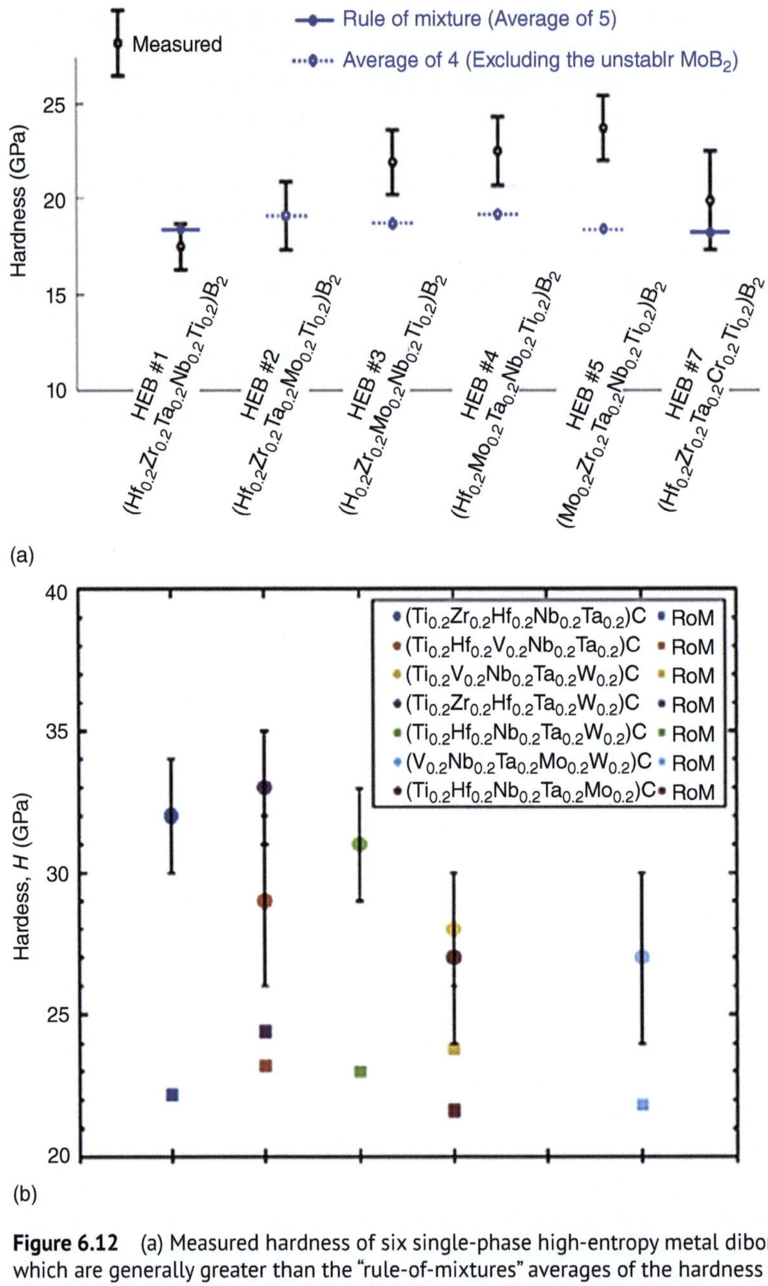


**Figure 6.12** (a) Measured hardness of six single-phase high-entropy metal diborides, which are generally greater than the "rule-of-mixtures" averages of the hardness values measured from individual metal diborides that were fabricated via the same route. Since $MoB_2$ is not in an equilibrium bulk phase below 1500 °C, the averages for HEB#2-HEB#5 were calculated without $MoB_2$. However, $MoB_2$ has a lower melting temperature and theoretical hardness than all other five other metal diborides in question; thus, the actual rule-of-mixtures averages should be even lower. Source: Glid et al. [22]/Springer Nature/CC BY-4.0. (b) Hardness from nanoindentation (circle markers) on seven high entropy carbide compositions along with rule of mixture (RoM) values (square markers) against valence electron concentration (VEC). Error bars are one standard deviation from 100 indents per sample. Source: Reproduced with permission from Harrington et al. [31]/Elsevier.

results in greater enhanced hardness [32–34]. In other words, those single-phase high-entropy carbides or diborides that are difficult to synthesize usually result in higher hardness. For example, Qin et al. [23] successfully synthesized five single-phase high-entropy diborides containing $WB_2$ and $MoB_2$ via reactive spark plasma sintering that could not been synthesized by other methods. In addition, high-entropy diborides containing softer $WB_2$ and/or $MoB_2$ components (difficult to synthesize with large lattice mismatch) are significantly harder than $(Hf_{0.2}Nb_{0.2}Ta_{0.2}Ti_{0.2}Zr_{0.2})B_2$ (easy to synthesize with small lattice mismatch). Before the efforts being made on preparing bulk high-entropy carbides or diborides, investigation on high-entropy carbide and diboride coatings has been done, where films with ultra-high hardness were obtained. For example, Gorban' et al. [35] prepared super hard $(HfNbTaTiVZr)C_6$ coatings by ion sputtering of a HfNbTaTiVZr alloy target in the plasma of compressed vacuum-arc discharge and produced a thin film with hardness of 43–48 GPa and friction coefficient of 0.14–0.16 at the loads 2.2–5.2 N; Mayrhofer et al. [36] employed physical vapor deposition and prepared a

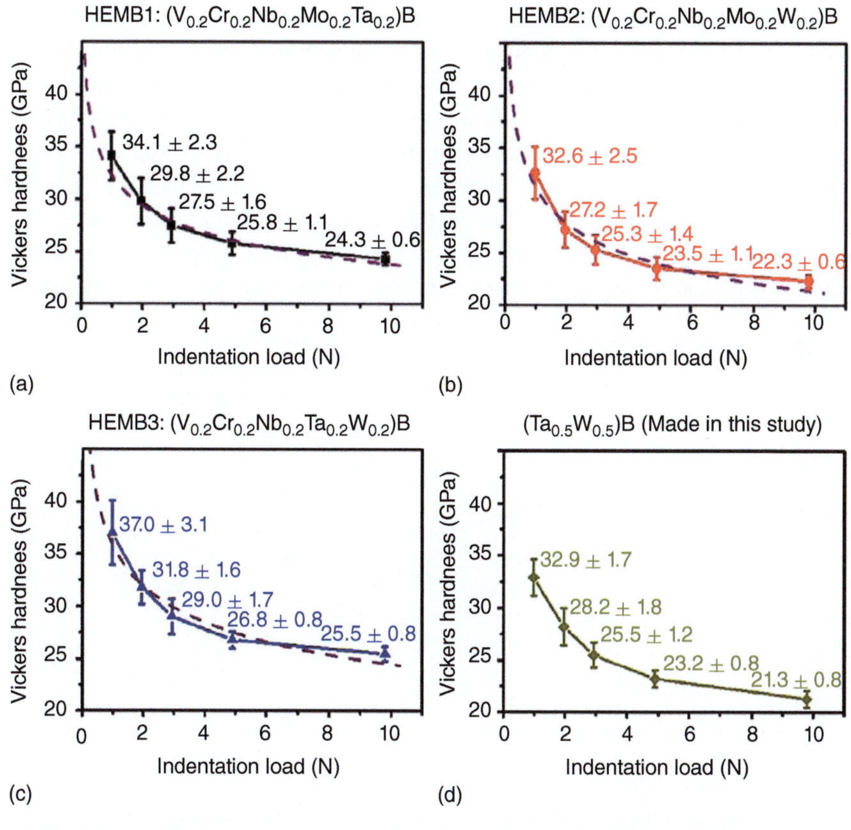


**Figure 6.13** Measured Vickers microhardness of (a) $(Cr_{0.2}Mo_{0.2}Nb_{0.2}Ta_{0.2}V_{0.2})B$, (b) $(Cr_{0.2}Mo_{0.2}Nb_{0.2}V_{0.2}W_{0.2})B$, and (c) $(Cr_{0.2}Nb_{0.2}Ta_{0.2}V_{0.2}W_{0.2})B$, along with (d) a reference $(Ta_{0.5}W_{0.5})B$ specimen made by the same procedure, under indentation loads ranging from 0.98 to 9.8 N. The purple dash lines represent the best-fitted microhardness versus indentation load curves by Meyer's law for each specimen. Source: Reproduced with permission from Qin et al. [37]/Elsevier.

$(Hf_{0.19}Ta_{0.24}Ti_{0.20}V_{0.14}Zr_{0.23})B_2$ thin film with improved hardness (47.2 GPa), which is higher than that of $ZrB_2$ film (43.2 GPa) and $Ti_{0.39}Zr_{0.61}B_2$ film (45.8 GPa).

Super hard bulk materials have also been developed under the concept of HEMs. For example, Qin et al. [37] developed single-phase high-entropy monoborides ($(Cr_{0.2}Mo_{0.2}Nb_{0.2}Ta_{0.2}V_{0.2})B$, $(Cr_{0.2}Mo_{0.2}Nb_{0.2}V_{0.2}W_{0.2})B$, and $(Cr_{0.2}Nb_{0.2}Ta_{0.2}V_{0.2}W_{0.2})B$) of the CrB prototype structure. The measured Vickers hardness was ~22–26 GPa at an indentation load of 9.8 N and ~32–37 GPa at 0.98 N. In particular, the load-dependent hardness of $(Cr_{0.2}Mo_{0.2}Nb_{0.2}V_{0.2}W_{0.2})B$ is higher than those of $(Ta_{0.5}W_{0.5})B$ (already considered as a super hard material) and the hardest reported high-entropy metal diborides, and on a par with the classical super hard boride $WB_4$, as shown in Figure 6.13. High hardness was also obtained by Zhao et al. [38], who designed a new high-entropy monoboride $(Cr_{0.2}Mo_{0.2}Ni_{0.2}Ta_{0.2}W_{0.2})B$ ceramic with a WB type structure, the hardness of which was 48.51 GPa at 0.49 N, higher than those high-entropy monoborides reported by Qin et al. [37].

Except for those transition metal carbides and borides, efforts have also been made to develop HEAs with improved hardness or hot hardness. In a review paper by Yeh and Lin [39], the authors showed some intriguing examples. As they suggested, HEAs may be used as a bond layer in gas turbines, molds and dies, bond medium or surface coating in cutting tools due to their improved hardness. For example, Hsu et al. [40] developed a high-entropy $AlCo_{0.6}Cr_{1.5}Fe_{0.2}NiSiTi_{0.2}$ to replace $Al_{15}Co_{33.5}Cr_{21}Ni_{30}Y_{0.5}$ (in wt%), where the density of HEA was 20% lower, room temperature hardness was 2.3 times higher, 1100 °C hardness was 4 times higher, oxidation resistance was the same, thermal conductivity was 33% lower, and thermal expansion was 17% lower [39]. These improved properties guarantee smaller centrifugal force, higher wear resistance, better heat resistance, and good

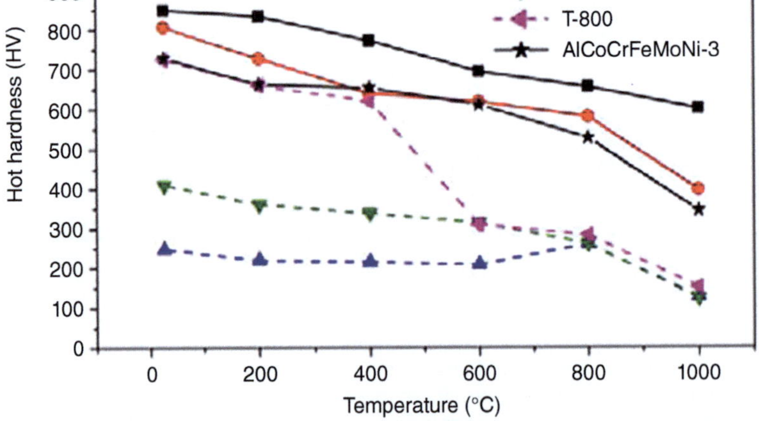


**Figure 6.14** Curves of hot hardness versus temperature for six alloys including three HEAs and commercial alloys IN718, IN718H, and T-800. Source: Reproduced with permission from Yeh et al. [39]/Springer Nature.

thermal expansion compatibility between the substrate and top coat. Then, the HEA may exhibit better performance in comparison to the current bond coating. Comparing to commercial hot alloys IN718, IN718H, and T-800, some HEAs reveals better hardness retention, e.g. at the temperature ranging from 400 to 1000 °C, as shown in Figure 6.14. As suggested by Yeh and Lin [39], these HEAs may be good candidates for hot working dies for steels, the softening temperature of which may be as high as 500 °C. Another example is that replacing the bond material Co in cutting tools by HEAs can result in both better hardness and fracture toughness, which are located at above the conventional scattering band and show their potential in cutting tools, as shown in Figure 6.15a. Figure 6.15c compares the flank wears between two commercial inserts and two inserts bonded with 15% and 20% HEAs after dry-cutting 304 stainless steel for a distance of 314 m at the cutting speed of

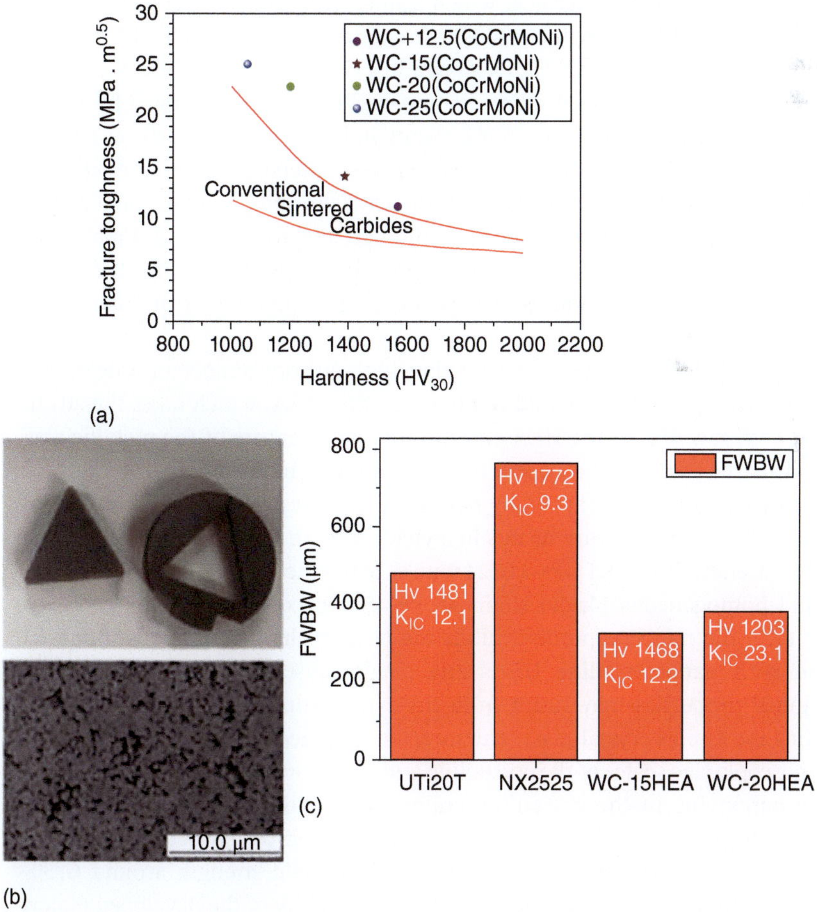


**Figure 6.15** (a) Typical scatter band of fracture toughness versus hardness for conventional sintered carbides; (b) pictures of sintered carbide bonded with 15 wt% HEA and its microstructure; and (c) comparison of flank wear band width between two commercial inserts and two inserts bonded with two different amounts of HEA. FWBW, flank wear band width. Source: Reproduced with permission from Yeh et al. [39]/Springer Nature.

94 m/s, the depth of cut 0.2 mm, and the feed rate of 0.3 mm/rev. It is apparent that high-entropy inserts have much lower flank wear, thus indicating that suitable high-entropy composition could improve the lifetime of a cutting tool.

## 6.5 More Examples on HEMs with Intriguing Mechanical Properties

In this part, more examples on HEMs with intriguing mechanical properties will be shown, where the underlying microstructure or mechanisms will also be introduced. Lei et al. [41] discovered an intriguing phenomenon in a HfNbTiZr HEA, where ordered oxygen complexes can enhance both the strength and ductility, thus breaking the long-standing strength–ductility trade-off. They doped a model HfNbTiZr HEA with 2 at.% oxygen, which lead to unprecedented enhancement in both strength and ductility with the tensile strength increased by $48.5 \pm 1.8\%$ and the ductility increased by $95.2 \pm 8.1\%$, as shown in Figure 6.16. Moreover, they found that the oxygen atoms formed ordered complexes that is characterized by enrichment of O, Ti, and Zr (see Figure 6.16), a state in between oxide particles and frequently occurring random interstitials. The ordered interstitial complexes changed the dislocation shear mode from planar slip to wavy slip and promoted double cross-slip and thus dislocation multiplication through the formation of Frank–Read sources during deformation. Their results provide a new insight into the role of interstitial solid solutions and associated ordering strengthening mechanisms in metallic materials.

Wu et al. [42] designed a novel crystal-glass high-entropy nanocomposite by doping glass-forming elements Si and B into CoCrFeNi HEA, which takes the advantages of both low SFE of the HEA and the compliant nature of the metallic glass phase at the nanoscale. The typical microstructure is nanocolumnar grains with high-density nanotwins wrapped by nanoscale glass regions at grain boundaries. The nanocomposite structure results in a yield strength approaching the theoretical shear strength limit (4.1 GPa $\approx G/24$ with $G$ being the shear modulus of a material) and homogeneous plastic strain above 45% in compression. As shown in Figure 6.17, the nanocomposite exhibits the best mechanical properties in comparison with nanocrystalline HEAs and single crystal CoCrNi. The superior mechanical properties may result from the synergistic strengthening hierarchy, including the duplex crystal-glass structure, the interface–dislocation interactions, the nanoscopic size scales of the crystalline and glassy building units, and the high-density nanotwins in the crystalline grains. Nano-pillars with ultra-high yield strength and good ductility have also been discovered by Zou et al. [43–45] in MoNbTaW HEAs, who obtained nano-pillars with yield strength around $G/50 \sim G/30$. In addition, they also checked the thermal stability of the developed materials, which revealed the high stability of the designed nanostructure. For example, no significant property degradation was observed when the material was annealed at 1100 °C for three days [44]. The stability of the microstructure and the retention

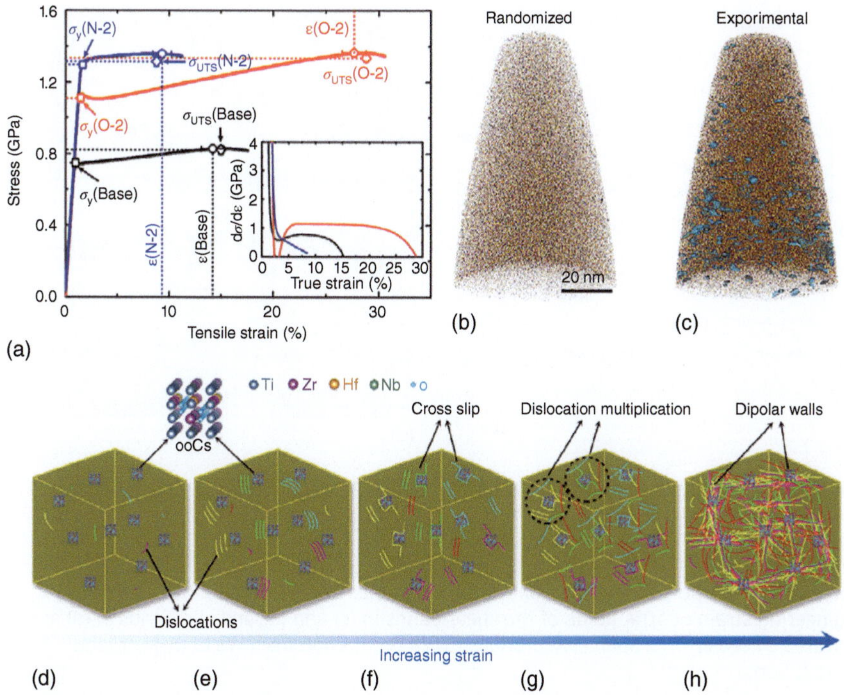


**Figure 6.16** (a) Room-temperature tensile stress–strain curves for the as-cast HfNbTiZr (denoted as base alloy), $(HfNbTiZr)_{98}O_2$ (denoted as O-2), and $(HfNbTiZr)_{98}N_2$ (denoted as N-2) HEAs. The inset shows the corresponding strain hardening response ($d\sigma/d\varepsilon$). Three-dimensional reconstruction of the O-2 HEA atom probe tomography dataset. Randomized (b) and experimental (c) datasets on which an iso-composition surface encompassing regions in the point cloud containing more than 3.0 at.% O was superimposed. The experimental dataset clearly shows evidence for OOCs (ordered oxygen complexes). (d–h) Schematic diagram illustrating the plastic deformation mechanism in the oxygen-rich alloy variant O-2 HEA.

of the superior properties to high temperatures may be highly correlated with grain boundary segregations [45].

Eutectic HEAs are a new set of HEAs that are typical examples on how microstructure modification can improve the mechanical properties of HEAs. Lu et al. [46] first proposed the designation concept of eutectic HEAs, which combines the high strength of BCC phase and the high ductility of FCC phase. They produced $AlCoCrFeNi_{2.1}$ eutectic HEAs with regular FCC($L1_2$)/BCC(B2) lamellar structure, which really exhibited a good combination of high strength and high ductility. Then, thermal–mechanical processes, including cold rolling, annealing, or directional solidification, were adopted to tune the microstructure (fine-grained, hierarchical structure, herringbone structure, etc.) and to optimize the mechanical properties [47, 48]. For example, Shi et al. [47] modified the microstructure of eutectic HEAs by thermal–mechanical processes and obtained dual-phase heterogeneous lamella microstructure, which exhibited significantly improved strength and ductility, as shown in Figure 6.18.

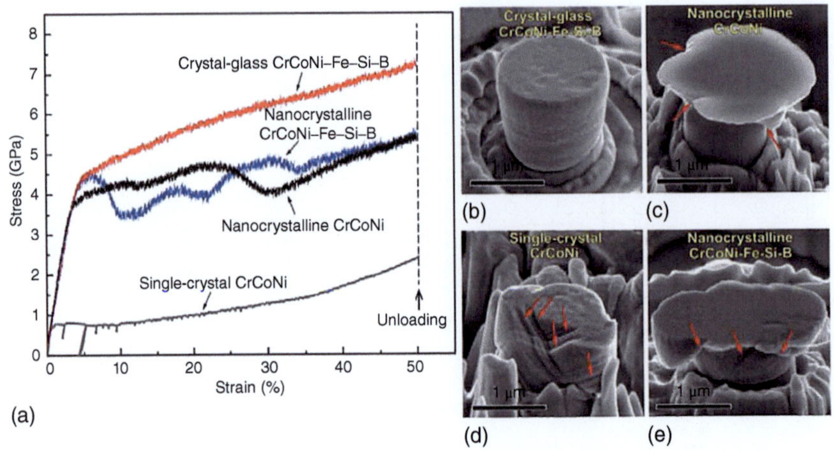


**Figure 6.17** Mechanical properties of the crystal-glass high-entropy nanocomposite. (a) Compressive engineering stress–strain curves of the crystal-glass CrCoNi–Fe–Si–B high-entropy nanocomposite, nanocrystalline CrCoNi–Fe–Si–B, nanocrystalline CrCoNi, and single-crystal CrCoNi pillar samples, each with the same diameter of 1 $\mu$m. (b–e) Scanning electron microscope (SEM) images of the same samples after compression to an identical engineering strain of 50%. Some of the shear bands in (c) and (e), and slip bands in (d) are indicated by red arrows. Source: Reproduced with permission from Wu et al. [42]/John Wiley & Sons.

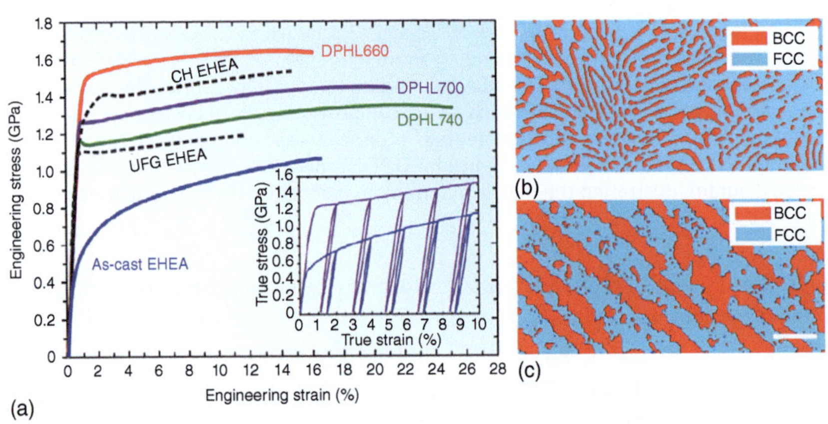


**Figure 6.18** (a) Tensile stress–strain curves of different eutectic HEAs. Ultrafine-grained (UFG) eutectic high-entropy alloy (EHEA) and CH EHEA refer to the ultrafine-grained eutectic HEA, and the complex and hierarchical eutectic HEA, dual-phase heterogeneous lamella (DPHL) means dual-phase heterogeneous lamella, respectively. The number following DPHL is annealing temperature. The inset is the loading-unloading-reloading behavior of the DPHL700 and the as-cast eutectic HEA. EBSD phase image of the (b) as-cast eutectic HEA and (c) DPHL eutectic HEA. Scale bars in (b) and (c) are the same in length in the figure, which represent 20 and 2 $\mu$m, respectively. Source: Shi et al. [47]/Springer Nature/CC BY-4.0.

## 6.6 Strengthen Mechanisms

### 6.6.1 Theory on Yield Strength

The yield strength of an alloy depends on its intrinsic dislocation properties, compositions, grain size (Hall–Petch effect), etc. In this part, only solid solution strengthening effect on yield strength is discussed, which is considered one of the key properties of HEMs. The solid solution models of HEAs were developed based on traditional solid solution models derived from dilute alloys. For more comprehensive discussion on dislocation hardening theory of HEAs, please refer to a well-organized review paper by George et al. [2]. Here, only a brief introduction is provided.

For FCC alloys, the yield is initialized by planer glide of dislocations. The solid solution strengthening effect depends on the interaction strength between solute atoms and dislocation cores, which is usually characterized by an interaction volume. Toda-Caraballo et al. [49–51] made great efforts to develop simplified models based on elastic theory to evaluate lattice distortions and to predict solid solution strengthening in HEAs. They [51] modified and extended models developed for binary alloys suitable for multicomponent systems and checked the generality of the model by applying it to both binary and multicomponent alloys. Similar works were done by Varvenne et al. [52, 53], who modeled a HEA by an average medium with fluctuations due to local chemical composition. The model developed by Varvenne et al. is parameter-free, which can predict the temperature-, composition-, and strain rate-dependence of the yield strength. In their model, the interaction energy between elements and dislocation cores is necessary, which can then be simplified by approximating the interaction energy as $-p\Delta V_i$ that is similar to the model of dilute alloys. Figure 6.19 illustrates the predictive power of Varvenne's model on temperature dependence and composition dependence of yield strength.

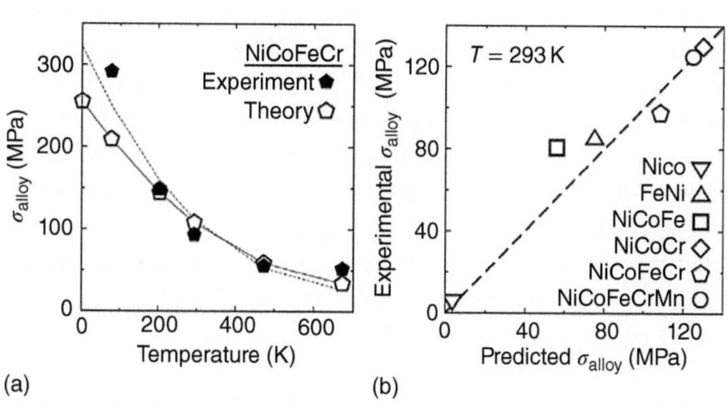


**Figure 6.19** (a) Solid solution (non-Hall–Petch) contribution to initial yield strength versus temperature for the FCC CoCrFeMnNi HEA, as predicted (open symbols) and as measured (solid symbols); the dashed line indicates predictions using a slightly different value of the dislocation line tension. (b) Non-Hall–Petch contribution to the initial yield strength, experimental versus predicted, for a range of equal composition alloys in the Co–Cr–Fe–Mn–Ni family at $T$ = 293 K. Source: Varvenne et al. [53]/With permission of Elsevier.

For BCC alloys, Suzuki's model suggested that the dominant strengthening was the inhibition of the glide of kink along a screw dislocation by the solutes. Rao et al. [54] adapted the Suzuki model to HEAs and revealed slightly overestimated yield strengths. One of the problems in BCC HEAs is that the dominating yield phenomenon may change due to the severe lattice distortions. Atomistic simulations have given different conclusions on the dominating effect. For example, Chen et al. [55] simulated dislocation glide in $Co_{16.67}Fe_{36.67}Ni_{16.67}Ti_{30}$ BCC-HEA and found that the controlling step for screw dislocation motion changed to kink spread on dislocation lines, while Yin et al. [56] simulated dislocation glide in NbTaMoW BCC-HEA and found that the rate-limiting process was still kink pair nucleation, which was similar to conventional BCC alloys. Moreover, contributions from edge dislocations can no longer be neglected in BCC HEAs, where the Peierls stress of dislocations may be comparable to that of screw dislocations. For example, a large fraction of nonscrew segments on the gliding dislocations in a BCC refractory medium-entropy alloy was reported by Wang et al. [57]. In contrast, contributions from edge dislocations are usually negligible in conventional BCC alloys, where the Peierls stress is very low. Thus, more studies are necessary to model solid solution strengthening in BCC HEAs.

### 6.6.2 Short Range Order

Originally, it is assumed that alloying elements in medium/high-entropy materials tend to occupy lattice sites randomly to enlarge configurational entropy. In a perfectly random alloy, the occupation of any site is equal to its overall concentration. However, it has been reported that short range order (SRO) may exist in many medium/high-entropy materials, where the occupations of neighboring sites are correlated depending on element interactions to reduce enthalpy. Today, there is increasing experimental evidence come out that SRO exists in medium/high-entropy materials, which may have strong influence on materials' mechanical properties, e.g. providing additional pinning strength to dislocations. For example, Zhang et al. [58] reported the observation of SRO in the medium-entropy alloy CoCrNi by using energy-filtered transmission electron microscopy, where the SRO may be driven by the disfavor of Cr–Cr bonds. In addition, they also revealed that increasing the amounts of SRO would give rise to both higher SFE and hardness. As mentioned above, the yield strength of CoCrNi is higher than that of the Cantor alloy. It is possible that the SRO has some contribution to the high yield strength. Similar SROs have also been observed in CoNiV by Chen et al. [59]. Chen et al. [59] discovered that the SRO could be best described in terms of the V occupancy, as shown in Figure 6.20. Specifically, two V-enriched $(\bar{3}11)$ planes sandwich one V-depleted $(\bar{3}11)$ plane. In other words, the V-enriched $(\bar{3}11)$ planes alternate with those enriched in Co and/or Ni. This alternating chemical occupancy extends across only a few $(\bar{3}11)$ planes, that is, a distance of less than 1 nm. In addition, Chen et al. [59] also revealed the interaction between SROs and dislocations, where they observed profuse dislocation tangles in the FCC phase, rather than planar slip, which may result from the trapping effect of SROs on moving dislocations. Except

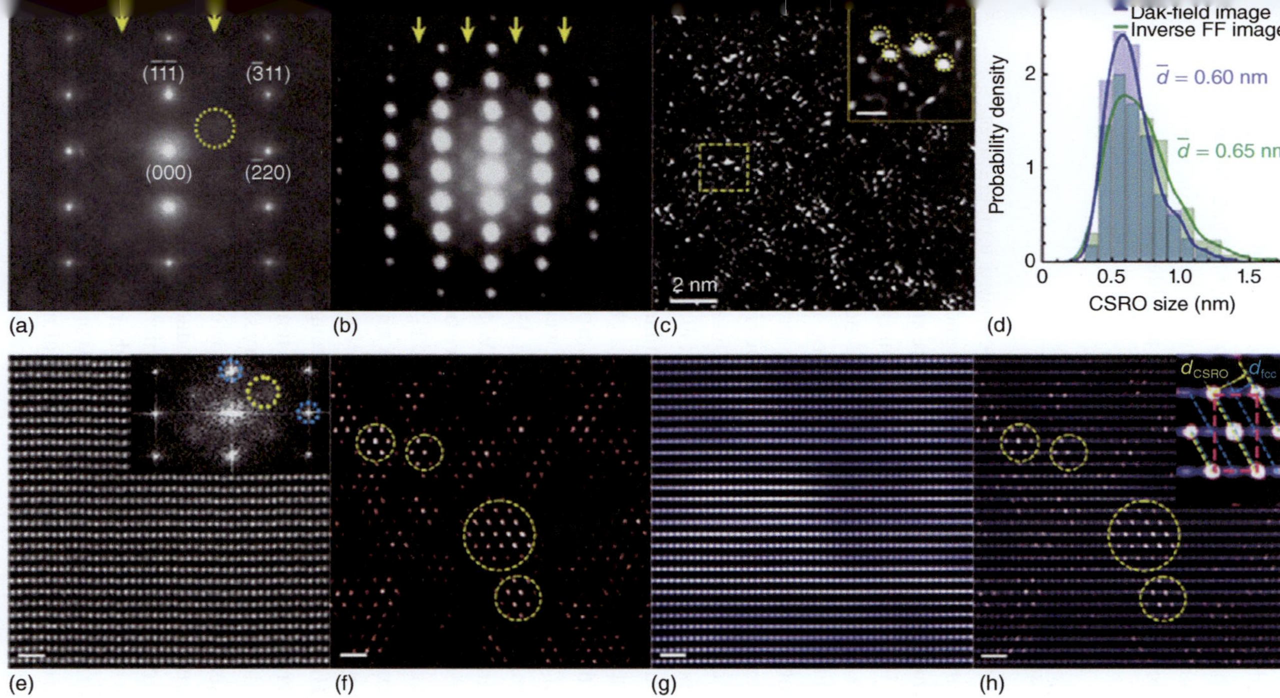


**Figure 6.20** Evidence of CSRO (chemical short range order) in the FCC CoNiV. The sample was deformed in tension to 18% plastic strain. (a) Selected-area EDP (electron diffraction patten) with the [112] zone axis. Two arrays of extra and diffuse disks (indicated by arrows) appearing at $\{\overline{3}\ 11\}/2$ positions (one example is inside the yellow circle). (b) Nano-beam EDP with the [112] zone axis. Arrays of superlattice reflections at $\{\overline{3}\ 11\}/2$ positions as indicated by arrows. (c) Energy-filtered dark-field transmission electron microscope (TEM) image taken using the diffuse reflections, with the inset showing a close-up view of the dashed square area, highlighting some coherently diffracting clusters corresponding to the local CSROs. The size distribution of these CSRO regions is shown in (d). The average size $\overline{d}$ is 0.60 nm and 0.65 nm, based on the dark-field TEM image and the inverse fast Fourier transform (FFT) image, respectively. (e) Lattice image of the FCC phase with the [112] zone axis and the corresponding FFT pattern (inset). This pattern again displays the extra diffuse reflections (see, for example, inside the yellow circle) at $\{\overline{3}\ 11\}/2$ positions, besides the sharper Braggs spots from the FCC phase (blue circles). (f and g) Inverse FFT image showing the CSRO regions (several are circled), and the FCC lattice, respectively. These two images are superimposed in (h), with a close-up view of a CSRO region in the inset. Overlapping in this way produces bright sites that highlight the extra CSRO lining up on $\{\overline{3}\ 11\}$ planes (yellow dashed lines). $d_{fcc}$ denotes the spacing of $\{\overline{3}\ 11\}$ planes in the normal FCC lattice, whereas $d_{CSRO}$ displays the spacing corresponding to the extra chemical order. The red dashed box outlines this unit period for the local CSRO configuration. All scale bars are 0.5 nm except in (c). Source: Chen et al. [59], Springer Nature.

for these experimental lines of evidence, atomistic simulations also suggested SROs may exist in HEAs [60, 61] (see Chapter 3 for more examples). It is possible that SROs usually exist in medium/high-entropy materials due to the driving force of enthalpy, especially in HEAs, since the first neighbor interaction dominates the enthalpy. Similar to the idea to design super hard HEMs, special attention on possible SROs should be paid to HEMs that are not easy to synthesize. In such cases, the entropy contribution to randomness and the enthalpy contribution to ordering are close to a critical point. Therefore, it is highly possible that SROs exist in these HEMs.

### 6.6.3 Grain Boundary Segregation

As mentioned in the section "superior performances at elevated temperatures," grain boundary segregation may be an intrinsic phenomenon in medium/high-entropy materials due to the multiprincipal element nature and the rugged PES at grain boundaries due to local distortions, bond breaking, and recombination. Here, element segregation formed during solidification is not considered. There are both experimental and theoretical evidence of grain boundary segregations in HEMs. For example, experimentally, Ming et al. [29] observed segregation and spinodal-like decomposition of grain boundaries in CoCrFeMnNi HEA. Sohn et al. [7] found grain boundary segregation of V in medium-entropy alloy CoNiV, even though they did not pay special attention to it. Zou et al. [45] reported segregation and compound formation in MoNbTaW grain boundaries, which may contribute to the improved stability nanostructure and strength retention at high temperatures (see Figure 6.21). The formation of the amorphous phase at grain boundaries in the work of Wu et al. [42] can also be viewed as grain boundary segregation of glass-forming elements Si and B. Wang et al. [30] reported synergic grain boundary segregation and precipitation in W- and W-Mo-containing high-entropy diborides. Theoretically, atomistic simulations have been adopted to study grain boundary segregations in HEMs, e.g. Hu and Luo [21] confirmed the grain boundary segregation phenomenon in CoCrFeMnNi HEA and trained a model to predict the amounts of segregation. Li et al. [28] found Nb segregation in grain boundaries of MoNbTaW HEA, which could improve the strength of MoNbTaW. Dai et al. [24] discovered general grain boundary segregation tendencies in medium/high-entropy carbides, which may enhance grain boundary strength and improve high temperature strengths of these medium/high-entropy carbides. On the one hand, grain boundary segregation may impede grain boundary grow, which guarantee uniform fine-grained microstructure, enhancing microstructural stability, and possibly improving mechanical properties. On the other hand, grain boundary segregation may significantly change grain boundary properties, which will then change material's performance, e.g. loss of ductility at medium temperature of CoCrFeMnNi HEA, improvement of strength retention at elevated temperatures of medium/high temperature strength.

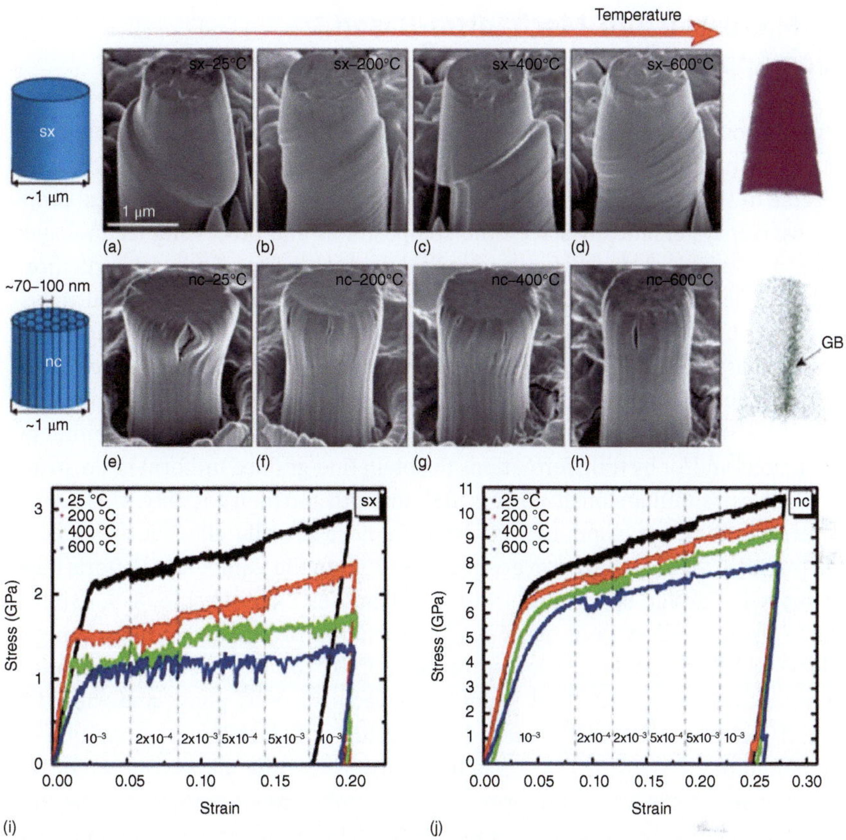


**Figure 6.21** Compression results for the sx-HEA and nc-HEA pillars from room temperature to 600 °C. Representative SEM images of the deformed sx-HEA pillars (a–d) and nanostructured columnar-grained HEA pillars (e–h). Corresponding engineering stress–strain curves of (i) the sx-HEA and (j) nc-HEA pillars, showing how flow stresses changes by temperature. Strain-rate jump tests are applied to measure the strain-rate sensitivity using initial and final strain rates of $10^{-3}$ $s^{-1}$ and four other strain rates of $2 \times 10^{-4}$, $2 \times 10^{-3}$, $5 \times 10^{-4}$, and $5 \times 10^{-3}$ $s^{-1}$. sx-HEA means single crystal HEA, nc-HEA means nanocrystal HEA. Source: Zou et al. [45], American Chemical Society.

Actually, when considering grain boundary segregation, not only those elements manually added (including impurities in the raw materials) should be considered, but also those elements possibly absorbed from environments during processing, storage, and service should also be considered. The complexity in structure and composition makes the investigation on grain boundary segregation a non-trivial task. Thus, studying on grain boundary segregation in HEMs is just in the initial stage. There is no doubt that more and more work on grain boundary segregation of HEMs will be done when high-performance HEMs are designed.

## 6.7 Microstructure-Mechanism-Based Design Approaches

It is well known that the mechanical properties of materials depend strongly on their microstructures and mechanisms that activated under deform, which is also applicable to HEMs. Therefore, to design a HEM with intriguing mechanical properties, we can systematically screen the compositions and processing technologies to achieve a desired microstructure and candidate mechanisms. The microstructure–mechanism-based design approaches have been well studied in conventional alloys, e.g. steels, Ti-based alloys, and Al-based alloys, where different strengthening mechanisms or plasticity enhancing mechanisms are designed. Some ideas in the microstructure–mechanism-based design approaches are now being transferred into the design of HEMs. For example, the most effective one is to reduce grain size either by deformations and/or by transformations to obtain fine-grained uniform microstructure, which can simultaneously possess high strength and high ductility. This microstructure-based design approach has been found especially effective in eutectic HEAs [47, 48, 62]. After the discovery of TWIP effect in the Cantor alloy, efforts have been applied to design better HEAs with TWIP and/or TRIP effect [6, 8, 9, 13]. For example, the CoCrNi medium-entropy alloy with improved solid solution strength and lowered SFE guarantees room temperature TWIP [6]. By systematic evaluation of compositional dependence of stability between the FCC phase and the HCP phase, HEAs with TRIP effect were designed to show excellent combination of strength and ductility [8, 9, 13]. It has been proven that the microstructure of Ni-based alloy, which consists of uniform distributed coherent solid solution phase and intermetallic compound, is a good candidate for applications in high temperatures. Based on the idea to develop super-alloy like microstructure, Feng et al. [63] used a calculation of phase diagrams (CALPHAD)-based high-throughput computational method to design light, strong, and low-cost HEAs for high-temperature applications. The crystal–glass nanocomposite with extremely good mechanical properties was first found in Mg-based alloys by Wu et al. [64], which is found also appliable to HEAs, as illustrated by Wu et al. [42]. Similar ideas can also be transferred to HECs, where Dai et al. [24] have contributed good high temperature strength retention of medium/high-entropy carbides as grain boundary segregation, which is also the dominating mechanism in $ZrB_2$ and ZrC ceramics with the addition of W or WC [65–67].

In a review paper, George et al. [1] provided an illustrative graph on microstructure–mechanism-oriented design approaches (Figure 6.22), which summarizes the mechanisms that have been discovered in conventional alloys and applied in the design of conventional materials, and may be transferred to the design of HEMs. We believe that such microstructure–mechanism-based approaches learned from conventional alloys will play more important roles in the design of high-performance HEMs in the future.

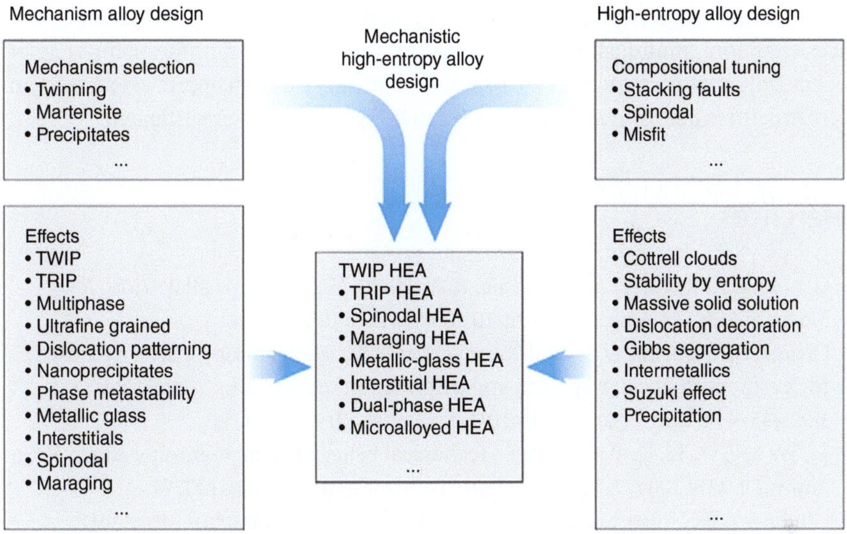


**Figure 6.22** Microstructure–mechanism-oriented design approaches to the design of HEAs. Source: George et al. [1]/Springer Nature.

## 6.8 Conclusions and Perspectives

In this chapter, mechanical properties of HEMs were briefly reviewed by some typical examples, where both the novel properties and the underlying microstructure and/or mechanisms were presented. These microstructures and/or mechanisms may serve as design goals, when one tries to tune the mechanical properties of a HEM or design a new HEM with intriguing mechanical properties. Nowadays, it is still a non-trivial task to inverse design a new material based on a given set of properties. Usually, a new material is still designed under "try and error" paradigm under the intuition of researchers, sometimes designed based on qualitative way, or rarely in a semi-quantitative way. Many examples introduced in this chapter were somehow designed on qualitative or semi-qualitative way under the microstructure- and/or mechanism-based approaches. For example, Sohn et al. [7] employed DFT-based methods to predict lattice distortions and designed a CoNiV medium-entropy alloy with high strength (yield strength ~1 GPa); Li et al. [13] predicted the relative stability of FCC and HCP phases versus compositions, and successfully designed HEAs with TRIP effect; Feng et al. [63] aimed to design super-alloy like microstructure, employed a CALPHAD-based high-throughput computational method to screen the composition, and successfully designed a light, strong, and low-cost HEAs for high-temperature applications. In these examples, only a single goal is considered, e.g. increasing lattice distortions to improve yield strength and tuning stability between FCC and HCP phases by varying composition. In practice, we need a material meet a combination of properties, e.g. elastic, strength, toughness, corrosion resistance, and price. Therefore, multitask design will be the future work. From the point view of mechanical properties, qualitative, or semi-quantitative design approaches based on microstructures and/or mechanisms need to be systematically investigated.

## References


1 George, E.P., Raabe, D., and Ritchie, R.O. (2019). High-entropy alloys. *Nat. Rev. Mater.* 4: 515–534. https://doi.org/10.1038/s41578-019-0121-4.

2 George, E.P., Curtin, W.A., and Tasan, C.C. (2020). High entropy alloys: a focused review of mechanical properties and deformation mechanisms. *Acta Mater.* 188: 435–474. https://doi.org/10.1016/j.actamat.2019.12.015.

3 Li, W., Xie, D., Li, D. et al. (2021). Mechanical behavior of high-entropy alloys. *Prog. Mater Sci.* 118: 100777. https://doi.org/10.1016/j.pmatsci.2021.100777.

4 Gludovatz, B., Hohenwarter, A., Catoor, D. et al. (2014). A fracture-resistant high-entropy alloy for cryogenic applications. *Science* 345: 1153–1158. https://doi.org/10.1126/science.1254581.

5 Cantor, B., Chang, I.T.H., Knight, P., and Vincent, A.J.B. (2004). Microstructural development in equiatomic multicomponent alloys. *Mater. Sci. Eng., A* 375–377: 213–218. https://doi.org/10.1016/j.msea.2003.10.257.

6 Gludovatz, B., Hohenwarter, A., Thurston, K.V.S. et al. (2016). Exceptional damage-tolerance of a medium-entropy alloy CrCoNi at cryogenic temperatures. *Nat. Commun.* 7: 10602. https://doi.org/10.1038/ncomms10602.

7 Sohn, S.S., Kwiatkowski da Silva, A., Ikeda, Y. et al. (2019). Ultrastrong medium-entropy single-phase alloys designed via severe lattice distortion. *Adv. Mater.* 31: 1807142. https://doi.org/10.1002/adma.201807142.

8 Li, Z., Pradeep, K.G., Deng, Y. et al. (2016). Metastable high-entropy dual-phase alloys overcome the strength–ductility trade-off. *Nature* 534: 227–230. https://doi.org/10.1038/nature17981.

9 Lu, W., Liebscher, C.H., Dehm, G. et al. (2018). Bidirectional transformation enables hierarchical nanolaminate dual-phase high-entropy alloys. *Adv. Mater.* 30: 1804727. https://doi.org/10.1002/adma.201804727.

10 Huang, S., Li, W., Lu, S. et al. (2015). Temperature dependent stacking fault energy of FeCrCoNiMn high entropy alloy. *Scr. Mater.* 108: 44–47. https://doi.org/10.1016/j.scriptamat.2015.05.041.

11 Zaddach, A.J., Niu, C., Koch, C.C., and Irving, D.L. (2013). Mechanical properties and stacking fault energies of NiFeCrCoMn high-entropy alloy. *JOM* 65: 1780–1789. https://doi.org/10.1007/s11837-013-0771-4.

12 Zhao, S., Stocks, G.M., and Zhang, Y. (2017). Stacking fault energies of face-centered cubic concentrated solid solution alloys. *Acta Mater.* 134: 334–345. https://doi.org/10.1016/j.actamat.2017.05.001.

13 Li, Z., Körmann, F., Grabowski, B. et al. (2017). Ab initio assisted design of quinary dual-phase high-entropy alloys with transformation-induced plasticity. *Acta Mater.* 136: 262–270. https://doi.org/10.1016/j.actamat.2017.07.023.

**14** Senkov, O.N., Wilks, G.B., Scott, J.M., and Miracle, D.B. (2011). Mechanical properties of $Nb_{25}Mo_{25}Ta_{25}W_{25}$ and $V_{20}Nb_{20}Mo_{20}Ta_{20}W_{20}$ refractory high entropy alloys. *Intermetallics* 19: 698–706. https://doi.org/10.1016/j.intermet.2011.01.004.

**15** Miracle, D.B. and Senkov, O.N. (2017). A critical review of high entropy alloys and related concepts. *Acta Mater.* 122: 448–511. https://doi.org/10.1016/j.actamat.2016.08.081.

**16** Senkov, O.N., Miracle, D.B., Chaput, K.J., and Couzinie, J.-P. (2018). Development and exploration of refractory high entropy alloys – a review. *J. Mater. Res.* 33: 3092–3128. https://doi.org/10.1557/jmr.2018.153.

**17** Senkov, O.N., Jensen, J.K., Pilchak, A.L. et al. (2018). Compositional variation effects on the microstructure and properties of a refractory high-entropy superalloy $AlMo_{0.5}NbTa_{0.5}TiZr$. *Mater. Des.* 139: 498–511. https://doi.org/10.1016/j.matdes.2017.11.033.

**18** Demirskyi, D., Borodianska, H., Suzuki, T.S. et al. (2019). High-temperature flexural strength performance of ternary high-entropy carbide consolidated via spark plasma sintering of TaC, ZrC and NbC. *Scr. Mater.* 164: 12–16. https://doi.org/10.1016/j.scriptamat.2019.01.024.

**19** Feng, L., Chen, W., Fahrenholtz, W.G., and Hilmas, G.E. (2021). Strength of single-phase high-entropy carbide ceramics up to 2300 °C. *J. Am. Ceram. Soc.* 104: 419–427. https://doi.org/10.1111/jace.17443.

**20** Yang, Q.-Q., Wang, X.-G., Wu, P. et al. (2022). Ultra-high strength medium-entropy (Ti,Zr,Ta)C ceramics at 1800 °C by consolidating a core-shell structured powder. *J. Am. Ceram. Soc.* 105: 823–829. https://doi.org/10.1111/jace.18197.

**21** Hu, C. and Luo, J. (2021). Data-driven prediction of grain boundary segregation and disordering in high-entropy alloys in a 5D space. *Mater. Horiz.* https://doi.org/10.1039/D1MH01204E.

**22** Gild, J., Zhang, Y., Harrington, T. et al. (2016). High-entropy metal diborides: a new class of high-entropy materials and a new type of ultrahigh temperature ceramics. *Sci. Rep.* 6: 37946. https://doi.org/10.1038/srep37946.

**23** Qin, M., Gild, J., Wang, H. et al. (2020). Dissolving and stabilizing soft $WB_2$ and $MoB_2$ phases into high-entropy borides via boron-metals reactive sintering to attain higher hardness. *J. Eur. Ceram. Soc.* 40: 4348–4353. https://doi.org/10.1016/j.jeurceramsoc.2020.03.063.

**24** Dai, F.-Z., Wen, B., Sun, Y., Ren, Y., Xiang, H., and Zhou, Y. (2022). Grain boundary segregation induced strong UHTCS at elevated temperatures: a universal mechanism from conventional UHTCS to high entropy UHTCS. *J. Mater. Sci. Technol.* 123: 26–33. https://doi.org/10.1016/j.jmst.2021.12.074.

**25** Wang, X.-F., Wang, X.-G., Yang, Q.-Q. et al. (2021). High-strength medium-entropy (Ti,Zr,Hf)C ceramics up to 1800 °C. *J. Am. Ceram. Soc.* 104: 2436–2441. https://doi.org/10.1111/jace.17677.

**26** Han, X., Girman, V., Sedlak, R. et al. (2020). Improved creep resistance of high entropy transition metal carbides. *J. Eur. Ceram. Soc.* 40: 2709–2715. https://doi.org/10.1016/j.jeurceramsoc.2019.12.036.

**27** Senkov, O.N. and Semiatin, S.L. (2015). Microstructure and properties of a refractory high-entropy alloy after cold working. *J. Alloys Compd.* 649: 1110–1123. https://doi.org/10.1016/j.jallcom.2015.07.209.

**28** Li, X.-G., Chen, C., Zheng, H. et al. (2020). Complex strengthening mechanisms in the NbMoTaW multi-principal element alloy. *npj Comput. Mater.* 6: 70. https://doi.org/10.1038/s41524-020-0339-0.

**29** Ming, K., Li, L., Li, Z. et al. (2019). Grain boundary decohesion by nanoclustering Ni and Cr separately in CrMnFeCoNi high-entropy alloys. *Sci. Adv.* 5: eaay0639. https://doi.org/10.1126/sciadv.aay0639.

**30** Wang, C., Qin, M., Lei, T. et al. (2021). Synergic grain boundary segregation and precipitation in W- and W-Mo-containing high-entropy borides. *J. Eur. Ceram. Soc.* 41: 5380–5387. https://doi.org/10.1016/j.jeurceramsoc.2021.04.004.

**31** Harrington, T.J., Gild, J., Sarker, P. et al. (2019). Phase stability and mechanical properties of novel high entropy transition metal carbides. *Acta Mater.* 166: 271–280. https://doi.org/10.1016/j.actamat.2018.12.054.

**32** Feng, L., Fahrenholtz, W.G., and Hilmas, G.E. (2020). Processing of dense high-entropy boride ceramics. *J. Eur. Ceram. Soc.* S0955221920302466. https://doi.org/10.1016/j.jeurceramsoc.2020.03.065.

**33** Gild, J., Wright, A., Quiambao-Tomko, K. et al. (2019). Thermal conductivity and hardness of three single-phase high-entropy metal diborides fabricated by borocarbothermal reduction and spark plasma sintering. *Ceram. Int.* S0272884219333784. https://doi.org/10.1016/j.ceramint.2019.11.186.

**34** Wang, Y., Csanádi, T., Zhang, H. et al. (2020). Enhanced hardness in high-entropy carbides through atomic randomness. *Adv. Theory Simul.* 3: 8.

**35** Gorban', V.F., Andreyev, A.A., Kartmazov, G.N. et al. (2017). Production and mechanical properties of high-entropic carbide based on the TiZrHfVNbTa multicomponent alloy. *J. Superhard Mater.* 39: 166–171. https://doi.org/10.3103/S1063457617030030.

**36** Mayrhofer, P.H., Kirnbauer, A., Ertelthaler, P., and Koller, C.M. (2018). High-entropy ceramic thin films: a case study on transition metal diborides. *Scr. Mater.* 149: 93–97. https://doi.org/10.1016/j.scriptamat.2018.02.008.

**37** Qin, M., Yan, Q., Wang, H. et al. (2020). High-entropy monoborides: towards superhard materials. *Scr. Mater.* 189: 101–105. https://doi.org/10.1016/j.scriptamat.2020.08.018.

**38** Zhao, P., Zhu, J., Zhang, Y. et al. (2020). A novel high-entropy monoboride ($Mo_{0.2}Ta_{0.2}Ni_{0.2}Cr_{0.2}W_{0.2}$)B with superhardness and low thermal conductivity. *Ceram. Int.* 46: 26626–26631. https://doi.org/10.1016/j.ceramint.2020.07.131.

**39** Yeh, J.-W. and Lin, S.-J. (2018). Breakthrough applications of high-entropy materials. *J. Mater. Res.* 33: 3129–3137. https://doi.org/10.1557/jmr.2018.283.

**40** Hsu, W.-L., Murakami, H., Yeh, J.-W. et al. (2016). A heat-resistant $NiCo_{0.6}Fe_{0.2}Cr_{1.5}SiAlTi_{0.2}$ overlay coating for high-temperature applications. *J. Electrochem. Soc.* 163: C752–C758. https://doi.org/10.1149/2.0821613jes.

**41** Lei, Z., Liu, X., Wu, Y. et al. (2018). Enhanced strength and ductility in a high-entropy alloy via ordered oxygen complexes. *Nature* 563: 546–550. https://doi.org/10.1038/s41586-018-0685-y.

**42** Wu, G., Balachandran, S., Gault, B. et al. (2020). Crystal–glass high-entropy nanocomposites with near theoretical compressive strength and large deformability. *Adv. Mater.* 32: 2002619. https://doi.org/10.1002/adma.202002619.

**43** Zou, Y., Maiti, S., Steurer, W., and Spolenak, R. (2014). Size-dependent plasticity in an $Nb_{25}Mo_{25}Ta_{25}W_{25}$ refractory high-entropy alloy. *Acta Mater.* 65: 85–97. https://doi.org/10.1016/j.actamat.2013.11.049.

**44** Zou, Y., Ma, H., and Spolenak, R. (2015). Ultrastrong ductile and stable high-entropy alloys at small scales. *Nat. Commun.* 6: 7748. https://doi.org/10.1038/ncomms8748.

**45** Zou, Y., Wheeler, J.M., Ma, H. et al. (2017). Nanocrystalline high-entropy alloys: a new paradigm in high-temperature strength and stability. *Nano Lett.* 17: 1569–1574. https://doi.org/10.1021/acs.nanolett.6b04716.

**46** Lu, Y., Dong, Y., Guo, S. et al. (2015). A promising new class of high-temperature alloys: eutectic high-entropy alloys. *Sci. Rep.* 4: 6200. https://doi.org/10.1038/srep06200.

**47** Shi, P., Ren, W., Zheng, T. et al. (2019). Enhanced strength–ductility synergy in ultrafine-grained eutectic high-entropy alloys by inheriting microstructural lamellae. *Nat. Commun.* 10: 489. https://doi.org/10.1038/s41467-019-08460-2.

**48** Shi, P., Li, R., Li, Y. et al. (2021). Hierarchical crack buffering triples ductility in eutectic herringbone high-entropy alloys. *Science* https://doi.org/10.1126/science.abf6986.

**49** Toda-Caraballo, I. and Rivera-Díaz-del-Castillo, P.E.J. (2016). A criterion for the formation of high entropy alloys based on lattice distortion. *Intermetallics* 71: 76–87. https://doi.org/10.1016/j.intermet.2015.12.011.

**50** Toda-Caraballo, I., Wróbel, J.S., Dudarev, S.L. et al. (2015). Interatomic spacing distribution in multicomponent alloys. *Acta Mater.* 97: 156–169. https://doi.org/10.1016/j.actamat.2015.07.010.

**51** Toda-Caraballo, I. (2017). A general formulation for solid solution hardening effect in multicomponent alloys. *Scr. Mater.* 127: 113–117. https://doi.org/10.1016/j.scriptamat.2016.09.009.

**52** Varvenne, C., Luque, A., and Curtin, W.A. (2016). Theory of strengthening in fcc high entropy alloys. *Acta Mater.* 118: 164–176. https://doi.org/10.1016/j.actamat.2016.07.040.

**53** Varvenne, C., Leyson, G.P.M., Ghazisaeidi, M., and Curtin, W.A. (2017). Solute strengthening in random alloys. *Acta Mater.* 124: 660–683. https://doi.org/10.1016/j.actamat.2016.09.046.

**54** Rao, S.I., Antillon, E., Woodward, C. et al. (2019). Solution hardening in body-centered cubic quaternary alloys interpreted using Suzuki's kink-solute interaction model. *Scr. Mater.* 165: 103–106. https://doi.org/10.1016/j.scriptamat.2019.02.012.

**55** Chen, B., Li, S., Zong, H. et al. (2020). Unusual activated processes controlling dislocation motion in body-centered-cubic high-entropy alloys. *Proc. Natl. Acad. Sci. U.S.A.* 117: 16199–16206. https://doi.org/10.1073/pnas.1919136117.

**56** Yin, S., Zuo, Y., Abu-Odeh, A. et al. (2021). Atomistic simulations of dislocation mobility in refractory high-entropy alloys and the effect of chemical short-range order. *Nat. Commun.* 12: 4873. https://doi.org/10.1038/s41467-021-25134-0.

**57** Wang, F., Balbus, G.H., Xu, S. et al. (2020). Multiplicity of dislocation pathways in a refractory multiprincipal element alloy. *Science* 370: 95–101.

**58** Zhang, R., Zhao, S., Ding, J. et al. (2020). Short-range order and its impact on the CrCoNi medium-entropy alloy. *Nature* 581: 283–287. https://doi.org/10.1038/s41586-020-2275-z.

**59** Chen, X., Wang, Q., Cheng, Z. et al. (2021). Direct observation of chemical short-range order in a medium-entropy alloy. *Nature* 592: 712–716. https://doi.org/10.1038/s41586-021-03428-z.

**60** Körmann, F., Ruban, A.V., and Sluiter, M.H.F. (2017). Long-ranged interactions in bcc NbMoTaW high-entropy alloys. *Mater. Res. Lett.* 5: 35–40. https://doi.org/10.1080/21663831.2016.1198837.

**61** Wang, Y., Yan, M., Zhu, Q. et al. (2018). Computation of entropies and phase equilibria in refractory V–Nb–Mo–Ta–W high-entropy alloys. *Acta Mater.* 143: 88–101. https://doi.org/10.1016/j.actamat.2017.10.017.

**62** Pan, Q., Zhang, L., Feng, R. et al. (2021). Gradient cell–structured high-entropy alloy with exceptional strength and ductility. *Science* https://doi.org/10.1126/science.abj8114.

**63** Feng, R., Zhang, C., Gao, M.C. et al. (2021). High-throughput design of high-performance lightweight high-entropy alloys. *Nat. Commun.* 12: 4329. https://doi.org/10.1038/s41467-021-24523-9.

**64** Wu, G., Chan, K.-C., Zhu, L. et al. (2017). Dual-phase nanostructuring as a route to high-strength magnesium alloys. *Nature* 545: 80–83. https://doi.org/10.1038/nature21691.

**65** Dai, F.-Z., Zhou, Y., and Sun, W. (2017). Segregation of solute atoms (Y, Nb, Ta, Mo and W) in $ZrB_2$ grain boundaries and their effects on grain boundary strengths: a first-principles investigation. *Acta Mater.* 127: 312–318. https://doi.org/10.1016/j.actamat.2017.01.048.

**66** Ma, H.-B., Zou, J., Zhu, J.-T. et al. (2018). Segregation of tungsten atoms at $ZrB_2$ grain boundaries in strong $ZrB_2$–SiC–WC ceramics. *Scr. Mater.* 157: 76–80. https://doi.org/10.1016/j.scriptamat.2018.07.038.

**67** Dai, F.-Z., Sun, Y., Ren, Y. et al. (2022). Segregation of solute atoms in ZrC grain boundaries and their effects on grain boundary strengths. *J. Mater. Sci. Technol.* 101: 234–241. https://doi.org/10.1016/j.jmst.2021.05.058.

# 7

# Functional Properties

The composition, structural features, and thermodynamics of high-entropy materials (HEMs) are quite different from those of conventional materials, resulting in modifications in the movement of atoms, electrons, phonons, etc. This movement modification reshapes the performance of HEMs in an unexpected but fascinating way due to the "cocktail effect." Moreover, the vast chemical space gives huge design freedom in demand-driven selection of properties of HEMs. Thus, it would be a scientifically interesting topic to review the properties of HEMs. In this chapter, the functional properties, including conductivity, resistance to oxidation and corrosion, dielectric, and magnetic behavior, of HEMs will be the focus.

## 7.1 Thermal Conductivity

Thermal conductivity evaluates the ability of a material to dissipate heat through the movement of electrons and phonons. Thus, the scattering process of electrons and phonons determines the thermal conductivity of materials. Compared to conventional materials, the structure of HEMs is severely distorted, and the chemical composition is complex, which leads to mass disorder and force constant fluctuations, as discussed in Chapters 2.2 and 2.3. Smearing in electronic structure and phonon dispersion resulting from this disorder are observed in both high-entropy alloys (HEAs) and high-entropy ceramics (HECs), implying more scattering channels for electrons and phonons. Hence, a significant reduction in thermal conductivity is anticipated for these distorted materials.

As expected, the thermal conductivity of HEAs shows a dramatic deterioration as compared to that of pure metals and solid solutions [1]. However, the response of thermal conductivity of HEAs to temperature contravenes that of pure metals, that is, the thermal conductivity $\kappa$ of HEAs keeps increasing in the investigated temperature range. This phenomenon has been observed in Ni-based HEAs, NiCoFeCrMn and NiCoFeCrPd at cryogenic temperatures [1, 2], $Al_xCoCrFeNi$ [3], AlCrFeMnNi, AlCrFeMnNiMo [4], $Y_xCoCrFeNi$, $Gd_xCoCrFeNiCu$, $Al_{0.3}CoCrFeNi$ [5], and

*High-Entropy Materials: From Basics to Applications,* First Edition.
Huimin Xiang, Fu-Zhi Dai, and Yanchun Zhou.

CuCrFeTiV [6]. Different interpretations have been proposed to understand the difference. Chou et al. believed that the augmentation of electronic mean free path induced by low sensitivity of phonon concentration with respect to temperatures and large linear thermal expansion in HEAs was responsible [3]. Lu et al. argued that the enhanced heat transfer at higher temperatures in the $Al_xCoCrFeNi$ alloys should be associated with the increased phonon mean free path at higher temperature due to the expansion of lattice [4]. Jin et al. separated the contribution of electrons and phonons to the total thermal conductivity of Ni-based HEAs according to the Wiedemann–Franz relationship [1]. They found that the lattice thermal conductivity of HEAs increased rapidly with temperature up to 50 K and showed weak temperature dependence thereafter, while the electronic contribution increased almost linearly with temperature. Accordingly, they considered that large and nearly constant resistivity of these alloys was the major reason. Up to now, no mature conclusion has been drawn on the increasing thermal conductivity of HEAs with temperature. Theoretical descriptions on the temperature dependence of $\kappa$ for disordered multicomponent alloys might be the solution, which is, however, quite a challenge due to the exhausting simulations on the electronic–electronic scattering, phonon–phonon scattering, and electronic–phonon scattering.

Another interesting phenomenon regarding the thermal conductivity of HEAs is that the contribution from lattice to the total thermal conductivity is comparable to that from electrons or even higher at cryogenic temperatures [1, 3, 7], while in pure metals, the majority of the heat is carried by electrons. Intriguingly, after separating the phonon and electron thermal conductivity using the Wiedemann–Franz relationship, the lattice thermal conductivity ($\kappa_L$) seems insensitive to the compositions and structure disorder, while that from electrons ($\kappa_e$) dropped dramatically with the increasing chemical complexity [1, 7]. Chemical disorder appears to scatter electrons more effectively than phonons in HEAs. Although no explanation has been proposed, some logical interpretations could be deduced by examining the electronic structures and phonon dispersions of HEAs. As discussed in Chapter 2.2.1, theoretical calculations on the electronic structure of HEAs have demonstrated remarkable smearing near the Fermi level throughout the whole Brillouin zone [8, 9], indicating severe scattering of these electrons even at cryogenic temperatures. Considering the heat is mainly carried by these electrons, it might not be surprising to observe insensitive temperature dependence of electrical conductivity and a sharp decrease in electronic thermal conductivity in HEAs. Moreover, this smearing is composition-dependent. As demonstrated by Zhang et al. [10], with the increase in chemical complexity, the degrees of disorder smearing increased, thereby inducing decrease in $\kappa_e$ along with chemical disorder. Regarding phonon dispersion, strong broadening is also observed with the introduction of chemical disorder; however, the broadening spreads in the mid- and high-frequency region (above ~4 THz), and the long wavelength limit (region near $\Gamma$) is unaffected, as shown by Körmann et al. [11]. For metals, the lattice thermal conductivity mainly comes from the long wavelength phonon modes due to their large group velocity and small scattering rate [12]. Since these modes are unaffected by chemical disorder, adding more elements does not necessarily further decrease the lattice thermal conductivity [10]. With the increase in temperature, the high-lying acoustic phonons are activated thermally. However, due to strong broadening, the contribution from these modes is limited, and the phonon thermal conductivity would be temperature insensitive in this temperature range. Thus, for materials with electrons as the main heat carrier, introducing chemical complexity is an efficient way to decrease the intrinsic thermal conductivity by massing the transportation of electrons.

It is worth mentioning that, in the pursue of $\kappa_L$ and $\kappa_e$, Wiedemann–Franz relationship is the most common and convenient method. In this relationship, the electronic thermal conductivity is related to electrical conductivity by Lorentz number:

$$\kappa_e = L\sigma T \tag{7.1}$$

where $L$ is the Lorentz number with the ideal value of $2.44 \times 10^{-8}\,\mathrm{W}/\Omega/\mathrm{K}^{-2}$, $\sigma$ is the electric conductivity, and $T$ is the temperature. The ideal value of Lorentz number is derived from the free electron theory, in which the electrons are allowed to move freely. In real HEAs, the movement of free electrons is modulated by metal atoms, especially by magnetic transition metals, resulting in the deviation of Lorentz number from its Sommerfeld value [12]. Thus, the use of Wiedemann–Franz relationship to separate $\kappa_L$ and $\kappa_e$ of HEAs should process with caution; since the contributions from these two parts are comparable, a slight change in Lorentz number may lead to different conclusions. Reasonable estimation of Lorentz number is feasible theoretically for HEAs; furthermore, exhausting and time-consuming calculations are needed.

The diminishment in thermal conductivity of HECs has also been observed in multiply material systems, as shown in Figure 7.1 [13]. Efforts have been made to interpret such deterioration. Braun et al. [14] investigated the thermal conductivity

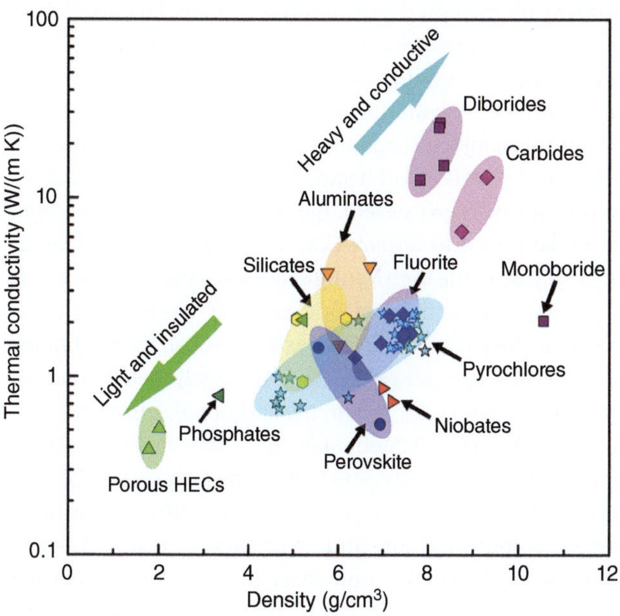


**Figure 7.1** Summarized thermal conductivity of some typical high-entropy oxides, carbides, and borides. Source: Xiang et al. [13]/Springer Nature/CC BY-4.0.

of (MgCoNiCuZn)O. They found that with the addition of the sixth cation (Sc, Sb, Sn, Cr, or Ge), 50% reduction in thermal conductivity could be achieved. The results from a virtual crystal approximation (VCA) model suggested that the amorphous or glass-like thermal conductivity was attributed to disorder in interatomic force constants (IFCs) rather than mass and size disorder. Wright et al. [15] investigated a series of 22 single-phase pyrochlores, including 18 compositionally complex (i.e. medium- and high-entropy) pyrochlore oxides and four additional "low-entropy" benchmark compositions. They found that size disorder parameter was a more effective descriptor to forecast reduce thermal conductivity than the ideal mixing configurational entropy, and severe lattice distortion was significant in reducing the thermal conductivity.

The behavior of thermal conductivity with temperature of conductive and insulative HECs is largely different. For conductive ceramics, the temperature dependence of thermal conductivity is similar to that of HEAs, i.e. increasing with temperature [16], while for insulative ceramics, the temperature dependency of $\kappa$ is rather weak [14, 17–21]. The difference originates from "free" electrons in conductive ceramics. Wen et al. [16] separated the thermal conductivity of phonons and electrons in the temperature range of 300–800 K for three HEMs, namely $(Hf_{0.2}Zr_{0.2}Ta_{0.2}Nb_{0.2}Ti_{0.2})C$, $(Hf_{0.2}Zr_{0.2}Ta_{0.2}Nb_{0.2}Ti_{0.2})N$, and $(Hf_{0.2}Zr_{0.2}Ta_{0.2}Nb_{0.2}Ti_{0.2})(N_{0.5}C_{0.5})$, using Wiedemann–Franz relationship. They found that the contributions from electrons for three materials increased with temperature, while the phonon thermal conductivity was insensitive to the temperature. Up to now, no explanation has been proposed since no fine electronic structure and phonon dispersion of HECs have been reported. However, the deduction used to understand the thermal response of heat conductivity of HEAs might be suitable to conductive HECs, that is, electrons near Fermi surface and acoustic phonons at the Brillouin zone boundary are heavily scattered. If so, the introduction of chemical disorder might be an efficient way to decrease the thermal conductivity of conductive HECs and insulative ones with thermal conductivity mainly from zone boundary acoustic phonons and low-lying optic phonons. Verification of this conjecture requires solid theoretical results.

The thermal conductivity of HEMs is also affected by the composition. As the content of Al increases in the $Al_xCoCrFeNi$ system, the structure of the alloy varies from single face-centered cubic (FCC) to duplex FCC–BCC, and finally to single body-centered cubic (BCC) [3]. Due to the high Young's modulus and phonon velocity, at each temperature, thermal conductivity of BCC phase is the highest, while that of dual phases is the lowest since more interfaces exist. Jin et al. found that the thermal conductivity of Ni-based single-phase HEAs was largely determined by magnetic elements, such as Cr and Mn [1, 2]. Due to the additional source of disorder scattering from mixed ferro/antiferro-magnetic state, the alloys containing these elements possessed lower thermal conductivity. The reduction effect on the thermal conductivity is also observed in HECs. Gild et al. [22] synthesized three dense high-entropy borides. They found that following substitution of Nb with heavy Mo atom or antiferromagnetic Cr atom, the thermal conductivity of high-entropy borides decreased efficiently.

Theoretical prediction is appealing in the investigation of thermal conductivity of HEMs due to its ability of modeling to reveal the micro-mechanism of transportation, screen material systems, and tune the thermal property. Considering the complexity in structural and chemical composition, molecular dynamics (MD) is always chosen in the theoretical works of HEMs. Caro et al. performed nonequilibrium MD calculations of phonon thermal conductivity in multiple component solid solutions represented by Lennard–Jones (LJ) potentials [23]. They found that the lattice thermal conductivity of a multicomponent alloy was coupled with the mass and atomic size of the components. Heavier element was more efficient in decreasing the thermal conductivity, and it had a smaller size. Lim et al. [24] constructed a cell containing 30 720 atoms and calculated the thermal conductivity of $(Mg_{0.2}Co_{0.2}Ni_{0.2}Cu_{0.2}Zn_{0.2})O$ as well as ones plus Sc, Sn, Cr, or Ge in equal-molar cation proportions by MD. The thermal conductivity was calculated using the Green–Kubo method, and the interatomic potential was modeled using a pair sum of an exponential-6 function and long-range Coulomb interactions. They found that phonon scattering from charge disorder was responsible for lowering of thermal conductivity with the addition of a sixth cation. The lack of high accurate interatomic potentials in classical MD leads to results without guarantee. To tackle this issue, Dai et al. [25, 26] employed a deep leaning potential to investigate lattice thermal conductivity of HE $(Zr_{0.2}Hf_{0.2}Ti_{0.2}Nb_{0.2}Ta_{0.2})C$ and $(Zr_{0.2}Hf_{0.2}Nb_{0.2}Ta_{0.2}Ti_{0.2})B_2$. They found that the predicted lattice thermal conductivity of these two HECs was rather low, which was consistence with experiments.

## 7.2 Thermal Expansion

Generally, the lattice parameters of binary solid solutions with the end members having the same structure follow Vegard's law, i.e. the lattice parameters of solid solution vary linearly with the content of one end member. In this case, the thermal expansion coefficient of the solution can be obtained using the rule of mixtures (ROMs) for the molar volume. Mostly, the results from ROMs agree reasonably with the experimental values [27, 28]. However, in HEMs, due to the cocktail effects, an unexpected response of lattice to temperature is observed. In the investigation of thermal expansion coefficients (TECs) of medium-entropy subsystems of the CrMnFeCoNi HEA, Laplanche et al. found systematic underestimation of thermal expansion for all investigated systems at low temperatures and complete deviation from the experiments for CoNi and FeCoNi [29]. They attributed the discrepancy to ferromagnetic nature of these two systems and the uneven structure of constituents at low temperatures.

In HECs, the situation is more complicated. Yan et al. found that the TEC of (TiZrHfNbTa)C at 25–300 °C did not show significant deviation from binary carbides [30]. In $(La_{0.2}Ce_{0.2}Nd_{0.2}Sm_{0.2}Eu_{0.2})PO_4$, $(Sm_{1/6}Eu_{1/6}Y_{1/6}Yb_{1/6}Lu_{1/6}Er_{1/6})_3(Nb_{1/2}Ta_{1/2})O_7$, $(Y_{0.2}Yb_{0.2}Er_{0.2}Lu_{0.2})_2(Zr_{0.5}Hf_{0.5})_2O_7$ systems, an obvious decrease in TEC was observed as compared to that in their single component counterparts [31–33].

The high entropy effect that hindered the oscillation amplitude of constituting atoms by severe lattice distortion was believed to be responsible for the decrease in TEC. Meanwhile, an increase in TECs of $(Y_{0.2}Nd_{0.2}Sm_{0.2}Eu_{0.2}Er_{0.2})AlO_3$, $(Yb_{0.25}Y_{0.25}Lu_{0.25}Er_{0.25})_2SiO_5$, $(Y_{0.2}Yb_{0.2}Lu_{0.2}Eu_{0.2}Er_{0.2})_3Al_5O_{12}$, $(Nd_{0.2}Sm_{0.2}Eu_{0.2}Y_{0.2}Yb_{0.2})_4Al_2O_9$ has also been reported [34–37]. The anisotropy of TECs is also disturbed by high entropy effects. Using high temperature powder X-ray diffraction, Zhao et al. and Chen et al. [34–37] explored the structure change of HECs with temperature, and they found that the anisotropy of thermal expansion was dependent on the material systems. For $(Nd_{0.2}Sm_{0.2}Eu_{0.2}Y_{0.2}Yb_{0.2})_4Al_2O_9$ and $(Yb_{0.25}Y_{0.25}Lu_{0.25}Er_{0.25})_2SiO_5$, the TECs were more anisotropic, while for $(Y_{0.2}Nd_{0.2}Sm_{0.2}Eu_{0.2}Er_{0.2})AlO_3$, a decrease in anisotropy of TECs was observed. A simple theory called chemical bonding theory (CBT) was employed to interpret their observation [37]. They found that the expansion behavior of high-entropy oxide was determined by the chemical environments around RE atoms due to their weak bond energies with oxygen. The RE-O bonds with large energy degradation were almost parallel with *ab* plane, leading to the increase in anisotropy in bonding strength between *ab* plane and *c* direction. In the investigation of anisotropic thermal expansion of X2-$(ScYDyErYb)_2SiO_5$, Ridley et al. [17] found that linear coefficient of thermal expansion (CTE) generally followed the ROM, while the *c*-axis CTE show larger differences. They attributed this deviation to the rigid Si–O tetrahedron in the structure interacted less with incorporated elements.

Thermodynamically, the expansion of a structure is related to the Grüneisen parameters by the following equations:

$$\alpha = \gamma \frac{BC_V}{V} \tag{7.2}$$

where $\alpha$ is the expansion, $\gamma$ is the Grüneisen parameters, $B$ is the bulk modulus, $C_V$ is the heat capacity, and $V$ is the mole volume. The overall Grüneisen parameters $\gamma$ is the average of the mode Grüneisen parameters $\gamma_{i,q}$ weighted with the contribution of each mode to lattice specific heat $C_i(q)$:

$$\gamma = \frac{\Sigma \gamma_{i,q} C_i(q)}{\Sigma C_i(q)} \tag{7.3}$$

$$C_i(q) = k_B \left( \frac{\hbar \omega_{i,q}}{k_B T} \right)^2 \frac{\exp(\hbar \omega_{i,q} / k_B T)}{\left[ \exp(\hbar \omega_{i,q} / k_B T) - 1 \right]^2} \tag{7.4}$$

Thus, the thermal response of lattice parameters is intrinsically determined by the anharmonicity of phonons, especially by acoustic modes, as demonstrated in the $\alpha$-$ZrW_2O_8$, $TiB_2$, $ZrB_2$, and $HfB_2$ [38–40]. A positive Grüneisen parameter contributes to positive and large thermal expansion, while a negative Grüneisen parameter leads to small or even negative expansion. By adjusting the composition, the mode Grüneisen parameter $\gamma_{i,q}$ and specific heat $C_i(q)$ of acoustic phonons in HEM change accordingly, leading to the variation in thermal expansion eventually. Ren et al. used this interpretation to explain the thermal behavior of $(YHoErYb)_2SiO_5$ [19] by examining the mode Grüneisen parameters of each constituent monosilicate [41]. However, no solid proof on the mode Grüneisen parameters of HEMs has been presented, and the effects of different element on the $\gamma_{i,q}$ of HEMs are still hidden in the dark due to the time-demanding calculations.

Although it is arduous to uncover the micro-mechanism behind the thermal behavior of HEMs, it is possible to estimate the TECs from the first-principles calculation and MD by auxiliary methods. Using the Debye–Grüneisen model, the thermal expansion of a material is conveniently obtained from their static-free energies at different volumes [42, 43]. This model has been used in the estimation of thermal expansion of Hf–Nb–Ta–Ti–Zr HEA [44], (TaNbHfTiZr)C, and (TaNbHfTiZr)N [45]. Due to the absence of volume-dependency of free energy in the model, the TECs estimated from this simple method is not always satisfactory [45]. Deep leaning potential has also been employed in the calculation of properties of (TaNbHfTiZr)C and $(TaNbHfTiZr)B_2$ [25, 26]. The lattice parameter and thermal properties of high entropy carbide agreed well with the experiment. For high entropy borides, unexpectedly large thermal expansion along *a* direction was obtained. The inaccurately fitted deep learning potential was believed to be responsible due to the chemical complexity of high entropy borides.

Up to now, no mature conclusions on the thermal response of lattice parameters of HEMs have been drawn; however, these preliminary findings have demonstrated the possibility of tailoring the thermal expansion of HEMs by composition design.

## 7.3 Oxidation Resistance

Considering the excellent mechanical properties at high temperatures, HEMs are promising as high-temperature structural components. Problems brought by the high temperature applications are the inevitable oxidation for HEAs and high-entropy non-oxide ceramics. The oxidation resistance of a material is largely determined by the oxidation products and adhering strength between the oxide layer and matrix, which is variable with the composition. According to the Ellingham diagram and data of the Pilling–Bedworth ratio (PBR) of metal oxidation, elements capable to form a dense and protective oxide layer are beneficial for the improvement of oxidation resistance of materials, such as Al, Cr, and Si. Thus, for HEMs with different chemical compositions, the oxidation behavior is largely different.

For Al-containing HEAs, good oxidation resistance was confirmed by Huang et al. wherein they proposed the concept of HEAs [46]. They observed that the weight gains of two HEAs, $AlSiTiCrFeCoNiMo_{0.5}$ and $AlSiTiCrFeNiMo_{0.5}$, approached a constant level after about 100 hours at high temperatures in air. The protection to oxidation was attributed to the formation of an inner chromium oxide layer. Good oxidation resistance of AlCuTiFeNiCr and $Al_{0.5}FeCoCrNi(Si_{0.2}, Ti_{0.5})$ HEAs has also been observed [47, 48]. By examining the microstructural evolution of oxidation process of two Al–Co–Cr–Ni–(Fe or Si)-based HEAs, Butler et al. vividly demonstrated the protection of a dense oxide layer on the matrix from oxidation [49]. They found that the formation of a discontinuous, external $Cr_2O_3$ scale along with an internal $Al_2O_3$ subscale and AlN precipitates could not efficiently protect low Al-contenting HEAs, $Al_{10}Cr_{22.5}Co_{22.5}Ni_{22.5}Fe_{22.5}$, from oxidation. The weight gain of this alloy increased markedly during testing. For the other two HEAs with Al content of 15 and 20 at.%, a continuous external $Al_2O_3$ scale with a small volume fraction of $Cr_2O_3$ was formed on the matrix, resulting in much smaller weight gain and a thinner oxidation scale, as shown in Figure 7.2. Further, they examined the oxidation resistance of a series of NiCoCrFe-based HEAs with different Al contents and confirmed that the increase in Al content enhanced oxidation resistances of HEAs by promoting the formation of a continuous $Al_2O_3$ scale [50]. Their works demonstrated that compositionally complex alloys still oxidized selectively without microstructure changing of the matrix beneath the oxide scale, and the existing oxide formation models were still suitable for these complex alloys.

The oxidation resistance of Al-containing HEAs is adjustable by alloying different elements. Dąbrowa et al. carefully examined the Cu content on high-temperature oxidation behavior of $AlCoCrCu_xFeNi$ HEAs [51]. They observed a deterioration in the adhesion of the oxide scale with the increase in Cu content; however, the mass gain and parabolic constant of alloys with high Cu content were much lower. The mechanism of oxidation was similar to that predicted by the Giggins–Pettit theory [52], with the scale formation occurring mostly via

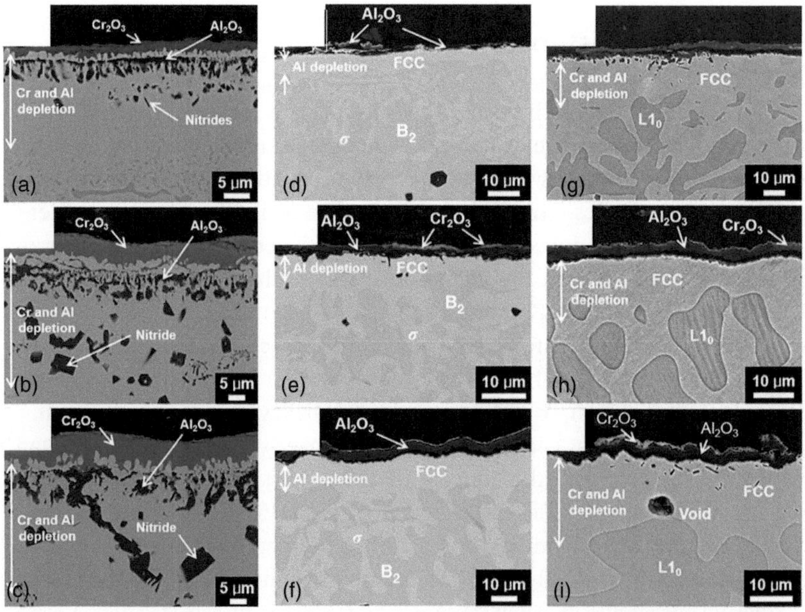


**Figure 7.2** Micostructure of the oxide scale and the underlying substrate of alloy containing 10% Al after (a) 5 hours, (b) 50 hours, (c) 100 hours, alloy containing 15% Al after (d) 20 hours, (e) 96 hours, (f) 500 hours, and alloy containing 20% Al after (g) 20 hours, (h) 96 hours, (i) 500 hours of oxidation. Source: Reproduced with permission from Butler et al. [49]/Springer Nature.

outward diffusion of Al. Meanwhile, phase composition and microstructure of the matrix beneath the oxide scale changed drastically after oxidation due to the depletion of Al and long-term annealing. They also suggested that alloying with highly active elements such as Y, Zr, or Hf could improve the adhesion of the scale and oxidation resistance of HEAs, which was verified by Lu et al. [53]. In their work, minor concentrations of Y (0.02 at.%) and Hf (0.02 at.%) were doped into the AlCoCrFeNi HEA. After oxidation at 1100 °C for 1000 hours, a uniform and continuous oxide scale developed on the HEA, and the oxide scale/substrate interface was smooth and clean. The oxide scale predominantly consisted of α-$Al_2O_3$ with minor $HfO_2$ and $Y_3Al_5O_{12}$ dispersion at the $Al_2O_3$ grain boundaries. The scale was strongly bonded to the substrate for even up to 1000 hours of oxidation at 1100 °C, as shown in Figure 7.3. The double-layer structure of the oxide scale consisting of an inner columnar and outer equiaxed grains could be clearly observed, which resulted from the concurrent inward oxygen diffusion and the outward aluminum diffusion. By examining the evolution of the double-layer structure of the oxide scale, the inward O diffusion was confirmed to be dominated in the growth of the scale. They attributed the slow oxidation rate of

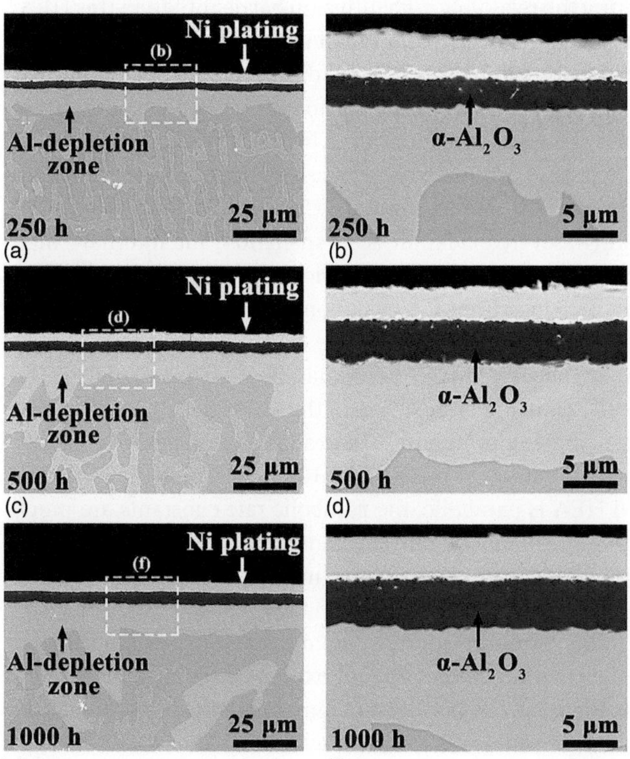


**Figure 7.3** Cross-sectional microstructure of the AlCoCrFeNiYHf HEA after isothermal oxidation at 1100 °C for (a and b) 250 hours, (c and d) 500 hours and (e and f) 1000 hours. Source: Reproduced with permission from Tsai et al. [53]/Elsevier.

Y/Hf-doped HEA to the fast establishment of an exclusive α-$Al_2O_3$ scale in the early oxidation stage, a beneficial RE-effect, and the large columnar grain size in the oxide scale for the HEA.

For Cantor alloy, MnCoCrFeNi system, on the one hand, large Mn content is detrimental to oxidation resistance since the outward diffusion rates of metal cations through the Mn-rich spinel is higher than that through a protective Al or Cr oxide scale [54]. On the other hand, sluggish diffusion of Mn in the CrMnFeCoNi HEA might retard the oxidation process of Mn [55]. To investigate the oxidation process of this alloy, Laplanche et al. performed isothermal oxidation tests in air between 500 and 900 °C for up to 100 hours on Cantor HEA [56]. Thermogravimetric analysis revealed an initial linear region at short oxidation times followed by parabolic oxidation behavior at longer times. The duplex oxide scale consisted of a Cr-rich inner layer and a thicker Mn-rich outer layer. A significant amount of porosity can be observed in the CrMnFeCoNi substrate close to the oxide-substrate interface, and the porosity increased with increasing oxidation temperature. They determined the oxidation activation energy of this alloy as $130 \pm 10$ kJ/mol, which was close to that of Mn diffusion in Mn oxides, indicating strongly that the outward diffusion of Mn through the oxide scale controlled the oxidation kinetics of the HEA. However, considering the relatively high diffusion barrier of Mn in this HEA, 288 kJ/mol, the pores beneath the oxide scale might accelerate the diffusion process of Mn in the substrate by providing short-circuit diffusion paths through vapor transport. Holcomb et al. designed a serial of alloys to investigate the composition dependency of oxidation resistance of Cantor alloy [57]. They found that for the alloy without Mn, CoCrFeNi, the oxidation resistance was excellent and mass gain was the least due to the thin oxide scale consisting of $Cr_2O_3$ primarily. While for the alloy without Cr, CoFeMnNi, severe oxidation, scale spallation, and internal oxidation were observed. It has been shown that the addition of Y, mainly to reduce the amount of S in the alloy, also improved the oxidation resistance of CrMnFeCoNi HEA, most likely by improving the adhesion of the oxide layer to the substrate. Based on these results, they suggested that the elimination of Mn element from the alloy should be considered before applying HEAs in the Mn–Co–Cr–Fe–Ni system in high-temperature oxidizing environments. The oxidation resistance of Cantor alloy is generally poorer than that of Al-containing HEAs. Although the oxidation kinetics of CrMnFeCoNi HEA is parabolic, the parabolic rate constants are much larger. Kim et al. measured the parabolic rate constant $k_p$ of CrMnFeCoNi HEA at 900, 1000, and 1100 °C [58]. The value is at least one magnitude larger than that of Al-containing HEAs at the same temperature.

In the category of HEC, high-entropy borides, carbides, and nitrides are vulnerable to oxidation due to their high temperature and oxygen-involving applications. Gild et al. synthesized six high-entropy borides and found that the oxidation performance of these high-entropy borides was much better than that of the individual metal diborides at 1000–1200 °C [59]. Detailed works on the high-entropy carbides have been conducted to reveal the oxidation mechanism of HECs [60–64]. Ye et al. found out that the oxidation of $(Hf_{0.2}Zr_{0.2}Ta_{0.2}Nb_{0.2}Ti_{0.2})C$ at 1073–1673 K followed a parabolic rate rule, indicating a diffusion-controlled oxidation process [60, 61]. Based on the oxidation activation energy, they pointed out that the oxidation was mainly controlled by the inward diffusion of the oxidant. By increasing the oxidation temperature to 1773 K, the oxidation kinetics obeyed a parabolic-linear rule, and the carbide was completely oxidized after isothermal oxidation for 60 minutes. On the other hand, Wang et al. observed a narrow Ti-depleted layer at the matrix-oxide interface [62]. Using thermodynamic calculations, they found that TiO was easily formed with large vapor pressure during high temperature oxidation. Based on these results, they attributed the formation of a Ti-depleted layer to the active oxidation of TiC and the thickening of the oxide layer to the outward diffusion of Ti-containing species. By removing element Ti from the high-entropy carbides, no Ti-depleted layer was observed at the matrix-oxide interface, and the dominant process for the oxidation of $(Hf_{0.25}Zr_{0.25}Ta_{0.25}Nb_{0.25})C$ was likely transformed to the inward diffusion of oxygen through the formed oxide layer [63]. This oxidation mechanism is confirmed by Wang et al. [64]. They observed an amorphous, thin, and dense oxycarbide interlayer, which acted as an effective diffusion barrier for oxygen. Thus, the oxidation rate was determined by the rate of oxygen diffused through this dense interlayer.

Although the oxidation mechanism is still debatable, the multilayer structure of the oxide scale is a common feature for high-entropy carbides [60–64]. Normally, the Hf/Zr-rich oxide formed the outer layer of the scale, while the inner layer consisted of Ta/Nb-rich oxide mainly based on the similarity solution principle. To provide a theoretical ground for this oxidation behavior of high-entropy carbides and borides, Backman et al. extended an analytical approach for binary systems to quantify the extent of preferential oxidation in high-entropy carbide systems through the development of a computational approach using free energy minimization [65]. They concluded that the tendency for preferential oxidation was highly sensitive to the difference in relative thermodynamic stabilities of the constituent oxides. The sequence of binding ability for the five transition metals for oxygen atoms was Hf > Zr > Ti > Ta > Nb. By examining the oxidation products of a dogbone specimen at 1700 °C in a resistive heating system, they proved the correctness of their calculations [66].

Due to the lack of oxidation resisting element, the oxidation of high-entropy carbides is much severer than that of HEAs at the same temperature, according to the reported experiment data. The thickness of oxide scale increased rapidly in a short duration time for HE carbides. Introducing an oxidation-resistant element, Al and/or Si, is an effective way to improve the oxidation resistance of HECs. Cheng et al. deposited (AlCrTaTiZr)–$Si_x$–N films with different silicon contents by dual-gun reactive RF magnetron co-sputtering [67]. After oxidation at 1000 °C for two hours, the solid solution nitride was still stable and remained unchanged. Using the same strategy, oxidation resistant nitrides, $(Al_{34}Cr_{22}Nb_{11}Si_{11}Ti_{22})N$, $(Al_{23.1}Cr_{30.8}Nb_{7.7}Si_{7.7}Ti_{30.7})N$, and $(Al_{29.1}Cr_{30.8}Nb_{11.2}Si_{7.7}Ti_{21.2})N$, had also been synthesized [68, 69]. For carbides and borides, introducing Al and Si into the structure is quite a challenge due to their quite different electronic configuration and chemical properties.

## 7.4 Molten Salt Corrosion Resistance

Considering their potential applications in nuclear reactors, concentrating solar power plants, and gas turbine engines, good resistance of HEAs in a hot molten salt is required. Tsao et al. designed two promising high-entropy superalloys based on the Ni–Co–Fe–Al–Cr–Ti system [70], and the corrosion behavior of these two alloys in 75% $Na_2SO_4$ + 25% NaCl at 900 °C for 100 hours was investigated. After the corrosion, they found that the alloy with higher Cr content was more resistant to molten salt corrosion. The weight loss of this alloy was only 10% of that of CM247LC, a typical Ni-based superalloy. The corrosion thickness was much thinner, as shown in Figure 7.4. They attributed good resistance of this HEA to the high Cr content by forming protective $Cr_2O_3$ layer against molten salt. Jalbuena et al. exposed $Al_{0.1}$CoCrFeNi HEA and Alloy 718 in a molten $Na_2SO_4$-NaCl eutectic salt mixture at 750 °C [71]. The dual oxide layer was found for the HEA after corrosion, with a dense outer chromium oxide layer and an inner aluminum oxide layer. Due to the existence of corrosion-resistant $Cr_2O_3$ layer and oxidation-resistant $Al_2O_3$ layer, the corrosion rate of HEA was lower than that of Alloy 718.

Elbakhshwan et al. examined the corrosion behavior of the FCC $Cr_{18}Mn_{27}Fe_{27.5}Ni_{27.5}$ HEA in molten FLiBe salt at 700 °C for 1000 hours [72]. They found that HEA lost a higher mass as compared to the reference 316H stainless steel due to the dissolution of Mn into the molten salt; however, the dissolution of Mn appeared to hinder the dissolution of Cr in the molten fluoride salts, which was recognized as the main mechanism of corrosion degradation of alloys. These results have clearly demonstrated that HEAs are more resistant to molten salt corrosion than traditional alloys.

Due to the excellent thermophysical, mechanical properties, and high-temperature stability, high-entropy oxides exhibit great potential as high-temperature thermal/environmental barrier coating materials. For these applications, attack from molten calcia–magnesia–aluminosilicate (CMAS) is of great concern. Dong et al.

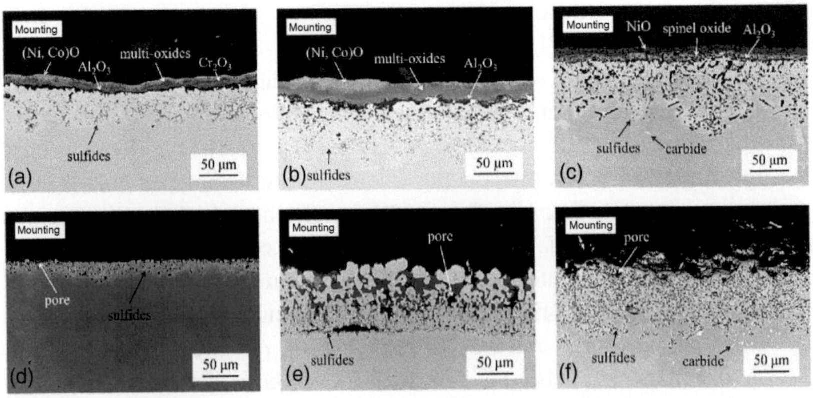


**Figure 7.4** Microstructure of samples after 900 °C/100 hours (five cycles) hot corrosion of (a and d) high Cr content HEA, (b and e) low Cr content alloy, and (c and f) CM247LC. Source: Tsao et al. [70], MDPI, CC BY 4.0.

and Ren et al. developed high-entropy disilicate ($(Yb_{0.2}Y_{0.2}Lu_{0.2}Sc_{0.2}Gd_{0.2})Si_2O_7$) and monosilicate ($(Y_{1/4}Ho_{1/4}Er_{1/4}Yb_{1/4})Si_2O_5$), respectively, which exhibited more desirable properties (thermophysical and corrosion resistance) compared with their single constituents [19, 73]. The improvement of good properties was attributed to the synergistic effects of the different kinds of RE elements. Sun et al. investigated the corrosion mechanism of high-entropy disilicate ($(Er_{0.25}Tm_{0.25}Yb_{0.25}Lu_{0.25})_2Si_2O_7$) in CMAS at 1500 °C [74]. The same mechanism that involved reaction and precipitation was observed in high-entropy disilicate, as in individual constituents. When contacting with molten CMAS, high-entropy disilicate dissolved and Ca–Si–RE containing apatite precipitated when the elements reached saturation. Nevertheless, the resistance of high-entropy disilicate to CMAS was significantly higher than that of single constituents, and these improvements could be attributed to the slower dissolution rate and enhanced grain boundary stability, which resulted from the sluggish diffusion effect and lattice distortion. A similar corrosion mechanism is also observed in the corrosion of high-entropy fluorite against CMAS [75]. Compared to the corrosion resistance of high-entropy fluorite, medium-entropy fluorites and 8YSZ, they found that high-entropy fluorite performed best against CMAS corrosion, and no grain boundary infiltration layer was found in the post-corrosion samples. Sun et al. tested CMAS corrosion resistance of two rare-earth bixbyite structured high-entropy oxides, $(Eu_{0.2}Er_{0.2}Lu_{0.2}Y_{0.2}Yb_{0.2})_2O_3$ and $(Sm_{0.2}Er_{0.2}Lu_{0.2}Y_{0.2}Yb_{0.2})_2O_3$, at 1250 and 1350 °C for four hours [76]. Compared to the other thermal barrier coating/environmental barrier coating (TBC/EBC) materials, the thickness of reaction layers for two high-entropy oxides were much thinner, as shown in Figure 7.5. The reaction mechanism was characterized by the diffusion of CMAS to HE $RE_2O_3$. Qiu et al. designed two medium-entropy oxides, $(Y, Ti)_{0.1}(Zr, Hf, Ce)_{0.9}O_2$ and $(Ta, Ti)_{0.1}(Zr, Hf, Ce)_{0.9}O_2$ [77]. Besides the good high-temperature thermal radiation shielding performance at the waveband of 1–5 μm and low thermal conductivity in the investigated temperature range, these two oxides exhibited remarkable CMAS resistance as the penetration layer was only 2 μm thick. Analysis on the distribution of elements suggested that no element diffusion occurred between the substrate and molten salt. They attributed the good CMAS resistance to the sluggish diffusion effect of solid solution.

High-entropy ceramics are also found to be resistant to molten salt. Qiu et al. synthesized a series of $(Ca,Sr,Ba)TMO_3$ [78]. After exposure to molten NaCl-KCl salts at 800 °C for two, five, and eight hours, the mass losses of $(Ca,Sr,Ba)(Zr,Hf)O_3$ and $(Ca,Sr,Ba)(Zr,Hf,Ta)O_3$ were significantly lower than $BaZrO_3$. In addition, the surfaces of these two HE ceramics after corrosion were plain and smooth, indicating their stability in the molten salt. Different results were obtained when ceramics were exposed to molten NaF–KF salts, since F was much smaller than Cl. Although the resistance of $(Ca,Sr,Ba)(Zr,Hf,Ta)O_3$ to molten salt was still better, the dissolution of CaO, SrO, BaO, $ZrO_2$, and $HfO_2$ into the salt was observed since $Ta_2O$ and $K_2Ta_2O_3F_6$ were formed on the corroded surface. The thickness of the formed corrosion layer was merely 8 μm.

Benefited from high-entropy effects, improved hot corrosion resistance is observed in both HEAs and HECs, guaranteeing their applications in harsh environments.

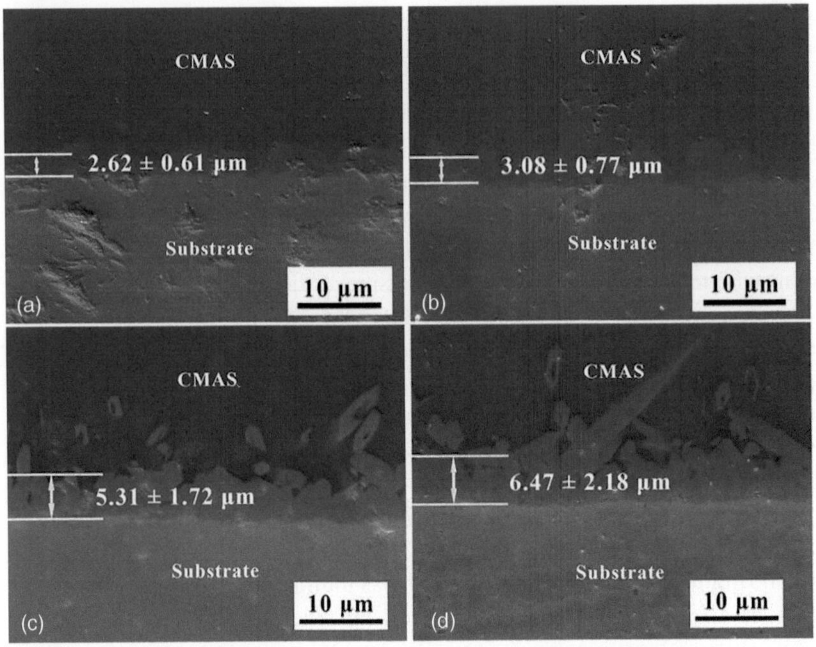


**Figure 7.5** Cross-section morphologies of (a and c) $(Eu_{0.2}Er_{0.2}Lu_{0.2}Y_{0.2}Yb_{0.2})_2O_3$ and (b and d) $(Sm_{0.2}Er_{0.2}Lu_{0.2}Y_{0.2}Yb_{0.2})_2O_3$ corroded by CMAS at 1250 and 1350 °C for four hours. Source: Sun et al. [76], Reproduced with permission from Springer Nature.

Moreover, considering the chemical and structural space of HEMs, the design of corrosion resistance materials with desired properties is feasible.

## 7.5 Irradiation Resistance

For materials under irradiation, severe damages such as void, swelling, atom segregation, irradiation hardening, phase instability, and amorphization could be detected. These damages depend not only on the incident particle but also on the characteristics of the material and the temperature of irradiation [79]. Searching for irradiation-resistant materials is an important topic for the applications involving nuclear reactors.

As speculated by Egami et al. [80], high atomic-level stresses originating from mixing of elements with different atomic sizes promoted the local amorphization of the alloy under the irradiation conditions; however, local melting and re-crystallization due to the thermal spikes would leave much less defects in HEAs. To support their speculation, they conducted electron irradiation tests on ZrHfNb alloy films and found remarkable structure stability of these alloys up to 10 dpa at 298 K and up to 50 dpa at 103 K. They also speculated that HEAs with the atomic-level volume strain close to 0.1 were self-healing and highly irradiation resistant. This "self-healing" mechanism could be illustrated by Figure 7.6 [81]. Good structure stability was also observed in CoCrFeNi-based and $Al_x$CoCrFeNi-based HEAs [81, 82].

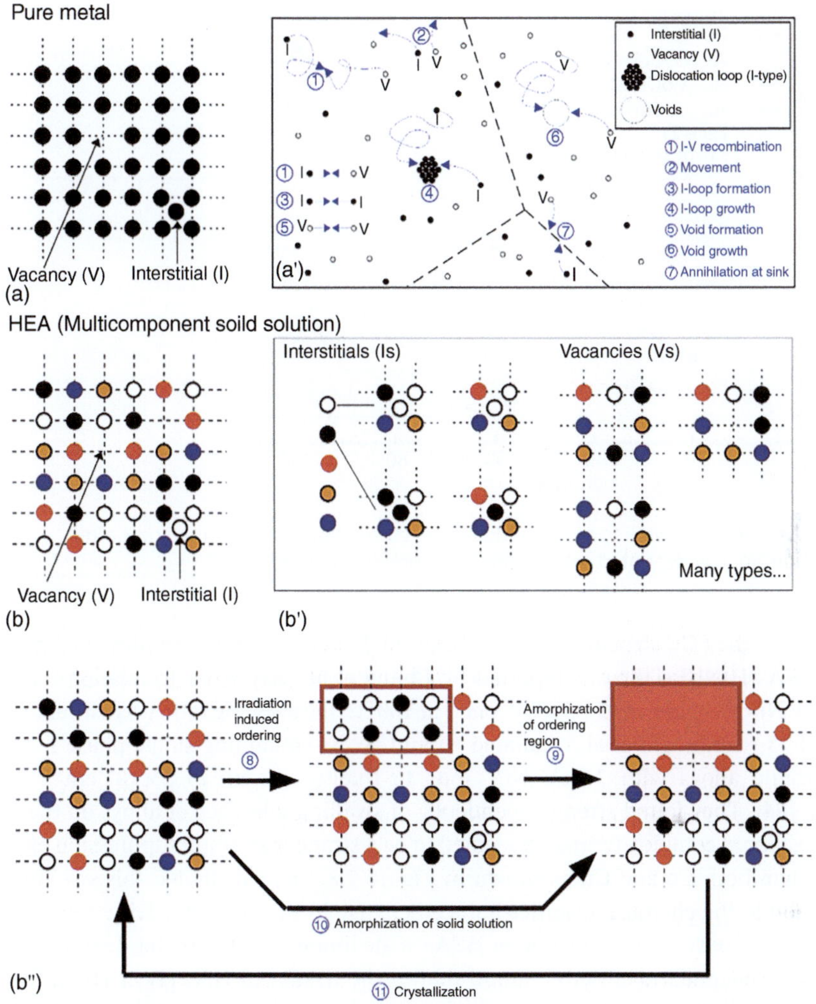


**Figure 7.6** Schematic illustration of the simplest type of defects introduced by the electron knock-on effect under MeV electron irradiation and defect recovery processes. (a) (a′) pure metal, (b) (b′) (b″) HEAs. (b″) is the schematic illustration of the defect recovery via the amorphization-crystallization process. Source: Reprinted with permission from Nagase et al. [81].

Besides the good phase stability under irradiation, HEAs seem more resistant to other irradiation-induced damages. Xia et al. exposed the $Al_xCoCrFeNi$ ($x = 0.1$, 0.75, 1.5) to the flux of 3 MeV Au-ions with fluences ranging from $1\times10^{14}$ to $1\times10^{16}\,cm^{-2}$ [82]. With the increase in Al content, the phase structure transformed from single FCC to primary FCC with minor BCC and dual BCC structure. After irradiation, the volume swelling of these HEAs was significantly lower than that of conventional nuclear materials at high dpa, as shown in Figure 7.7. They also observed unusually good irradiation resistance of FCC phase over BCC structure, which was contrast to the general impression that the BCC structure was more

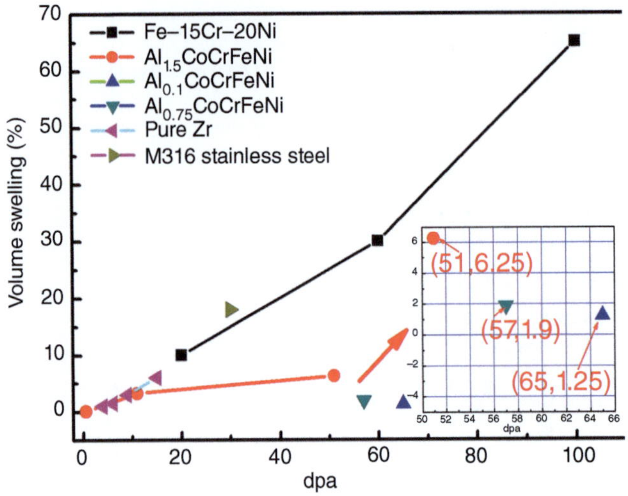


**Figure 7.7** Volume swelling of $Al_xCoCrFeNi$ and conventional nuclear materials after irradiation. Source: Reprinted with permission from Xia et al. [82].

stable than the FCC structure under irradiation. Better swelling resistance of refractory HEAs HfNbTaTiZr was reported by Chang et al. [83]. After irradiation with 300 keV $Ni^+$ to a fluence of $1.5 \times 10^{16}\,cm^{-2}$ at 100 °C, the swelling ratio was estimated to be $1.23 \pm 0.03\%$. Jin et al. irradiated Ni and six Ni-containing single-phase FCC equiatomic alloys with 3 MeV $Ni^{2+}$ ions to fluences of up to $5 \times 10^{16}\,cm^{-2}$ at 500 °C [84]. They found strong dependency of swelling tolerance of alloys on their compositional complexity, and Fe seemed to be a stronger swelling suppression element than both Co and Cr, as shown in Figure 7.8. They attributed this swelling reduction to the chemical disorder-induced change in vacancy diffusivity. Another commonly observed phenomenon in HEAs is the limited irradiation hardening. For instance, Sadeghilaridjani et al. found that in the irradiated HfTaTiVZr HEA, the magnitude of hardening (13%) was significantly smaller than that for SS304 (50%) [85]. They attributed the limited hardening to the reduced mobility of point defects and self-healing ability in HEA. Similar results were observed in TiVCrTa, WTaCrV, $Ti_2ZrHfV_{0.5}Mo_{0.2}$, and TiZrNbHfTa [86–89]. Some exceptions have also been reported. Patel et al. observed an unexpected decrease in the hardness of post-irradiated $V_{2.5}Cr_{1.2}WMoCo_{0.04}$ [90]. While significant hardening, comparable to that in 304 stainless steel and conventional austenitic Fe–Cr–Ni, was found in Fe–Ni–Mn–Cr and CrMnFeCoNi HEAs [91, 92].

Normally, the superior irradiation resistance of HEAs was attributed to the "self-healing" ability induced from severe lattice distortion and sluggish atomic diffusion [80, 81, 84, 86]. To verify such mechanism, Zhang et al. used the ab initio Korringa–Kohn–Rostoker coherent-potential-approximation (KKR–CPA) electronic structure method to explore the dependency of an electronic structure on the alloy complexity [10]. Substantial reduction in electron and phonon mean free path was observed, leading to a significant decrease in electron and lattice thermal conductivity.

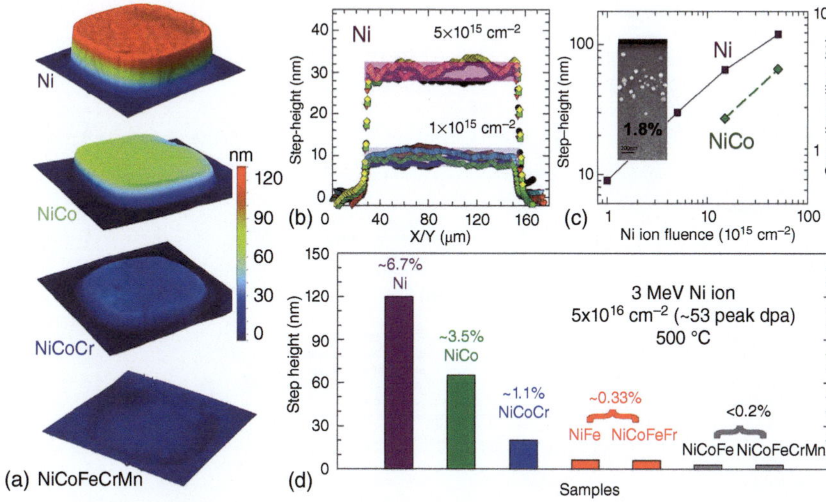


**Figure 7.8** (a) Surface step measurements of Ni, NiCo, NiCoCr, and NiCoFeCrMn after ion irradiations. (b) Step-height profiles for Ni under low-fluence irradiations. (c) Overall swelling from step-height as a function of ion fluence in Ni and NiCo. (d) Comparison of step-height and overall swelling among the seven materials. Source: Jin et al. [84], Reproduced with permission from Elsevier.

Such localized electron–electron interactions led to a prolonged thermal spike and an increase in electronic temperature along the ion path and in the vicinity of the collision cascades, and therefore have a substantial impact on defect dynamics. Granberg et al. investigated the resistant to radiation damage of equiatomic metal alloys and pure metals by combined experiments and simulation [93, 94]. They concluded that the substantial reduction of damage accumulation under prolonged irradiation in single-phase Ni-based alloys could be explained by reduced defect mobility and altered migration paths on a modified energy landscape, resulting from alloy complexity, which could modify defect dynamics to promote annihilation of radiation damage that ultimately enhance radiation tolerance, as shown in Figure 7.9.

Inspired by the achievement in HEAs, the irradiation effects of HECs were also investigated. After 120 keV helium ion irradiation at 25 °C, the coalescence of helium bubbles was significantly suppressed in $(Hf_{0.2}Zr_{0.2}Ta_{0.2}Nb_{0.2}Ti_{0.2})C$ as compared to the obvious structure swell in ZrC [95]. The irradiation damage of the high-entropy carbide $(Zr_{0.25}Ta_{0.25}Nb_{0.25}Ti_{0.25})C$ was explored systematically by Wang et al. [96]. After irradiation of 3 MeV Zr ions to 20 dpa at 25, 300, and 500 °C, the rock-salt structure was maintained without any phase transformation or amorphous formation, indicating a good phase stability of HEC. The lattice expanded by ~0.2%. The irradiation-induced microstructures comprised defect clusters with the diameters of several nanometers, without void formation or radiation-induced segregation at grain boundaries. Zhang et al. designed a novel thermal neutron and gamma-ray shielding ceramic: $(La_{0.2}Ce_{0.2}Gd_{0.2}Er_{0.2}Tm_{0.2})_2(WO_4)_3$ based on the relatively high neutron capture cross section of rare-earth elements [97]. The thermal

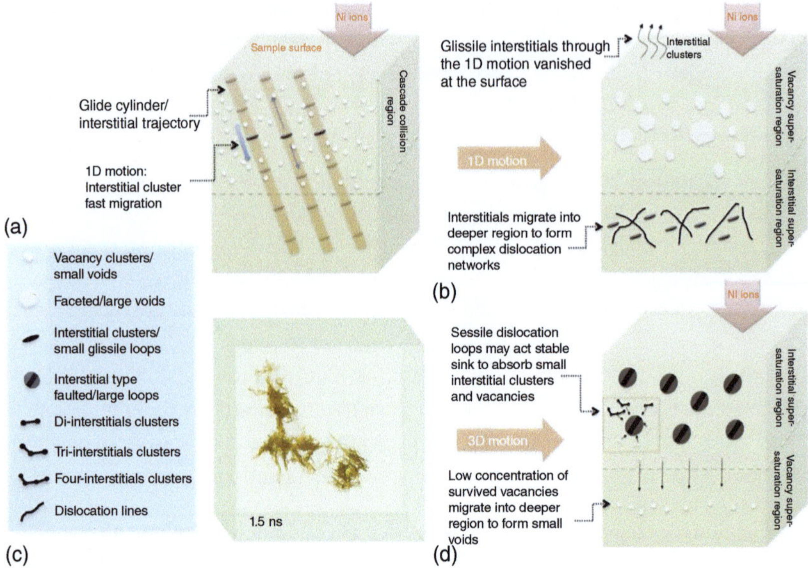


**Figure 7.9** (a) Schematic illustration of 1D motion in nickel and NiCo. Interstitial clusters migrated fast along the glide cylinder. (b) Schematic sketch of defect evolution and distribution in nickel and NiCo as a result of (a). (c) MD simulation result of the trajectory of the center of a four-interstitial cluster in NiFe showing a 3D migration mode. (d) Schematic sketch of defect evolution and distribution in NiFe, NiCoFe, NiCoFeCr, and NiCoFeCrMn as a result of (c). Source: Lu et al. [94], Springer Nature, CC BY 4.0.

neutron absorption rate and γ-ray shielding performance of this HEC was quite satisfactory, which is a promising material in the application of radiation protection field. Unlike HEAs, the underlying mechanisms responsible for radiation resistance of HECs have not yet been revealed.

## 7.6 Electronic and Ionic Conductivity

Two distinct features regarding the electronic conductivity of HEAs and HECs are both related to the severe structure distortion. The first one is the drastic decrease in electric conductivity. As demonstrated by Chou et al. [3], the electric conductivity of $Al_x$CoCrFeNi varied with temperature and Al content. The electrical resistivity was approximately a linear function of temperature with a positive slope, while that fluctuated with the content of Al. The electrical conductivity in the duplex phase region was smaller than that in the single-phase region. Nevertheless, a reduction in electrical conductivity was observed for all alloys with different compositions. This phenomenon is more obvious for Ni-based alloys, where the resistance of Ni-based HEAs was almost two magnitudes higher than that of metal Ni [1, 10]. Regarding HECs, the experimentally measured electrical conductivity of (CrMnFeCoMo)B ($(0.49 \pm 0.04) \times 10^6$ S/m) was much lower than that of CrB ($0.3 \times 10^7$ S/m), FeB ($0.5 \times 10^7$ S/m), and $ZrB_2$ ($1.0 \times 10^7$ S/m) [98]. The electrical resistance of high-entropy carbides and nitrides was also much higher than that of the binary carbides and nitrides [16, 99].

Another feature is the weak temperature dependency of resistance for HEAs and HECs. Kao et al. systematically investigated the resistance of as-cast, homogenized, and deformed $Al_x$CoCrFeNi HEAs [100]. Besides the large resistivity at 4.2 K for these alloys, compared to traditional crystalline elemental metals, the temperature coefficient of resistivity (TCR, defined as $(\rho_{300}-\rho_{150})/[\rho_{150}(300-150)]$) of these alloys was also quite low. The lowest one was for the homogenized alloys with Al content of 2.0, which was 82.5 ppm/K. Chen and Kao reduced the TCR of Al-containing HEAs further by increasing the Al content to form $Al_{2.08}$CoCrFeNi HEA [101]. The averaged TCR of this alloy from 4.2 to 360 K was only 72 ppm/K, which was much lower than that of manganin. In Cantor alloys, a similar phenomenon was also observed [1, 10]. One direct consequence of such weak temperature dependence of resistance was the almost linear temperature dependence of thermal conductivity of electrons with a positive slope for HEAs and HECs [1, 16]. We have discussed the possible mechanism for high resistance and weak temperature dependence of HEAs in Chapter 7.1, that is, remarkable smearing near the Fermi level throughout the whole Brillouin zone [8–10], indicating severe scattering of these electrons even at cryogenic temperatures.

Besides the normal conducting behaviors, superconductivity is also found in HEMs. For all reported superconducting HEAs, the superconducting transition temperatures ($T_c$) are lower than 10 K. Koželj et al. reported the first superconducting HEA with the composition of $Ta_{34}Nb_{33}Hf_8Zr_{14}Ti_{11}$ [102]. By measuring the electrical resistivity, magnetization, magnetic susceptibility, and specific heat, they confirmed that this alloy was a type II superconductor with a $T_c$ of 7.3 K, as shown in Figure 7.10. The transition temperature was much higher than the one calculated

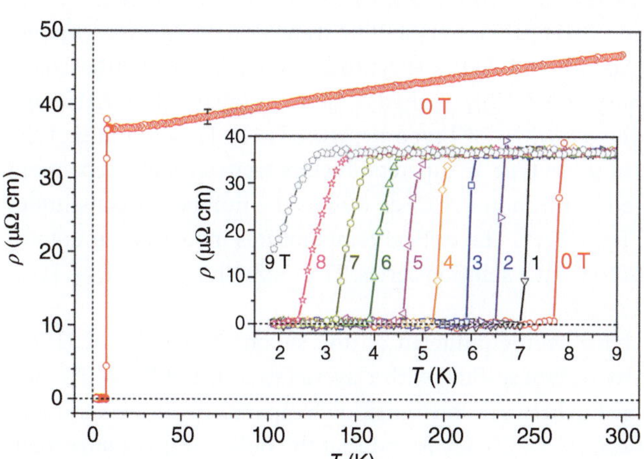


**Figure 7.10** Electrical resistivity in the zero magnetic field between 300 and 2 K. Magnetic-field dependence of the resistivity in the region of the superconductivity transition for fields up to 9 T is shown in the inset. Source: Reproduced with permission from Koželj et al. [102].

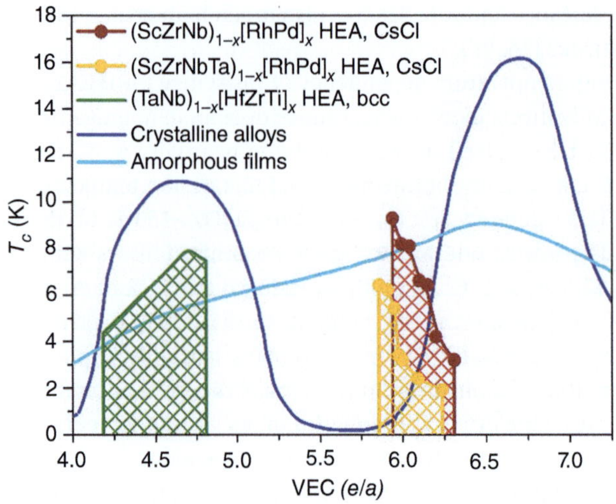


**Figure 7.11** Valence electron concentration dependency of the superconducting transition temperatures of $(ScZrNb)_{1-x}(RhPd)_x$ HEA with $x = 0.45, 0.42, 0.40$, and 0.39–0.35 and $(ScZrNbTa)_{1-x}(RhPd)_x$ with $x = 0.43, 0.38, 0.35, 0.33, 0.328, 0.32, 0.315$, and 0.31 compared to the $T_c$ versus VEC trend lines of $(TaNb)_{1-x}(ZrHfTi)_x$ HEA superconductor, crystalline binary alloys, and amorphous alloy films. Source: Reproduced with permission from Stolze et al. [104].

by the ROMs (5.5 K), indicating a strong cocktail effect. Inspired by this work, von Rohr et al. synthesized a series of $(TaNb)_{1-x}(ZrHfTi)_x$ HEAs by carefully adjust the compositions and investigated the superconducting properties of these HEAs [103]. The valence electron dependence of the transition temperatures for $(TaNb)_{1-x}(ZrHfTi)_x$ was clearly less pronounced than that seen for crystalline alloys. However, a maximum (8.03 K) was reached around *e/a* (*d*-electron) = 4.7, which was an essential feature of the Matthias rule for crystalline transition metal superconductors. Stolze et al. introduced Rh and Pd in the synthesis of HEAs and synthesized a series of new pentanary $(ScZrNb)_{1-x}(RhPd)_x$ and hexanary $(ScZrNbTa)_{1-x}(RhPd)_x$ HEA superconductors with CsCl-type structure [104]. They found that the transition temperatures of these type-II superconductors were strongly dependent on the valence electron concentration (VEC) as well as the number of constituent elements, as shown in Figure 7.11. The critical temperatures increased monotonically with decreasing VEC within a solution series. The $(ScZrNb)_{0.65}(RhPd)_{0.35}$ HEA exhibited the highest $T_c \approx 9.3$ K.

Few research studies have been conducted on the superconductivity of HECs. Sogabe et al. synthesized two sets of sulfides with a layered structure of $REO_{0.5}F_{0.5}BiS_2$ and examined the relationship between the transition temperatures of these sulfides and the mixing entropy $\Delta S_{mix}$ [105]. They found that the bulk nature of superconductivity of these sulfides was enhanced by increase in the $\Delta S_{mix}$, which clearly demonstrated that the increase in mixing entropy at the blocking layer could positively affect the emergence of bulk superconductivity in the conducting layer, while these layers could be regarded as spatially separated. The highest transition temperature for these sulfides was 3.97 K. Wang et al. designed four rare-earth based

YBCO ($YBa_2Cu_3O_{7-\delta}$) high-entropy oxides [106]. However, their results revealed that the there was no obvious correlation observed between the irreversible fields, superconducting transition temperatures, and the effective magnetic moments of the four samples. The mixing entropy did not affect the transition temperatures of these oxides, and the highest transition temperature of these materials was 92 K.

Up to now, superconductivity in HEMs has been reported, and more and more new superconducting ones will be discovered; however, the mechanism that underpins such phenomenon has not yet been fully uncovered. Theoretical studies on the electronic structures and phonon spectra of HEM superconductors are rare. The real factors that determine the superconducting properties of HEMs when mixing entropy and different elements are introduced in are still open questions.

One of the breakthroughs for HECs in ion conductivity is the high Li-ion conductivity. Most of the works focus on the Li-ion conductivity of the rock-salt Mg–Co–Ni–Cu–Zn-based oxides. These works were initialed by Bérardan et al. [107]. They reported a $Li^+$ conductivity of $>10^{-3}$ S/cm in $(MgCoNiCuZn)_{1-x-y}Ga_yA_xO$ (A = Li, Na, K) at ambient temperature, which was two orders of magnitude higher than that of LiPON. The oxygen vacancies created by lattice distortion and charge compensation were attributed to the high ionic conductivity. Good cycling stability of Mg–Co–Ni–Cu–Zn-based oxides has also been found and the entropy stabilization mechanism and sluggish diffusion were believed responsible [108, 109]. Wang et al. incorporated fluorine and chlorine into (MgCoCuNiZn)O to form high-entropy oxyhalides, based on the higher electrochemical potential of metal halides. The overall discharge potential of 3.4 V was superior to most disordered rock-salt oxyfluoride materials reported. An increase in the discharge potential was attributed to the change in the overall (de)lithiation mechanism from a conversion to an insertion type reaction and an increase in the theoretical potential of the redox reaction due to the increase in the Gibbs free energy of the system. Besides the rock-salt oxides, other HECs with different structures, such as spinel structure [110], perovskite [111], have also been explored for lithium-ion battery applications, and good capacity and cycling stability were found. These results have clearly demonstrated the promising potential of HEOs in Li-ion batteries for future commercialization.

## 7.7 Dielectric Properties

Due to the electrical conductivity nature, HEAs are not good dielectric materials. Thus, most of the investigations concentrate on dielectric properties of high-entropy oxides. Usually, in high-entropy oxides, there are more than one sublattices in the structure where at least one is occupied by multi-elements. This sublattice exhibits not only long-range periodicity with the distortion that influences the behaviors of electrons, dipoles and band structure but also compositional disorder that reduces short-range order resulting from the randomly distributed multielements [13]. The abundant choices for the composition and structure of HEOs have provided more flexible ways to tailor the dielectric properties of these materials.

Bérandan et al. reported colossal dielectric constants in rock-salt structured HEOs [112]. For instance, the dielectric constant (the relative permittivity) for $(Mg,Co,Ni,Cu,Zn)_{0.95}Li_{0.05}O$ measured at 300 K and 20 Hz with an LCR bridge method was close to $1\times10^3$. Large dielectric constants were obtained for all HECs in this material family, regardless of the substitution elements. The possible origins of this large relative permittivity included dipolar fluctuations in nano-size domains, interface effects, electronic phase separation, defect dipoles, dipoles relaxations or hopping transport, which needed further verification.

Zhou et al. synthesized a series of high entropy $Ba(Zr_{0.2}Ti_{0.2}Sn_{0.2}Hf_{0.2}Me_{0.2})O_3$ (Me = Y, Nb, Ta, V, Mo, W) perovskite oxides using a solid state reaction method [113]. Among these oxides, only those containing Nb and Ta were capable to form dense, single-phase bulks. The permittivity and dielectric loss as functions of frequency and temperature of these two oxides revealed that the permittivity possessed excellent stability at a wide temperature window, and loss tangent was quite low (<0.002) in the frequency range of 20 Hz to 2 MHz. The relatively low permittivity was attributed to the minor concentration of titanium in the B-site sublattice and small grain size. They also examined the effect of multielement doping in the A-site on the dielectric properties of perovskite [114]. Their results showed that the dielectric peak of $(Na_{0.2}Bi_{0.2}Ba_{0.2}Sr_{0.2}Ca_{0.2})TiO_3$ was broad and shifts toward higher temperature with the increasing frequency, which was similar to the results for the relaxation of other ferroelectric perovskite structures. The maximum discharge energy density for this HEO was determined as 1.02 J/cm$^3$ under the electric strength of 145 kV/cm, which exceeded those of many single-phase perovskite ceramics. With the institution of Sr by K, a decrease in discharge energy storage density was observed in $(Bi_{0.2}Na_{0.2}K_{0.2}Ba_{0.2}Ca_{0.2})TiO_3$ with the value of 0.684 J/cm$^3$ [115].

In spite of good dielectric properties of high-entropy oxides, the mechanism that responsible for good performance is still untouched, which would be the main obstacle for our understanding and tuning the dielectric properties of HECs.

## 7.8 Magnetic Properties

For most HEAs, excessive metallic elements such as ferromagnetic Ni, Co, and Fe are the main constituents. Thus, the magnetic property of HEAs is an attractive and scientific interesting topic. Since the simple cubic structure and random atomic distribution lead to low magnetic anisotropy, most HEAs exhibit soft magnetic behavior. Compared to conventional soft magnetic alloys, HEAs with near-equal atomic ratios are expected to have a reduced saturation magnetization $M_s$ due to the lower concentration of elements with the highest net magnetic moments and increased intrinsic coercivity $H_c$ due to the difficulty in magnetic reordering of distorted structure [116]. For the application of magnetic HEAs, a high value of magnetic saturation is desirable. Considering the ferromagnetic nature of Ni, Co, and Fe, medium entropy alloys constituted by these three elements are expected to have large saturation magnetization, which confirmed by experiments that $M_s$ of this alloy exceeded 1 T [117, 118]. With the introduction of different elements into the medium-entropy alloy CoFeNi, the magnetic behavior is quite different. Zhang et al. added Al and Si, simultaneously, to achieve an excellent combination of mechanical, electrical, and magnetic properties of $FeCoNi(AlSi)_x$ alloys [117]. Considering the non-magnetic nature of Al and Si, they observed a near-monotonic decrease in saturation magnetization with an increase in Al and Si content. For coercivity and electrical resistance, their dependence on the composition was more complex. They found that the alloy with $x = 0.2$ had the best combination of $M_s$ (1.5 T), $H_c$ (1400 A/m), $\rho$ (69.5 μΩ·cm), and $\varepsilon_p$ (>50%) for applications as soft magnetic materials. Similar results have also been observed when Al, Si, Mn, and Cu were added to CoFeNi [118–120].

On the other hand, when antiferromagnetic Cr is introduced, the situation is different. Kao et al. systematically examined the magnetic properties of $Al_xCoCrFeNi$ HEAs ($x = 0$–$2$) [100]. They found that all alloys were ferromagnetic at low temperatures (5 K, 50 K) with very low $M_s$. For instance, the $M_s$ of CoCrFeNi was 0.2 T at 50 K, and that of $Al_{0.5}CoCrFeNi$ was 0.143 T. The magnetic nature of CoCrFeNi even transformed to paramagnetic at room temperature. The possible reason for this was the anti-parallel magnetic moment of Cr with Fe/Co/Ni (i.e. anti-parallel magnetic coupling), such that magnetization was canceled out in Cr-containing HEAs [117]. When Cr was removed from the alloy, the $M_s$ of $Al_{0.5}CoFeNi$ increased to 0.992 T [118]. Similar results were reported in the Cantor alloy CoCrFeMnNi. Ji et al. synthesized CoCrFeMnNi alloy powders by mechanical alloying and consolidated milled powders by spark plasma sintering (SPS) [121]. The as-milled powder exhibited a similar superparamagnetic property due to their ultra-fine grain size. After consolidation, the bulk of CoCrFeMnNi HEA showed a paramagnetic nature with $M_s$ of 32.49 emu/g and $H_c$ of 0 Oe under 20 000 Oe magnetic field. Schneeweiss et al. carefully examined the magnetic property of CoCrFeMnNi with temperature [122], and they found that the system of magnetic moments of this alloy was in a paramagnetic state down to temperature 93 K and transited into the (super)spin glass state. They also observed a second magnetic transformation of a ferromagnetic type at 38 K.

Superior soft magnets require low magnetocrystalline anisotropy and low magnetostriction [123]. These two parameters are closely related to the alloying elements, phase structure, and microstructure. As stated by Zhang, while designing magnetic HEAs with high $M_s$, one should (i) stay away from Cr, (ii) use more Fe and Co and less nonmagnetic components, and (iii) stabilize the FCC structure. Regarding the coercivity of magnetic materials, controlling the movement of the magnetic domain by microstructure design and element doping would be effective.

Besides soft magnetic HEAs, hard magnetic ones have also been reported recently. Feng et al. synthesized a Nd-based hard magnetic HEA, $Nd_{20}Pr_{20}La_{20}Fe_{20}Co_{10}Al_{10}$ [124]. This HEA consisted of an amorphous phase, HCP Nd (Pr) phase, HCP La phase, and δ-phase. The intrinsic coercivity and saturation magnetization were 280 k·A/m and 0.117 T, respectively, which were slightly smaller than those of Nd-based bulk amorphous alloy. The high coercivity of the alloy may originate from the combined effect of exchange coupling and domain wall pinning. Na et al. synthesized a hard magnetic HEA without rare-earth elements [125]. They found that the addition of Cu and Ti together in FeCoNiAl-based HEAs was an effective way to enhance coercivity and maintain the hard magnetic nature of these alloys. With the addition of $Cu_{0.4}Ti_{0.4}$, the alloy showed promising hard magnetic properties as an isotropic cast magnet, with an $H_c$ of 1078 Oe and the maximum energy product of 2.06 MGOe, which were slightly better than those of Alnico 2 magnets. These results clearly demonstrated the promising applications of HEAs as hard magnetic materials.

Compared to HEAs, investigations on magnetic properties of HECs are still in their nascent stage. Mao et al. found that $(CoCrFeMnNi)_3O_4$ powders exhibited a typical ferromagnetic behavior with a hysteresis loop at room temperature [126]. The saturation magnetization and coercivity of this HEC varied in the range of 4.85–15.98 emu/g and 176.55–21.39 Oe, respectively, depending on the synthesis temperature. They also examined the magnetic properties of rock-salt $(Co_{0.2}Cu_{0.2}Mg_{0.2}Ni_{0.2}Zn_{0.2})O$ HEC. These nanocrystalline powder exhibited long-range antiferromagnetic behavior below Néel temperature ($T_N$ = 106 K), which could be well understood by the super-exchange interactions in the rock-salt high-entropy oxide (HEO). This long-range magnetic ordering in rock-salt HEC was confirmed by Jimenez-Segura et al. where they found antiferromagnetic order below 120 K in spite of the structural disorder of randomly distributed magnetic ions [127]. Meisenheimer et al. examined the role of disorder and composition on the exchange anisotropy of permalloy/$(Mg_{0.25(1-x)}Co_xNi_{0.25(1-x)}Cu_{0.25(1-x)}Zn_{0.25(1-x)})O$ heterostructures [128]. They grew an exchange bias heterostructure consisting of a ferromagnetic layer (Py, 3 nm), an antiferromagnetic layer, and an entropy-stabilized oxide layer with 80 nm by pulsed laser deposition. The disorder, exchange field, and magnetic anisotropy could be tuned by changing the composition of the oxide. the exchange bias observed at low temperature could reach 10 times of that observed in a CoO/Py control sample.

Besides rock-salt HEOs, the magnetic properties of perovskite-type high-entropy oxides have also been explored, and the relationship between magnetic behavior of these HECs and the composition was established [129–132].

## References

1 Jin, K., Sales, B.C., Stocks, G.M. et al. (2016). Tailoring the physical properties of Ni-based single-phase equiatomic alloys by modifying the chemical complexity. *Sci. Rep.* 6: 20159.

2 Jin, K., Mu, S., An, K. et al. (2016). Thermophysical properties of Ni-containing single-phase concentrated solid solution alloys. *Mater. Des.* 117: 185–192.

3 Chou, H., Chang, Y., Chen, S. et al. (2009). Microstructure, thermophysical and electrical properties in $Al_xCoCrFeNi$ ($0 \le x \le 2$) high-entropy alloys. *Mater. Sci. Eng., B* 163: 184–189.

4 Lu, C., Lu, S., Ye, J. et al. (2013). Thermal expansion and enhanced heat transfer in high-entropy alloys. *J. Appl. Crystallogr.* 46: 736–739.

5 Dong, W., Zhou, Z., Zhang, L. et al. (2018). Effects of Y, GdCu, and Al addition on the thermoelectric behavior of CoCrFeNi high entropy alloys. *Metals* 8: 781.

**6** Antao, F., Dias, M., Correia, J.B. et al. (2020). The effects of mechanical alloying on the physical and thermal properties of CuCrFeTiV alloy. *Mater. Sci. Eng., B* 263: 114805.

**7** Lee, J.I., Oh, H.S., and Park, E.S. (2016). Manipulation of $\sigma_y/\kappa$ ratio in single phase FCC solid-solutions. *J. Appl. Phys.* 109: 061906.

**8** Troparevsky, M.C., Morris, J.R., Daene, M. et al. (2015). Beyond atomic sizes and Hume–Rothery rules: understanding and predicting high-entropy alloys. *JOM* 67: 2350–2363.

**9** Odbadrakh, K., Enkhtor, L., Amartaivan, T. et al. (2019). Electronic structure and atomic level complexity in $Al_{0.5}TiZrPdCuNi$ high-entropy alloy in glass phase. *J. Appl. Phys.* 126: 095104.

**10** Zhang, Y., Stocks, G.M., Jin, K. et al. (2015). Influence of chemical disorder on energy dissipation and defect evolution in concentrated solid solution alloys. *Nat. Commun.* 6: 8736.

**11** Körmann, F., Ikeda, Y., Grabowski, B. et al. (2017). Phonon broadening in high entropy alloys. *npj Comput. Mater.* 3: 36.

**12** Tong, Z., Li, S., Ruan, X. et al. (2019). Comprehensive first-principles analysis of phonon thermal conductivity and electron–phonon coupling in different metals. *Phys. Rev. B* 100: 144306.

**13** Xiang, H., Xing, Y., Dai, F. et al. (2021). High-entropy ceramics: present status, challenges, and a look forward. *J. Adv. Ceram.* 10: 385–441.

**14** Braun, J.L., Rost, C.M., Lim, M. et al. (2018). Charge-induced disorder controls the thermal conductivity of entropy-stabilized oxides. *Adv. Mater.* 30: 1805004.

**15** Wright, A.J., Wang, Q., Ko, S.T. et al. (2020). Size disorder as a descriptor for predicting reduced thermal conductivity in medium- and high-entropy pyrochlore oxides. *Scr. Mater.* 181: 76–81.

**16** Wen, T., Ye, B., Nguyen, M.C. et al. (2020). Thermophysical and mechanical properties of novel high-entropy metal nitride-carbides. *J. Am. Ceram. Soc.* 103: 6475–6489.

**17** Ridley, M., Gaskins, J., Hopkins, P. et al. (2020). Tailoring thermal properties of multi-component rare earth monosilicates. *Acta Mater.* 195: 698–707.

**18** Li, F., Zhou, L., Liu, J. et al. (2019). High-entropy pyrochlores with low thermal conductivity for thermal barrier coating materials. *J. Adv. Ceram.* 8: 576–582.

**19** Ren, X., Tian, Z., Zhang, J. et al. (2019). Equiatomic quaternary $(Y_{1/4}Ho_{1/4}Er_{1/4}Yb_{1/4})_2SiO_5$ silicate: a perspective multifunctional thermal and environmental barrier coating material. *Scr. Mater.* 168: 47–50.

**20** Ren, K., Wang, Q., Shao, G. et al. (2020). Multicomponent high-entropy zirconates with comprehensive properties for advanced thermal barrier coating. *Scr. Mater.* 178: 382–386.

**21** Liu, Y., Jia, D., Zhou, Y. et al. (2020). $Zn_{0.1}Ca_{0.1}Sr_{0.4}Ba_{0.4}ZrO_3$: a non-equimolar multicomponent perovskite ceramic with low thermal conductivity. *J. Eur. Ceram. Soc.* 40: 6272–6277.

**22** Gild, J., Wright, A., Quiambao-Tomko, K. et al. (2020). Thermal conductivity and hardness of three single-phase high-entropy metal diborides fabricated by borocarbothermal reduction and spark plasma sintering. *Ceram. Int.* 46: 6906–6913.

**23** Caro, M., Béland, L.K., Samolyuk, G.D. et al. (2015). Lattice thermal conductivity of multi-component alloys. *J. Alloys Compd.* 648: 408–413.

**24** Lim, M., Rak, Z., Braun, J.L. et al. (2019). Influence of mass and charge disorder on the phonon thermal conductivity of entropy stabilized oxides determined by molecular dynamics simulations. *J. Appl. Phys.* 125: 055105.

**25** Dai, F., Wen, B., Sun, Y. et al. (2020). Theoretical prediction on thermal and mechanical properties of high entropy $(Zr_{0.2}Hf_{0.2}Ti_{0.2}Nb_{0.2}Ta_{0.2})C$ by deep learning potential. *J. Mater. Sci. Technol.* 43: 168–174.

**26** Dai, F., Sun, Y., Wen, B. et al. (2021). Temperature dependent thermal and elastic properties of high entropy $(Ti_{0.2}Zr_{0.2}Hf_{0.2}Nb_{0.2}Ta_{0.2})B_2$: molecular dynamics simulation by deep learning potential. *J. Mater. Sci. Technol.* 72: 8–15.

**27** Mogilevskyw, P., Boakye, E.E., and Hay, R.S. (2007). Solid solubility and thermal expansion in a $LaPO_4$–$YPO_4$ system. *J. Am. Ceram. Soc.* 90: 1899–1907.

**28** Cedillos-Barraza, O., Grasso, S., Nasiri, N.A. et al. (2016). Sintering behaviour, solid solution formation and characterisation of TaC, HfC and TaC-HfC fabricated by spark plasma sintering. *J. Eur. Ceram. Soc.* 36: 1539–1548.

**29** Laplanche, G., Gadaud, P., Bärsch, C. et al. (2018). Elastic moduli and thermal expansion coefficients of medium-entropy subsystems of the CrMnFeCoNi high-entropy alloy. *J. Alloys Compd.* 746: 244–255.

**30** Yan, X., Constantin, L., Lu, Y. et al. (2018). $(Hf_{0.2}Zr_{0.2}Ta_{0.2}Nb_{0.2}Ti_{0.2})C$ high-entropy ceramics with low thermal conductivity. *J. Am. Ceram. Soc.* 101: 4486–4491.

**31** Zhao, Z., Chen, H., Xiang, H. et al. (2020). High entropy defective fluorite structured rare-earth niobates and tantalates for thermal barrier applications. *J. Adv. Ceram.* 9: 303–311.

**32** Zhao, Z., Chen, H., Xiang, H. et al. (2019). $(La_{0.2}Ce_{0.2}Nd_{0.2}Sm_{0.2}Eu_{0.2})PO_4$: a high-entropy rare-earth phosphate monazite ceramic with low thermal conductivity and good compatibility with $Al_2O_3$. *J. Mater. Sci. Technol.* 35: 2892–2896.

**33** Zhao, Z., Chen, H., Xiang, H. et al. (2020). $(Y_{0.25}Yb_{0.25}Er_{0.25}Lu_{0.25})_2(Zr_{0.5}Hf_{0.5})_2O_7$: a defective fluorite structured high entropy ceramic with low thermal conductivity and close thermal expansion coefficient to $Al_2O_3$. *J. Mater. Sci. Technol.* 39: 167–172.

**34** Chen, H., Xiang, H., Dai, F. et al. (2020). High entropy $(Yb_{0.25}Y_{0.25}Lu_{0.25}Er_{0.25})_2SiO_5$ with strong anisotropy in thermal expansion. *J. Mater. Sci. Technol.* 36: 134–139.

**35** Zhao, Z., Chen, H., Xiang, H. et al. (2020). High-entropy $(Y_{0.2}Nd_{0.2}Sm_{0.2}Eu_{0.2}Er_{0.2})AlO_3$: a promising thermal/environmental barrier material for oxide/oxide composites. *J. Mater. Sci. Technol.* 47: 45–51.

**36** Chen, H., Zhao, Z., Xiang, H. et al. (2020). High entropy $(Y_{0.2}Yb_{0.2}Lu_{0.2}Eu_{0.2}Er_{0.2})_3Al_5O_{12}$: a novel high temperature stable thermal barrier material. *J. Mater. Sci. Technol.* 48: 57–62.

**37** Zhao, Z., Chen, H., Xiang, H. et al. (2020). High-entropy $(Nd_{0.2}Sm_{0.2}Eu_{0.2}Y_{0.2}Yb_{0.2})_4Al_2O_9$ with good high temperature stability, low thermal conductivity and anisotropic thermal expansivity. *J. Adv. Ceram.* 9: 595–605.

**38** Gava, V., Martinotto, A.L., and Perottoni, C.A. (2012). First-principles mode Grüneisen parameters and negative thermal expansion in $\alpha$-$ZrW_2O_8$. *Phys. Rev. Lett.* 109: 195503.

**39** Xiang, H., Feng, Z., Li, Z. et al. (2015). Temperature-dependence of structural and mechanical properties of $TiB_2$: a first principle investigation. *J. Appl. Phys.* 117: 225902.

**40** Xiang, H., Feng, Z., Li, Z. et al. (2017). First-principles investigations on elevated temperature elastic and thermodynamic properties of $ZrB_2$ and $HfB_2$. *J. Am. Ceram. Soc.* 100: 3662–3672.

**41** Li, Y., Luo, Y., Tian, Z. et al. (2018). Theoretical exploration of the abnormal trend in lattice thermal conductivity for monosilicates $RE_2SiO_5$ (RE = Dy, Ho, Er, Tm, Yb and Lu). *J. Eur. Ceram. Soc.* 38: 3539–3546.

**42** Otero-de-la-Roza, A. and Luaña, V. (2011). Gibbs2: a new version of the quasi-harmonic model. I. Robust treatment of the static data. *Comput. Phys. Commun.* 182: 1708–1720.

**43** Otero-de-la-Roza, A., Abbasi-Pérez, D., and Luaña, V. (2011). Gibbs2: a new version of the quasi-harmonic model. II. Implementation and test cases. *Comput. Phys. Commun.* 182: 2232–2248.

**44** Song, H., Tian, F., and Wang, D. (2016). Thermodynamic properties of refractory high entropy alloys. *J. Alloys Compd.* 682: 773–777.

**45** Yang, Y., Ma, L., Gan, G. et al. (2019). Investigation of thermodynamic properties of high entropy (TaNbHfTiZr)C and (TaNbHfTiZr)N. *J. Alloys Compd.* 788: 1076–1083.

**46** Huang, P., Yeh, J., Shun, T. et al. (2004). Multi-principal-element alloys with improved oxidation and wear resistance for thermal spray coating. *Adv. Eng. Mater.* 6: 74–78.

**47** Jiang, J. and Luo, X. (2013). High temperature oxidation behaviour of AlCuTiFeNiCr high-entropy alloy. *Adv. Mater. Res.* 652–654: 1115–1118.

**48** Zhang, H., Wang, Q., Tang, Q. et al. (2013). High temperature oxidation property of $Al_{0.5}FeCoCrNi(Si_{0.2}, Ti_{0.5})$ high entropy alloys. *Corr. Prot.* 34 (7): 561–565.

**49** Butler, T.M., Alfano, J.P., Martens, R.L. et al. (2015). High-temperature oxidation behavior of Al–Co–Cr–Ni–(Fe or Si) multicomponent high-entropy alloys. *JOM* 67: 246–259.

**50** Butler, T.M. and Weaver, M.L. (2016). Oxidation behavior of arc melted AlCoCrFeNi multi-component high-entropy alloys. *J. Alloys Compd.* 674: 229–244.

**51** Dąbrowa, J., Cieślak, G., Stygar, M. et al. (2017). Influence of Cu content on high temperature oxidation behavior of $AlCoCrCu_xFeNi$ high entropy alloys ($x = 0$; 0.5; 1). *Intermetallics* 84: 52–61.

**52** Giggins, C.S. and Pettit, F.S. (1971). Oxidation of Ni–Cr–Al alloys between 1000 and 1200°C. *J. Electrochem. Soc.* 118: 1782–1790.

**53** Lu, J., Chen, Y., Zhang, H. et al. (2020). Y/Hf-doped AlCoCrFeNi high-entropy alloy with ultra oxidation and spallation resistance. *Corros. Sci.* 166: 108426.

**54** Stott, F.H., Wei, F.I., and Enahoro, C.A. (1989). The influence of manganese on the high-temperature oxidation of iron-chromium alloys. *Mater. Corros.* 40: 198–205.

**55** Tsai, K.Y., Tsai, M.H., and Yeh, J. (2013). Sluggish diffusion in Co–Cr–Fe–Mn–Ni high-entropy alloys. *Acta Mater.* 61: 4887–4897.

**56** Laplanche, G., Volkert, U.F., Eggeler, G. et al. (2016). Oxidation behavior of the CrMnFeCoNi high-entropy alloy. *Oxid. Met.* 85: 629–645.

57 Holcomb, G.R., Tylczak, J., and Carney, C. (2015). Oxidation of CoCrFeMnNi high entropy alloys. *JOM* 67: 2326–2339.

58 Kim, Y., Joo, Y., Kim, H. et al. (2018). High temperature oxidation behavior of Cr–Mn–Fe–Co–Ni high entropy alloy. *Intermetallics* 98: 45–53.

59 Gild, J., Zhang, Y., Harrington, T. et al. (2016). High-entropy metal diborides: a new class of high-entropy materials and a new type of ultrahigh temperature ceramics. *Sci. Rep.* 6: 37946.

60 Ye, B.L., Wen, T.Q., Liu, D. et al. (2019). Oxidation behavior of $(Hf_{0.2}Zr_{0.2}Ta_{0.2}Nb_{0.2}Ti_{0.2})C$ high-entropy ceramics at 1073–1473 K in air. *Corros. Sci.* 153: 327–332.

61 Ye, B.L., Wen, T.Q., and Chu, Y.H. (2020). High-temperature oxidation behavior of $(Hf_{0.2}Zr_{0.2}Ta_{0.2}Nb_{0.2}Ti_{0.2})C$ high-entropy ceramics in air. *J. Am. Ceram. Soc.* 103: 500–507.

62 Wang, H.X., Han, X., Liu, W. et al. (2021). Oxidation behavior of high-entropy carbide $(Hf_{0.2}Ta_{0.2}Zr_{0.2}Ti_{0.2}Nb_{0.2})C$ at 1400–1600 °C. *Ceram. Int.* 47: 10848–10854.

63 Wang, H.X., Wang, S.Y., Cao, Y.J. et al. (2021). Oxidation behaviors of $(Hf_{0.25}Zr_{0.25}Ta_{0.25}Nb_{0.25})C$ and $(Hf_{0.25}Zr_{0.25}Ta_{0.25}Nb_{0.25})C$-SiC at 1300–1500 °C. *J. Mater. Sci. Technol.* 60: 147–155.

64 Wang, Y.C., Zhang, R.Z., Zhang, B.H. et al. (2020). The role of multi-elements and interlayer on the oxidation behaviour of (Hf–Ta–Zr–Nb)C high entropy ceramics. *Corros. Sci.* 176: 109019.

65 Backman, L., Gild, J., Luo, J. et al. (2020). Part I: theoretical predictions of preferential oxidation in refractory high entropy materials. *Acta Mater.* 197: 20–27.

66 Backman, L., Gild, J., Luo, J. et al. (2020). Part II: Experimental verification of computationally predicted preferential oxidation of refractory high entropy ultra-high temperature ceramics. *Acta Mater.* 197: 81–90.

67 Cheng, K.H., Tsai, C.W., Lin, S.J. et al. (2011). Effects of silicon content on the structure and mechanical properties of (AlCrTaTiZr)–$Si_x$–N coatings by reactive RF magnetron sputtering. *J. Phys. D: Appl. Phys.* 44: 205405.

68 Shen, W.J., Tsai, M.H., Tsai, K.Y. et al. (2013). Superior oxidation resistance of $(Al_{0.34}Cr_{0.22}Nb_{0.11}Si_{0.11}Ti_{0.22})_{50}N_{50}$ high-entropy nitride. *J. Electrochem. Soc.* 160: C531–C535.

69 Hsieh, M., Tsai, M., Shen, M. et al. (2013). Structure and properties of two Al–Cr–Nb–Si–Ti high-entropy nitride coatings. *Surf. Coat. Technol.* 221: 118–123.

70 Tsao, T., Yeh, A., Kuo, C. et al. (2016). High temperature oxidation and corrosion properties of high entropy superalloys. *Entropy* 18: 62.

71 Jalbuena, A.A., Ury, N., Bae, J. et al. (2019). Corrosion of $Al_{0.1}CoCrFeNi$ high entropy alloy in a molten eutectic salt. *J. Electrochem. Soc.* 166: C3488–C3492.

72 Elbakhshwan, M., Doniger, W., Falconer, C. et al. (2019). Corrosion and thermal stability of CrMnFeNi high entropy alloy in molten FLiBe salt. *Sci. Rep.* 9: 18993.

73 Dong, Y., Ren, K., Lu, Y. et al. (2019). High-entropy environmental barrier coating for the ceramic matrix composites. *J. Eur. Ceram. Soc.* 39: 2574–2579.

74 Sun, L., Luo, Y., Tian, Z. et al. (2020). High temperature corrosion of $(Er_{0.25}Tm_{0.25}Yb_{0.25}Lu_{0.25})_2Si_2O_7$ environmental barrier coating material subjected to water vapor and molten calcium-magnesium-aluminosilicate (CMAS). *Corros. Sci.* 175: 108881.

**75** Wright, A.J., Huang, C.Y., Walock, M.J. et al. (2021). Sand corrosion, thermal expansion, and ablation of medium- and high-entropy compositionally complex fluorite oxides. *J. Am. Ceram. Soc.* 104: 448–462.
**76** Sun, Y., Xiang, H., Dai, F. et al. (2021). Preparation and properties of CMAS resistant bixbyite structured high-entropy oxides $RE_2O_3$ (RE = Sm, Eu, Er, Lu, Y, and Yb): promising environmental barrier coating materials for $Al_2O_{3f}/Al_2O_3$ composites. *J. Adv. Ceram.* 10: 596–613.
**77** Qiu, S., Xiang, H., Dai, F. et al. (2022). Medium-entropy $(Me,Ti)_{0.1}(Zr,Hf,Ce)_{0.9}O_2$ (Me = Y and Ta): promising thermal barrier materials for high-temperature thermal radiation shielding and CMAS blocking. *J. Mater. Sci. Technol.* 123: 144–153.
**78** Qiu, S., Xiang, H., Dai, F. et al. (2021). Preparation and molten salts corrosion behavior of high-entropy $(Ca,Sr,Ba)TMO_3$ (TM = Zr, Hf and Ta): promising protecting materials of reference electrode for metal smelting through fused salts electrolysis. *Corr. Commun.* 3: 10–22.
**79** Pickering, E.J., Carruthers, A.W., Barron, P.J. et al. (2021). High-entropy alloys for advanced nuclear applications. *Entropy* 23: 98.
**80** Egami, T., Guo, W., Rack, P.D. et al. (2014). Irradiation resistance of multicomponent alloys. *Metall. Mater. Trans. A* 45: 180–183.
**81** Nagase, T., Rack, P.D., Noh, J.H. et al. (2015). In-situ TEM observation of structural changes in nano-crystalline CoCrCuFeNi multicomponent high-entropy alloy (HEA) under fast electron irradiation by high voltage electron microscopy (HVEM). *Intermetallics* 59: 32–42.
**82** Xia, S., Yang, X., Yang, T. et al. (2015). Irradiation resistance in $Al_x$CoCrFeNi high entropy alloys. *JOM* 67: 2340–2344.
**83** Chang, S., Tseng, K., Yang, T. et al. (2020). Irradiation-induced swelling and hardening in HfNbTaTiZr refractory high-entropy alloy. *Mater. Lett.* 272: 127832.
**84** Jin, K., Lu, C., Wang, L.M. et al. (2016). Effects of compositional complexity on the ion-irradiation induced swelling and hardening in Ni-containing equiatomic alloys. *Scr. Mater.* 119: 65–70.
**85** Sadeghilaridjani, M., Ayyagari, A., Muskeri, S. et al. (2020). Ion irradiation response and mechanical behavior of reduced activity high entropy alloy. *J. Nucl. Mater.* 529: 151955.
**86** Kareer, A., Waite, J., Li, B. et al. (2019). Short communication: 'low activation, refractory, high entropy alloys for nuclear applications'. *J. Nucl. Mater.* 526: 151744.
**87** El-Atwani, O., Li, N., Li, M. et al. (2019). Outstanding radiation resistance of tungsten-based high-entropy alloys. *Sci. Adv.* 5: 1–10.
**88** Lu, Y., Huang, H., Gao, X. et al. (2019). A promising new class of irradiation tolerant materials: $Ti_2ZrHfV_{0.5}Mo_{0.2}$ high-entropy alloy. *J. Mater. Sci. Technol.* 35: 369–373.
**89** Moschetti, M., Xu, A., Schuh, B. et al. (2020). On the room-temperature mechanical properties of an ion-irradiated TiZrNbHfTa refractory high entropy alloy. *JOM* 72: 130–138.

**90** Patel, D., Richardson, M.D., Jim, B. et al. (2020). Radiation damage tolerance of a novel metastable refractory high entropy alloy $V_{2.5}Cr_{1.2}WMoCo_{0.04}$. *J. Nucl. Mater.* 531: 152005.

**91** Kumar, N., Li, C., Leonard, K. et al. (2016). Microstructural stability and mechanical behavior of FeNiMnCr high entropy alloy under ion irradiation. *Acta Mater.* 113: 230–244.

**92** Yang, L., Ge, H., Zhang, J. et al. (2019). High He-ion irradiation resistance of CrMnFeCoNi high-entropy alloy revealed by comparison study with Ni and 304SS. *J. Mater. Sci. Technol.* 35: 300–305.

**93** Granberg, F., Nordlund, K., Ullah, M.W. et al. (2016). Mechanism of radiation damage reduction in equiatomic multicomponent single phase alloys. *Phys. Rev. Lett.* 116: 135504.

**94** Lu, C., Niu, L., Chen, N. et al. (2016). Enhancing radiation tolerance by controlling defect mobility and migration pathways in multicomponent single-phase alloys. *Nat. Commun.* 7: 13564.

**95** Wang, F., Yan, X., Shao, L. et al. (2019). Irradiation damage behavior in novel high-entropy carbide ceramics. *Trans. Am. Nucl. Soc.* 120: 327.

**96** Wang, F., Yan, X.L., Wang, T.Y. et al. (2020). Irradiation damage in $(Zr_{0.25}Ta_{0.25}Nb_{0.25}Ti_{0.25})C$ high-entropy carbide ceramics. *Acta Mater.* 195: 739–749.

**97** Zhang, X., Li, Y., Li, C. et al. (2021). A novel $(La_{0.2}Ce_{0.2}Gd_{0.2}Er_{0.2}Tm_{0.2})_2(WO_4)_3$ high-entropy ceramic material for thermal neutron and gamma-ray shielding. *Mater. Des.* 205: 109722.

**98** Zhang, H., Zhao, B., Dai, F. et al. (2021). $(Cr_{0.2}Mn_{0.2}Fe_{0.2}Co_{0.2}Mo_{0.2})B$: a novel high-entropy monoboride with good electromagnetic interference shielding performance in K-band. *J. Mater. Sci. Technol.* 77: 58–65.

**99** Willams, W.S. (1997). Transition metal carbides, nitrides, and borides for electronic applications. *JOM* 49: 38–42.

**100** Kao, Y., Chen, S., Chen, T. et al. (2011). Electrical, magnetic, and hall properties of $Al_xCoCrFeNi$ high-entropy alloys. *J. Alloys Compd.* 509: 1607–1614.

**101** Chen, S. and Kao, Y. (2012). Near-constant resistivity in 4.2–360 K in a $B_2$ $Al_{2.08}CoCrFeNi$. *AIP Adv.* 2: 012111.

**102** Koželj, P., Vrtnik, S., Jelen, A. et al. (2014). Discovery of a superconducting high-entropy alloy. *Phys. Rev. Lett.* 113: 107001.

**103** von Rohr, F., Winiarski, M.J., Tao, J. et al. (2016). Effect of electron count and chemical complexity in the Ta–Nb–Hf–Zr–Ti high-entropy alloy superconductor. *Proc. Natl. Acad. Sci. U.S.A.* 113: E7144–E7150.

**104** Stolze, K., Tao, J., von Rohr, F. et al. (2018). Sc–Zr–Nb–Rh–Pd and Sc–Zr–Nb–Ta–Rh–Pd high-entropy alloy superconductors on a CsCl-type lattice. *Chem. Mater.* 30: 906–914.

**105** Sogabe, R., Goto, Y., Abe, T. et al. (2019). Improvement of superconducting properties by high mixing entropy at blocking layers in $BiS_2$-based superconductor $REO_{0.5}F_{0.5}BiS_2$. *Solid State Commun.* 295: 43–49.

**106** Wang, K., Hou, Q., Pal, A. et al. (2021). Structural and physical properties of high-entropy $REBa_2Cu_3O_{7-\delta}$ oxide superconductors. *J. Supercond. Novel Magn.* 34: 1379–1385.

**107** Bérardan, D., Franger, S., Meena, A. et al. (2016). Room temperature lithium superionic conductivity in high entropy oxides. *J. Mater. Chem. A* 4: 9536–9541.

**108** Sarkar, A., Velasco, L., Wang, D. et al. (2018). High entropy oxides for reversible energy storage. *Nat. Commun.* 9: 3400.

**109** Qiu, N., Chen, H., Yang, Z. et al. (2019). A high entropy oxide ($Mg_{0.2}Co_{0.2}Ni_{0.2}Cu_{0.2}Zn_{0.2}O$) with superior lithium storage performance. *J. Alloys Compd.* 777: 767–774.

**110** Chen, H., Qiu, N., Wu, B. et al. (2020). A new spinel high-entropy oxide $(Mg_{0.2}Ti_{0.2}Zn_{0.2}Cu_{0.2}Fe_{0.2})_3O_4$ with fast reaction kinetics and excellent stability as an anode material for lithium ion batteries. *RCS Adv.* 10: 9736–9744.

**111** Yan, Y., Wang, D., Zhang, X. et al. (2020). A high-entropy perovskite titanate lithium-ion battery anode. *J. Mater. Sci.* 55: 6942–6951.

**112** Bérandan, D., Franger, S., Dragoe, D. et al. (2016). Colossal dielectric constant in high entropy oxides. *Phys. Status Solidi RRL* 10: 328–333.

**113** Zhou, S., Pu, Y., Zhang, Q. et al. (2020). Microstructure and dielectric properties of high entropy $Ba(Zr_{0.2}Ti_{0.2}Sn_{0.2}Hf_{0.2}Me_{0.2})O_3$ perovskite oxides. *Ceram. Int.* 46: 7430–7437.

**114** Pu, Y., Zhang, Q., Li, R. et al. (2019). Dielectric properties and electrocaloric effect of high-entropy $(Na_{0.2}Bi_{0.2}Ba_{0.2}Sr_{0.2}Ca_{0.2})TiO_3$ ceramic. *Appl. Phys. Lett.* 115: 223901.

**115** Liu, J., Ren, K., Ma, C. et al. (2020). Dielectric and energy storage properties of flash-sintered high-entropy $(Bi_{0.2}Na_{0.2}K_{0.2}Ba_{0.2}Ca_{0.2})TiO_3$ ceramic. *Ceram. Int.* 46: 20576–20581.

**116** Gao, M.C., Miracle, D.B., Maurice, D. et al. (2018). High-entropy functional materials. *J. Mater. Res.* 33: 3138–3155.

**117** Zhang, Y., Zuo, T.T., Cheng, Y.Q. et al. (2013). High-entropy alloys with high saturation magnetization, electrical resistivity, and malleability. *Sci. Rep.* 3: 1455.

**118** Zuo, T.T., Li, R.B., Ren, X.J. et al. (2014). Effects of Al and Si addition on the structure and properties of CoFeNi equal atomic ratio alloy. *J. Magn. Magn. Mater.* 371: 60–68.

**119** Li, P., Wang, A., and Liu, C. (2017). Composition dependence of structure, physical and mechanical properties of $FeCoNi(MnAl)_x$ high entropy alloys. *Intermetallics* 87: 21–26.

**120** Zhang, Q., Xu, H., Tan, X. et al. (2017). The effects of phase constitution on magnetic and mechanical properties of $FeCoNi(CuAl)_x$ ($x = 0–1.2$) high-entropy alloys. *J. Alloys Compd.* 693: 1061.

**121** Ji, W., Wang, W., Wang, H. et al. (2015). Alloying behavior and novel properties of CoCrFeNiMn high-entropy alloy fabricated by mechanical alloying and spark plasma sintering. *Intermetallics* 56: 24–27.

**122** Schneeweiss, O., Friák, M., Dudová, M. et al. (2017). Magnetic properties of the CrMnFeCoNi high-entropy alloy. *Phys. Rev. B* 96: 014437.

**123** Herzer, G. (2013). Modern soft magnets: amorphous and nanocrystalline materials. *Acta Mater.* 61: 718–734.

**124** Feng, X., Zheng, R., Wu, Z. et al. (2021). Study on a new high-entropy alloy $Nd_{20}Pr_{20}La_{20}Fe_{20}Co_{10}Al_{10}$ with hard magnetic properties. *J. Alloys Compd.* 882: 160640.

**125** Na, S., Lambert, P.K., and Jones, N.J. (2021). Hard magnetic properties of $FeCoNiAlCu_XTi_X$ based high entropy alloys. *AIP Adv.* 11: 015210.

**126** Mao, A.Q., Quan, F., Xiang, H.Z. et al. (2019). Facile synthesis and ferrimagnetic property of spinel $(CoCrFeMnNi)_3O_4$ high-entropy oxide nanocrystalline powder. *J. Mol. Struct.* 1194: 11–18.

**127** Jimenez-Segura, M.P., Takayama, T., Bérardan, D. et al. (2019). Long-range magnetic ordering in rocksalt-type high-entropy oxides. *Appl. Phys. Lett.* 114: 122401.

**128** Meisenheimer, P.B., Kratofil, T.J., and Heron, J.T. (2017). Giant enhancement of exchange coupling in entropy-stabilized oxide heterostructures. *Sci. Rep.* 7: 13344.

**129** Witte, R., Sarkar, A., Velasco, L. et al. (2020). Magnetic properties of rare-earth and transition metal based perovskite type high entropy oxides. *J. Appl. Phys.* 127: 185109.

**130** Vinnik, D.A., Trofimov, E.A., Zhivulin, V.E. et al. (2019). High-entropy oxide phases with magnetoplumbite structure. *Ceram. Int.* 45: 12942–12948.

**131** Witte, R., Sarkar, A., Kruk, R. et al. (2019). High-entropy oxides: An emerging prospect for magnetic rare-earth transition metal perovskites. *Phys. Rev. Mater.* 3: 034406.

**132** Ma, J., Zhao, B., Xiang, H. et al. (2022). High-entropy spinel ferrites $MFe_2O_4$ (M = Mg, Mn, Fe, Co, Ni, Cu, Zn) with tunable electromagnetic properties and strong microwave absorption. *J. Adv. Ceram.* 11: 754–768.

# 8

# Applications of High-Entropy Materials

## 8.1 Introduction

The multi-principal element nature of high-entropy materials (HEMs) and the corresponding vast chemical space make people believe that the properties of HEMs can be tailored in large ranges by composition design, and HEMs with unprecedented properties may be discovered. Indeed, a lot of HEMs with intriguing properties have been reported in literatures, including extremely high fracture resistance at cryogenic temperature [1], promising retention of strength at elevated temperatures [2], excellent resistance to irradiation [3], colossal dielectric constant [4], super ionic conductivity [5], amorphous-like thermal conductivity [6], and good combination of mechanical and soft magnetic properties [7]. These promising results inspire people to explore new HEMs in different fields. The high-entropy design concept can be applied to any solid-state materials, especially for inorganic materials. Therefore, HEMs may find potential application in many industrial fields. The application of HEMs is a too big topic to be well discussed in a short chapter. Any of the potential application fields may need a long space for a detailed discussion. Moreover, the application of a material depends on its properties. Properties of HEMs have been comprehensively reviewed in Chapters 6 and 7, where the potential applications are more or less mentioned. Thus, we will just give a brief introduction on the potential applications of HEMs in this chapter. There are some excellent review papers on the applications of HEMs, e.g. Yeh and Lin [8] introduced breakthrough applications of high-entropy alloys (HEAs), Xiang et al. [9] gave a comprehensive review on high-entropy ceramics (HECs), Chen et al. [10] reviewed the application of HEMs in rechargeable batteries, Sun and Dai [11] reviewed applications of HEMs in catalysis, etc. Readers who need an in-depth understanding on applications of HEMs can refer to these review papers.

*High-Entropy Materials: From Basics to Applications,* First Edition.
Huimin Xiang, Fu-Zhi Dai, and Yanchun Zhou.

## 8.2 Structural Applications

The mechanical properties of HEMs, including HEAs and HECs, are the most extensively investigated properties. Various mechanical properties have been reported, which enable possible application of HEMs in many fields, such as automotive, aerospace, nuclear energy, energy storage, and molding. Here, we only introduce some typical examples where HEMs exhibit excellent properties that may not be attenable by conventional materials. For example, it is well known that the Cantor alloy CrMnFeCoNi exhibits good toughness at cryogenic temperatures [1], exceeding its ambient temperature toughness value, which makes it fairly suitable for applications under cryogenic conditions, e.g. storage tanks of liquid hydrogen and oxygen. Further research studies found that the Cantor alloy also exhibits improved hydrogen embrittlement resistance due to high-hydrogen solubility, easy twin formation, and hydrogen trapping by twin boundaries [12], which confirms the potential application of HEAs as storage tanks for liquid hydrogen. HEAs with excellent combination of strength and ductility were further developed by similar TWIP (twinning-induced plasticity) or TRIP (transformation-induced plasticity) mechanisms, where the mechanical properties can be tuned by composition design. On the other hand, HEMs, both HEAs and HECs, with outstanding mechanical properties at elevated temperatures have been reported. Senkov et al. [2, 13] found that high-entropy MoNbTaW, MoNbTaVW, and $AlMo_{0.5}NbTa_{0.5}TiZr$ alloys show promising strength retention at high temperatures (>1000 °C or even 1200 °C), better than the state-of-the-art superalloys. These HEAs are believed as the next-generation superalloys, which can be used as blades to increase the working temperature, and to enhance the thrust of aero-engines. Readers who are interested in these two sets of HEMs can refer to the Chapter 6 "Mechanical Properties of High Entropy Materials," where more comprehensive review is presented.

In nuclear energy field, materials with good irradiation resistance are in urgent demand. Zhang et al. [3] reported that suppressed damage accumulation was observed by increasing the configurational entropy from pure Ni to binary and to more complex quaternary solid solutions, as shown in Figure 8.1. They contributed the improved irradiation property to the slow energy dissipation related to reduced electrical and thermal conductivity. The results reveal that understanding and controlling energy dissipation and defect dynamics by altering alloy complexity may pave the way for new design principles of irradiation-tolerant structural alloys for energy applications.

In comparison to conventional materials, HEMs usually show enhanced hardness, corrosion, and oxidation resistance and can be adopted as wear and corrosion-resistant coatings, such as in molds, dies, and cutting tools. Yeh and Lin [8] compared the performance of cutting tools coated with high-entropy nitride $(Al_{0.34}Cr_{0.22}Nb_{0.11}Si_{0.11}Ti_{0.22})_{50}N_{50}$, TiN, and TiAlN, as shown in Figure 8.2. After dry-cutting the workpiece of 304 stainless steel for a length of 361.5 m at the high cutting speed of 321.6 m/s, the depth of cut of 0.1 mm, and the feed rate of 0.28 mm/rev, only the high-entropy nitride $(Al_{0.34}Cr_{0.22}Nb_{0.11}Si_{0.11}Ti_{0.22})_{50}N_{50}$-coated insert could produce

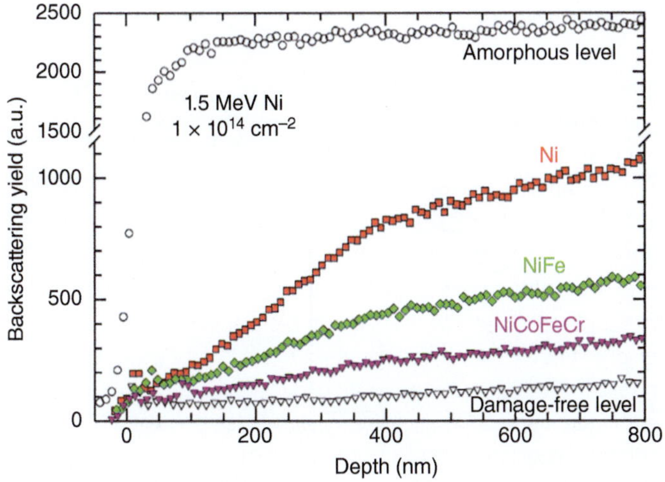


**Figure 8.1** Rutherford backscattering spectra show different irradiation response in Ni and Ni-based alloys. Result reveals that the alloy with good irradiation tolerant is the one with large complexity in chemistry. Source: Zhang et al. [3]/Springer Nature/CC BY-4.0.

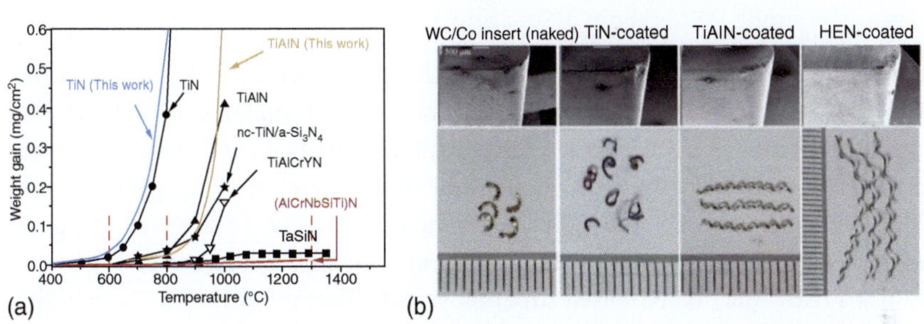


**Figure 8.2** (a) Weight gains of high-entropy $(Al_{0.34}Cr_{0.22}Nb_{0.11}Si_{0.11}Ti_{0.22})_{50}N_{50}$, TiN, and TiAlN as a function of temperature, shown as three colored curves. Black curves are the reported data for comparison; (b) chips produced after cutting 304 stainless steel for 361.5 m by the naked insert, TiN-coated insert, TiAlN-coated insert, and high-entropy $(Al_{0.34}Cr_{0.22}Nb_{0.11}Si_{0.11}Ti_{0.22})_{50}N_{50}$-coated insert. HEN, high-entropy nitride. Source: Reproduced with permission from Yeh and Lin [8]/Springer Nature.

long curled chips as shown in Figure 8.2b. The results revealed that high-entropy nitride coating is very promising to improve the lifetime of cutting tools. The enhanced cutting performance may result from the improved hardness and oxidation resistance, since the oxides formed during cutting can be easily removed by friction.

## 8.3 Thermal Protection and Management

One of the most studied candidate application fields of HEMs is thermal/environmental barrier coatings (T/EBCs). In these applications, the candidate materials studied include high-temperature oxides with fluorite structure, pyrochlore

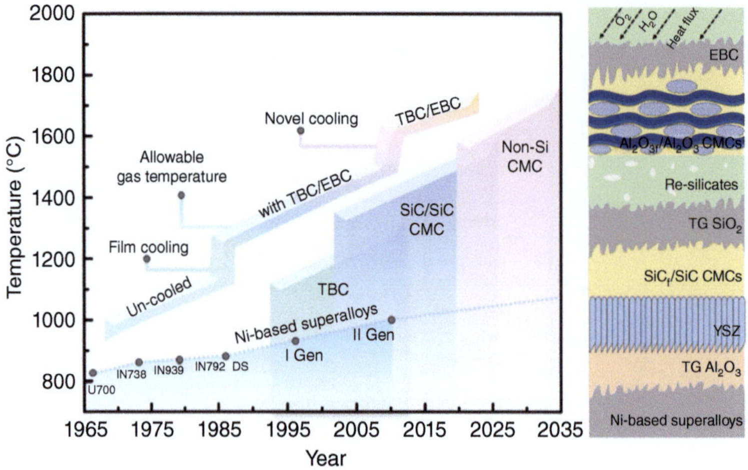


**Figure 8.3** Roadmap for the development of engine materials and schematic structures for the corresponding T/EBCs to protect Ni-based superalloys, $SiC_f/SiC$, and $Al_2O_{3f}/Al_2O_3$ composites. CMC, ceramic matrix composite. Source: Xiang et al. [9]/Springer Nature/ CC BY-4.0.

structure, or perovskite structure (e.g. rare-earth zirconates) [14–23], rare-earth aluminates [24–26], rare-earth silicates [27–29], rare-earth tantalates or niobates [16, 30], and phosphates [31, 32]). These HECs exhibit a superior combination of properties that are required by T/EBCs, such as low thermal conductivity, good thermal stability, low grain growth rate, good water vapor corrosion resistance, and good molten CMAS (calcia–magnesia–aluminosilicate) corrosion resistance. The substrate of a T/EBC may be Ni-based superalloys, SiC fiber-reinforced SiC ($SiC_f/SiC$) composites, or even $Al_2O_3$ fiber-reinforced $Al_2O_3$ ($Al_2O_{3f}/Al_2O_3$) composites. Ni-based superalloys are the most used materials for blades nowadays, $SiC_f/SiC$ composites are the new generation of materials for blades, and the $Al_2O_{3f}/Al_2O_3$ composites are believed to be the next generation of materials for blades. Figure 8.3 illustrates the roadmap for the development of aero-engine materials and schematic structures of the corresponding T/EBCs on different substrates [9]. Thermal matching between a T/EBC and a substrate is crucial for the service lifetime of the T/EBC. Xiang et al. [9] summarized the thermal conductivities and thermal expansion coefficients of the candidate T/EBC HECs and those of SiC, $Al_2O_3$, and YSZ ($Y_2O_3$-stabilized $ZrO_2$ state-of-the-art T/EBC materials used in the Ni-based superalloy blade of aero-engines), as shown in Figure 8.4. The figure suggests that high-entropy $REAlO_3$, $RE_3Al_5O_{12}$, $REPO_4$, rare earth zirconates, tantalates, and niobates are promising candidate T/EBCs for Ni-based superalloys or $Al_2O_{3f}/Al_2O_3$ composites, while high-entropy $RE_2Si_2O_7$ and $TMP_2O_7$ are more suitable T/EBCs for $SiC_f/SiC$ composites.

Rare-earth zirconates have long been proposed as potential T/EBCs for Ni-based superalloys due to their matching thermal expansion coefficient, good thermal stability, and low thermal conductivity. However, poor sintering resistance is the main

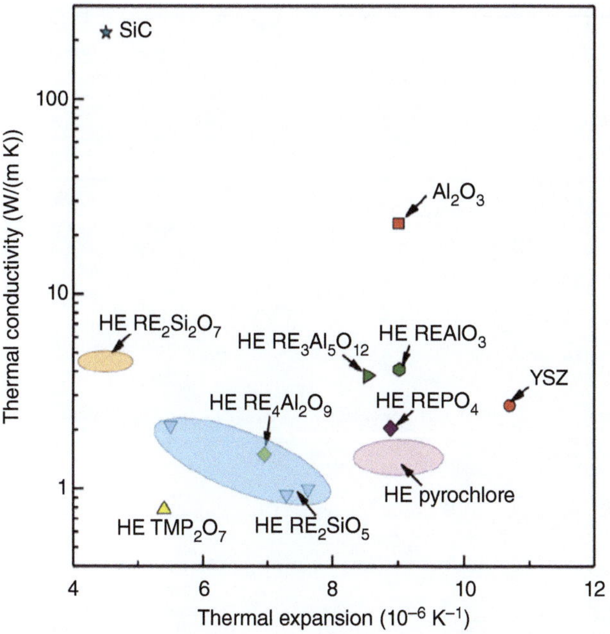


**Figure 8.4** Thermal conductivity and thermal expansion of high-entropy oxides. Source: Xiang et al. [9]/Springer Nature/CC BY-4.0.

drawback that limits their practical application. Zhao et al. [15] revealed that great benefits can be achieved by using high-entropy rare-earth zirconates (HE-REZs). They reported that high-entropy $(La_{0.2}Ce_{0.2}Nd_{0.2}Sm_{0.2}Eu_{0.2})_2Zr_2O_7$ exhibits lower thermal conductivity due to the enhanced scattering of phonons and much slower grain growth rate that may result from sluggish diffusion effect or grain boundary pinning due to segregation. Figure 8.5 compares the microstructure evolution of high-entropy $(La_{0.2}Ce_{0.2}Nd_{0.2}Sm_{0.2}Eu_{0.2})_2Zr_2O_7$ and $La_2Zr_2O_7$ after annealing at 1500 °C for different time periods, which reveals that the grain growth rate of high-entropy $(La_{0.2}Ce_{0.2}Nd_{0.2}Sm_{0.2}Eu_{0.2})_2Zr_2O_7$ is much lower. For high-temperature applications, microstructure stability is of critical importance for long time service, which means that the shortcomings of rare-earth zirconates can be overcome by high-entropy design. In addition, they further reported slower grain growth rate of high-entropy $(Y_{0.25}Yb_{0.25}Er_{0.25}Lu_{0.25})_2(Zr_{0.5}Hf_{0.5})_2O_7$ and $(Y_{0.2}Yb_{0.2}Lu_{0.2}Eu_{0.2}Er_{0.2})_3Al_5O_{12}$ [17, 25].

Another appealing example of applying HECs as T/EBCs is reported by Zhou et al. [18], who prepared double-ceramic-layer (DCL) thermal barrier coating (TBC) with HE-REZ $(La_{0.2}Nd_{0.2}Sm_{0.2}Eu_{0.2}Gd_{0.2})_2Zr_2O_7$ as the top layer and YSZ as the inner layer sprayed on Ni-based superalloy by atmospheric plasma spraying. For comparison, they also prepared DCL TBC with $La_2Zr_2O_7$ as the top layer and tested by thermal cycling. The results demonstrated that the coating with high-entropy $(La_{0.2}Nd_{0.2}Sm_{0.2}Eu_{0.2}Gd_{0.2})_2Zr_2O_7$ as the top layer exhibited improved thermal stability when compared with the one with $La_2Zr_2O_7$ as the top layer, as shown in Figure 8.6. They attributed the

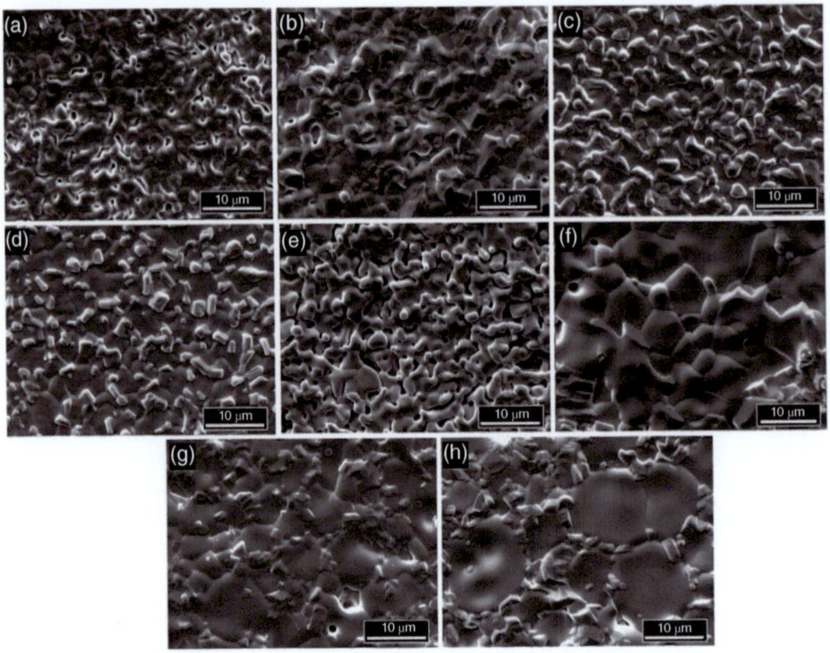


**Figure 8.5** Microstructures of $(La_{0.2}Ce_{0.2}Nd_{0.2}Sm_{0.2}Eu_{0.2})_2Zr_2O_7$ and $La_2Zr_2O_7$ compacts after annealing at 1500 °C for 1–18 hours in air: (a) 1 hour, (b): 6 hour, (c) 12 hour, (d) 18 hour for $(La_{0.2}Ce_{0.2}Nd_{0.2}Sm_{0.2}Eu_{0.2})_2Zr_2O_7$ and (e) 1 hour, (f) 6 hours, (g) 12 hours, (h) 18 hours for $(La_2Zr_2O_7)$. Source: Reproduced with permission Zhao et al. [15]/Elsevier.

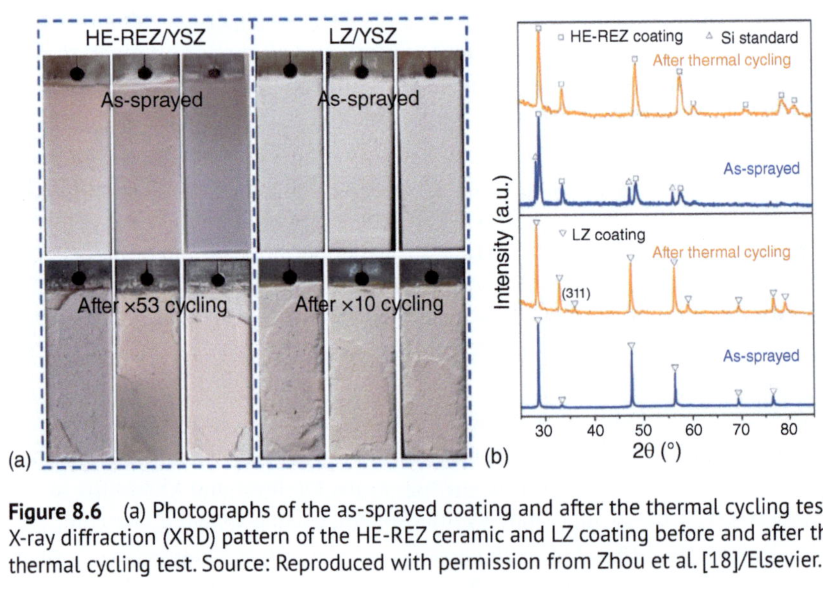


**Figure 8.6** (a) Photographs of the as-sprayed coating and after the thermal cycling test. (b) X-ray diffraction (XRD) pattern of the HE-REZ ceramic and LZ coating before and after the thermal cycling test. Source: Reproduced with permission from Zhou et al. [18]/Elsevier.

improvement of the thermal shock resistance to the improved anti-sinter ability and better thermal matching of the high-entropy $(La_{0.2}Nd_{0.2}Sm_{0.2}Eu_{0.2}Gd_{0.2})_2Zr_2O_7$.

Rare-earth tantalates and niobates display similar properties to YSZ, and they are also believed to be promising candidates as T/EBCs for Ni-based superalloys. One of the most appealing properties of rare-earth tantalates or niobates is ferro-elastic induced toughness, which is beneficial to the performance and lifetime of T/EBCs. Wang et al. [33] prepared a high-entropy rare-earth tantalate $(Y_{0.2}Ce_{0.2}Sm_{0.2}Gd_{0.2}Dy_{0.2})TaO_4$ and reported a higher fracture toughness and lower brittleness index than YSZ, which is induced by its ferro-elasticity. Zhao et al. [16] also studied high-entropy tantalates and niobates and found that they are thermally matching and chemically compatible with $Al_2O_3$ (in Ni-based superalloys, there is one layer of thermally grown $Al_2O_3$).

Another important requirement of T/EBC materials is good resistance to water vapor and CMAS corrosion. Poor resistance to water vapor and CMAS corrosion is one of the main challenges for $SiC_f/SiC$ composite blades. Based on thermal matching, rare-earth aluminates ($RE_4Al_2O_9$), and silicates ($RE_2SiO_5$, $RE_2Si_2O_7$) are considered promising candidate EBC materials for $SiC_f/SiC$ composites. However, the main restrictions for their practical applications are low phase stability and high thermal conductivity. With the high entropy design concept, there is no doubt that thermal conductivity can be reduced. It is also reported that the phase stability can also be improved. As shown by Zhao et al. [26], high-entropy $(Nd_{0.2}Sm_{0.2}Eu_{0.2}Y_{0.2}Yb_{0.2})_4Al_2O_9$ exhibits comparable thermal expansion coefficient with mullite and good phase stability from 300 to 1473 K. However, they also reported that thermal expansion coefficients are highly anisotropic in high-entropy $(Nd_{0.2}Sm_{0.2}Eu_{0.2}Y_{0.2}Yb_{0.2})_4Al_2O_9$, which may lead to risk of failure caused by thermal stress. Under the design concept of high entropy, water vapor and CMAS resistance can also be improved. For example, Sun et al. [29] reported that multicomponent $(Er_{0.25}Tm_{0.25}Yb_{0.25}Lu_{0.25})_2Si_2O_7$ shows superior water vapor corrosion resistance at 1400 °C and significantly enhanced CMAS resistance at 1500 °C as compared to single principal rare-earth disilicates. The results highlighted the merit of high-entropy engineering in advanced EBCs for the improvement of crucial high-temperature corrosion resistance in aero-engine applications.

For designing T/EBCs for $Al_2O_{3f}/Al_2O_3$ composites, the key requirements include good phase stability, low thermal conductivity, matching thermal expansion coefficient and chemical compatibility between T/EBCs and $Al_2O_3$, resistance to water vapor and CMAS corrosion, etc. Except for thermal conductivities and thermal expansion coefficients shown in Figure 8.4, Zhao et al. [24] also checked the chemical compatibility of high-entropy $(Y_{0.2}Nd_{0.2}Sm_{0.2}Eu_{0.2}Er_{0.2})AlO_3$ and $Al_2O_3$, where a high-entropy $(Y_{0.2}Nd_{0.2}Sm_{0.2}Eu_{0.2}Er_{0.2})_3Al_5O_{12}$ buffer layer was formed that might improve the bonding between them. In another work, Zhao et al. [32] checked the compatibility between high-entropy $(La_{0.2}Ce_{0.2}Nd_{0.2}Sm_{0.2}Eu_{0.2})PO_4$ and $Al_2O_3$, where no reaction was found, indicating good chemical compatibility between them.

Based on the results mentioned earlier, under the design scenario of high-entropy (or multi-principal element), rare-earth zirconates, tantalates, niobates, aluminates, silicates, phosphates, etc. exhibit lower thermal conductivity, better thermal expansion coefficient matching and chemical compatibility with substrate materials (or buffer layers) of blades, improved water vapor and CMAS corrosion resistance, slower grain growth rate, and possible ferro-elasticity, all of which are beneficial to design the new generation of T/EBCs. Though most works done are in the initial stage, it can be expected that more and more works will come forth in the future due to the promising potential of HEMs as T/EBCs.

Except for T/EBCs, the low thermal conductivity nature of HEMs guarantees the candidate application in other thermal applications that need low thermal conductivity. For example, thermal protection and insulation in aerospace applications. In the development of new-generation hypersonic vehicles, high-performance thermal insulation and thermal protection systems at ultra-high temperatures are necessary. Porous high-entropy ultra-high temperature ceramics (HE-UHTCs), including transition metal carbides and borides, have been designed for such applications [34–36]. Chen et al. [35] reported a porous $(Zr_{0.2}Hf_{0.2}Ti_{0.2}Nb_{0.2}Ta_{0.2})C$ with a porosity of 80.99% exhibiting an ultralow thermal conductivity of 0.39 W/(m K) and a low density of 1.79 g/cm$^3$. In addition, they also tested the microstructure stability by reheating the sample to 1850 °C, where no significant change in porosity was found, as shown in Figure 8.7. Except for HE-UHTCs, porous high-entropy oxides have also been designed as thermal insulation. For example, Yang et al. [37] prepared rare-earth niobate HEC foams by a 20 wt% slurry sintered at 1500 °C, which exhibited

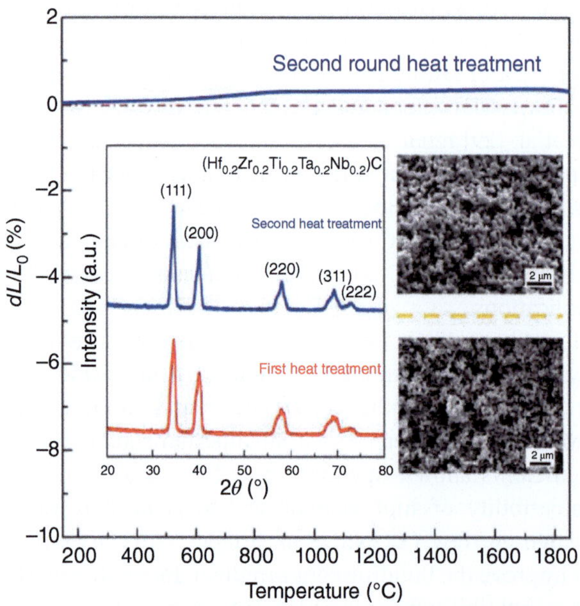


**Figure 8.7** Linear shrinkage curve of porous high-entropy $(Zr_{0.2}Hf_{0.2}Ti_{0.2}Nb_{0.2}Ta_{0.2})C$ during the second-round heat treatment. Inset showing the XRD patterns and scanning electron microscope (SEM) images of as-prepared porous high-entropy $(Zr_{0.2}Hf_{0.2}Ti_{0.2}Nb_{0.2}Ta_{0.2})C$ and the sample after second-round heat treatment at 1850 °C. Source: Xiang et al. [9] and Chen et al. [35], Springer Nature, CC BY 4.0.

high porosity of 96.12% and extremely low thermal conductivity of 0.0343 W/(m K). There was a temperature difference of over 800 °C between frontside and backside of the sample under fire resistance test, indicating that the as-prepared HEC foams are expected to be promising thermal insulation materials. When a porous material is designed with a high density of through-holes, it can be applied in active thermal protection, e.g. transpiration cooling [36].

## 8.4 Thermoelectricity

In thermoelectric materials, the efficiency of energy conversion is characterized by the dimensionless $ZT$ value:

$$ZT = \frac{\sigma S^2}{\kappa_{ph} + \kappa_{el}} T \tag{8.1}$$

where $\sigma$ is the electrical conductivity, $S$ is the Seebeck coefficient, $T$ is the absolute temperature, and $\kappa_{ph}$ and $\kappa_{el}$ are thermal conductivities contributed by phonons and electrons, respectively. High-performance thermoelectric property needs high power factor ($\sigma S^2$) and low thermal conductivity. Unfortunately, carrier concentration-associated coupling of $\sigma$, $S$, and $\kappa_{el}$ leads to difficulty in regulating every single parameter independently. Even though $\kappa_{ph}$ is not directly coupled with other parameters, reducing $\kappa_{ph}$ by scattering of the solid solution usually simultaneously enhances scattering of electrons and holes and usually reduces electronic transport properties. Thus, simultaneously improving power factor and reducing thermal conductivity are hard to achieve. Nevertheless, due to the fact that high-entropy design can usually reduce thermal conductivity, efforts have been paid to whether HEMs with enhanced thermoelectric property could be developed. Except for the enhanced scattering effect, the solid solution effect may change the band structure, which then changes the electronic transport properties, e.g. increasing the density of carrier and improving electric conductivity and Seebeck coefficient. For example, Roychowdhury et al. [38] showed that the carrier type and carrier concentration can be tailored by doping, a typical method in semiconductor industry, which is also applicable in HEMs. They introduced $AgBiSe_2$ into GeSe and synthesized a high-entropy selenide $(GeSe)_{0.5}(AgBiSe_2)_{0.5}$, changing GeSe from a p-type to n-type semiconductor and increased the carrier concentration from $1.20\times10^{17}$ to $3.29\times10^{18}\,cm^{-3}$. In combination of the reduced thermal conductivity, it indicates that high-performance thermoelectric material can be developed by high-entropy engineering.

Xiang et al. [9] summarized the reported temperature-dependent $ZT$ values of high-entropy thermoelectric materials, as shown in Figure 8.8. Among these works, an intriguing work was reported by Qiu et al. [39], who significantly improved the thermoelectric properties by high-entropy engineering. They showed that the phase transformation temperature can be reduced from more than 660 K in undoped GeTe to around 523 K in $Ge_{0.84}In_{0.01}Pb_{0.1}Sb_{0.05}Te_{0.997}I_{0.003}$. The high configuration-entropy samples possess an increased structural symmetry, larger effective masses, and higher Seebeck coefficients, leading to a record-high power factor of ~23 mW/(cm K$^2$) at

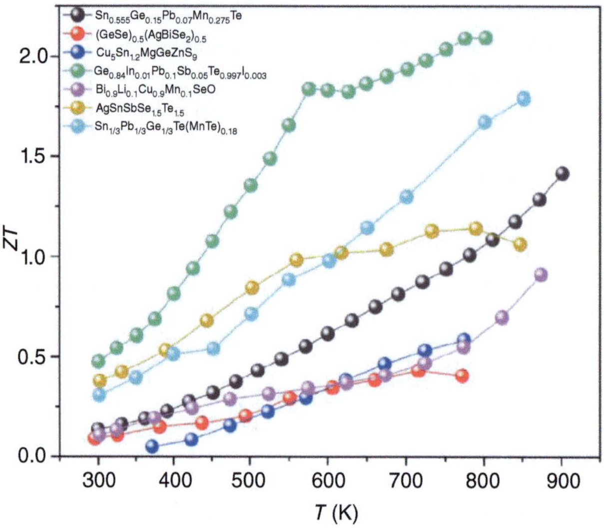


**Figure 8.8** Temperature-dependent *ZT* values of high-entropy materials reported in literatures. Source: Xiang et al. [9]/Springer Nature/CC BY-4.0.

300 K. In addition, the phonon thermal conductivity was reduced to 0.4 W/(m K). Combined with an outstanding power factor and low thermal conductivity, a maximum *ZT* value of ~2.1 at 800 K was achieved in $Ge_{0.84}In_{0.01}Pb_{0.1}Sb_{0.05}Te_{0.997}I_{0.003}$.

## 8.5 Electromagnetic Wave (EMW) Absorption

High-performance electromagnetic wave (EMW) absorption materials are in high demand due to the increase in electromagnetic interference and pollution or the need for electromagnetic stealth in defense. It has come to light that those materials with a good combination of both large permittivity and permeability exhibit enhanced EMW absorption. Therefore, great efforts have been made to develop composites of conductive and magnetic components. However, manufacturing composites with a uniform dispersion of different nanometer-sized phases and well-controlled interfaces is challenging. In addition, the stability problem is another issue that limits the application of EMW absorption composites to high temperatures. Thus, searching for single-phase materials with both high permittivity and permeability is appealing. Zhou et al. have done some pioneer work on developing HECs for EMW absorptions [40–46]. They reported that the electromagnetic parameters, i.e. complex permittivity and permeability, can be tuned by composition design, which may realize enhanced impedance matching and good EMW absorption by coupling dielectric/magnetic loss. Very impressive EMW absorption results were obtained in high-entropy diborides and carbides with the addition of Cr [42, 46], where effective absorption bandwidth (EAB) >10 GHz was achieved.

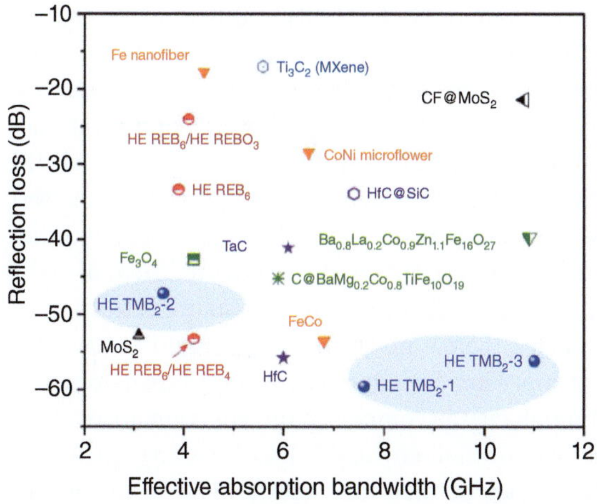


**Figure 8.9** Comparison of optimal optimized EAB and $RL_{min}$ values of high-entropy materials and other EMW absorbers. Source: Zhang et al. [42]/Springer Nature/CC BY-4.0.

Zhang et al. [41] compared EAB and the minimum reflection loss ($RL_{min}$) of HEMs and composites reported in literatures. As shown in Figure 8.9, high-entropy $(Cr_{0.2}Zr_{0.2}Hf_{0.2}Nb_{0.2}Ta_{0.2})B_2$ exhibit a good combination of EMW absorption properties with $RL_{min}$ of −56.2 dB (at 8.48 GHz with film thickness of 2.63 mm) and the EAB of 11.0 GHz (film thickness of 2.2 mm), which is superior to most other EMW-absorbing materials. Later, they also found that adding Cr into high-entropy carbides can achieve high-performance EMW absorption. In contrast, the addition of Ti would degenerate EMW absorption properties. The enhanced EMW absorption by introducing Cr is caused by the additional magnetic loss effect, where an extra magnetic loss peak around 10 GHz was observed both in high-entropy $(Cr_{0.2}Zr_{0.2}Hf_{0.2}Nb_{0.2}Ta_{0.2})B_2$ and $Cr_{0.2}Zr_{0.2}Hf_{0.2}Nb_{0.2}Ta_{0.2}C$ [42, 46]. Moreover, first-principles calculations have shown that the ground magnetic state of $CrB_2$ is ferro-magnetic. Compared to composites made of nano-materials, where specialize nanostructures (e.g. core–shell structure and nanowire) are designed, HECs are much easier to be fabricated and with much better high temperature stability and oxidation resistance. Thus, HECs may be promising candidates for high-temperature EMW absorption materials. In the future, their EMW absorption properties at elevated temperatures need to be measured.

## 8.6 Rechargeable Batteries

Similar to other fields, HEMs have also raised much attention for energy-related applications, especially for rechargeable batteries, e.g. Li-ion batteries. An interesting application of HEMs in Li-ion batteries is the rock-salt structure high-entropy oxides, which are believed as a promising candidate for solid-state electrolytes, anodes, and cathodes.

Electrolytes in Li-ion batteries are composed of volatile and flammable organic solvents, corrosive lithium salts, and additives. They are the most vulnerable parts in the battery, which usually cause safety problems. Developing solid-state electrolytes is believed the most promising method to solve this safety issue. Ionic conductivity is one of the key properties of solid-state electrolytes. Bérardan et al. [5] reported that rock-salt high-entropy $(MgCoNiCuZn)_{1-x}A_xO$ (with A = Li, Na) exhibit super-ionic conductivity. For example, $Li^+$ conductivity increases with the amount of $Li^+$ ions for $x$ in the range of 0.02–0.33 and reaches $10^{-3}$ S/cm at 20 °C with 30% $Li^+$ ions, which is approaching the best value of solid-state electrolytes. In rock-salt structure high-entropy oxides, there is no layered or channel-like ion pathway. Therefore, the mechanism of super-ionic conductivity is contributed from oxygen vacancies that are formed through charge compensation. For comparison, they also measured the ionic conductivity of high-entropy $(MgCoNiCuZn)_{1-2x}Ga_xLi_xO$ and $(MgCoNiCuZn)_{1-x}K_xO$, where both of them are nearly ionic insulators. In $(MgCoNiCuZn)_{1-2x}Ga_xLi_xO$, with Li and Ga co-doping, no oxygen vacancy is necessary for charge compensation. Therefore, no super-ionic conductivity can be observed. In $(MgCoNiCuZn)_{1-x}K_xO$, the $K^+$ might be too large to transfer through oxygen vacancy. These results confirm that super-ionic conductivity of high-entropy $(MgCoNiCuZn)_{1-x}A_xO$ (with A = Li, Na) results from the transition of $Li^+$ or $Na^+$ from oxygen vacancies.

State-of-the-art anode material in Li-ion batteries is graphite, which exhibits a good combination of merits. However, the limited capacity (~372 mAh/g) of graphite anode has significantly impeded the development of Li-ion batteries with high specific energy. Transition metal oxides exhibit higher specific capacities, which are investigated to replace graphite. However, transition metal oxides usually suffer from poor structure stability. Sarkar [47] proposed that the concept of high-entropy may help to solve the stability problem of transition metal oxide anodes. They developed high-entropy oxides under the Mg–Co–Ni–Cu–Zn–O rock-salt oxide family. They found that the stabilization effect of entropy brings significant benefits for the storage capacity retention of high-entropy oxides and greatly improves the cycling stability. Additionally, it is observed that the electrochemical behavior of the high-entropy oxides depends on each of the metal cations present (as shown in Figure 8.10), thus providing the opportunity to tailor the electrochemical properties by simply changing the elemental composition. In a later work, Qiu et al. [48] synthesized nano-particles of $(Mg_{0.2}Co_{0.2}Ni_{0.2}Cu_{0.2}Zn_{0.2})O$, which delivers a high reversible capacity of 920 mAh/g after 300 cycles.

Rock-salt high-entropy $(Co_{0.2}Cu_{0.2}Mg_{0.2}Ni_{0.2}Zn_{0.2})O$ can not only be used as solid-state electrolyte and the precursor of anode in Li-ion batteries, but it can also be used as a cathode in Li-ion batteries. Wang et al. [49] designed high-entropy $Li_x(Co_{0.2}Cu_{0.2}Mg_{0.2}Ni_{0.2}Zn_{0.2})OF_x$, which exhibits a working potential of 3.4 V versus $Li^+/Li$, thus enabling its use as a cathode active material. They compared LiNiOF and $Li_x(Co_{0.2}Cu_{0.2}Mg_{0.2}Ni_{0.2}Zn_{0.2})OF_x$ and found that $Li_x(Co_{0.2}Cu_{0.2}Mg_{0.2}Ni_{0.2}Zn_{0.2})OF_x$ has higher specific capacity and better columbic efficiency. Further, Lun et al. [50] performed a systematic comparison on a group of rock-salt structure high-entropy cathodes $Li_{1.3}TM_xO_{1.7}F_{0.3}$ containing two (TM2), four (TM4), or six (TM6)

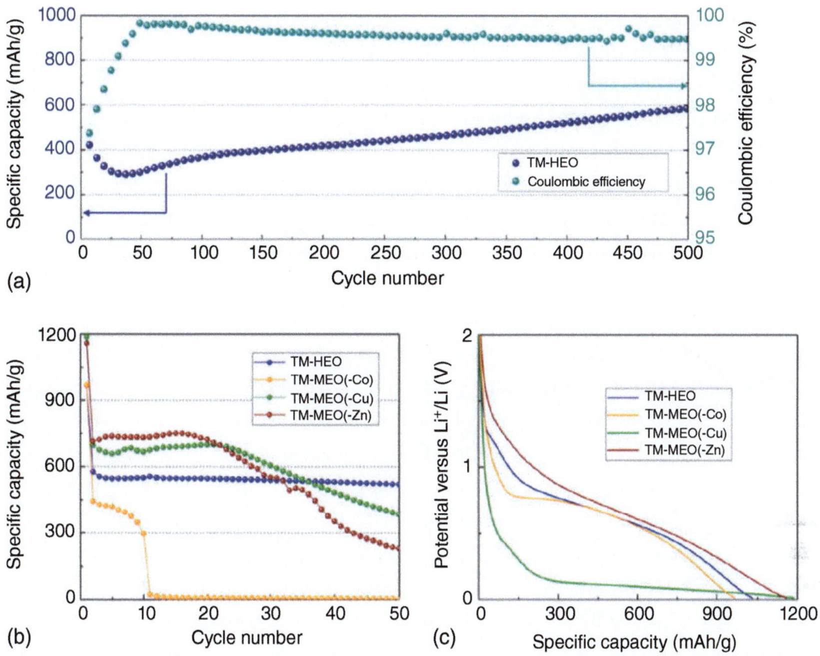


**Figure 8.10** Specific capacities and lithiation profiles of TM-HEO and TM-medium-entropy oxide (MEO). Comparison of the different medium and high-entropy oxides under investigation. (a) Long-term cycling stability of the calcined TM-HEO at 200 mA/g together with the corresponding Coulombic efficiency. (b) The TM-HEO shows stable capacity retention at 50 mA/g, while the materials without Zn and Cu reveal severe capacity degradation. The material without Co fails completely after 10 cycles. (c) Discharge (lithiation) profiles of the first cycle for the different compounds. The material without Cu shows a significantly lower discharge potential, and it might be interesting as an anode material for primary batteries. Source: Sarkar [47]/Springer Nature/CC BY-4.0.

transition metals. When more cation species are mixed, the energy density and rate capability are systematically improved. When cycled between 1.5 and 4.7 V at the rate of 20 mA/g, TM2 delivers a capacity (specific energy) of 220 mAh/g (704 Wh/kg) and TM4 delivers 269 mAh/g (849 Wh/kg), which further increases to 307 mAh/g (955 Wh/kg) for TM6 (Figure 8.11a–c). With the increase in the TM variety, the rate performance will also be improved, as shown in Figure 8.11d–f. For TM2, the discharge capacity decreases from 220 mAh/g when cycled at 20 mA/g to 58 mAh/g at 2000 mA/g, corresponding to a 74% capacity loss at the high rate. For a similar rate increase, the capacity loss is reduced to 58% for TM4 and to 45% for TM6. The rate capability of the TM6 compound is remarkable, with a discharge capacity of more than 170 mAh/g delivered at the very high cycling rate of 2000 mA/g, corresponding to a 7-minute (dis)charge process.

Except for rock-salt structure high-entropy oxides, spinel-type high-entropy oxides, high-entropy perovskite oxides, and high-entropy sulfides, etc. have also been designed for applications in Li-ion batteries [51–55]. As illustrated, HEMs can be applied in solid-state electrolytes, anodes, and cathodes in rechargeable batteries,

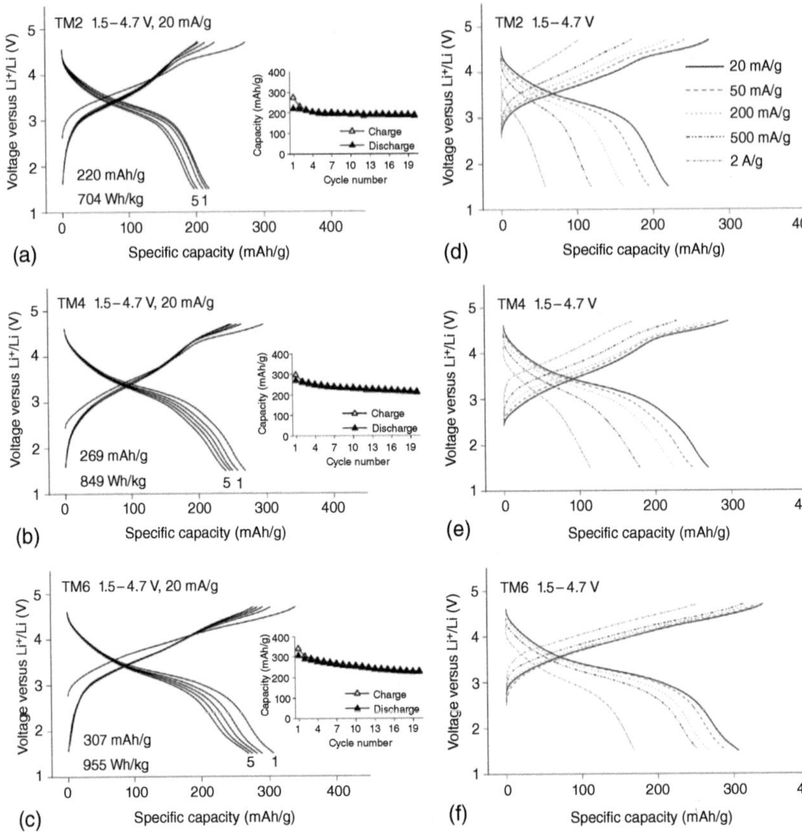


**Figure 8.11** Electrochemical performance of the three compounds. (a)–(c) Voltage profiles of TM2 (a), TM4 (b), and TM6 (c) within the voltage window of 1.5–4.7V at 20 mA/g and room temperature. The corresponding capacity retention plots are shown as insets. (d)–(f) Rate capability of TM2 (d), TM4 (e), and TM6 (f) showing the first-cycle voltage profiles of all three compounds when cycled between 1.5 and 4.7V at 20, 50, 200, 500, and 2000 mA/g. The legend in d also applies to (e) and (f). Source: Lun et al. [50]/Springer Nature.

and related works are emerging. Though HEMs are in their nascent beginning in rechargeable battery applications, the vast combination of elements gives rise to tailorable properties, affording fresh opportunities into the pursuit of high-performance batteries.

## 8.7 Other Applications

In principle, the idea of high entropy, which is the extension of alloying or doping, can be applied in any solid-state material. Thus, any solid-state material may be redesigned by employing the high-entropy concept. Except for applications listed earlier, there are still enormous candidate fields in which HEMs may be applied.

In microelectronics, high-entropy films have been designed as novel buffer layers to prevent the interdiffusion and reaction between Cu and Si. Jiang et al. [56] reported a bilayer AlCrTaTiZr/AlCrTaTiZr-N high-entropy film for a Cu/Si connect system, where no Cu–Si intermetallic compound formed after annealing at 600–900 °C. They suggested that the amorphous structure of the AlCrTaTiZr layer had a lower driving force to form intermetallic compounds. Moreover, the severe lattice distortion and the reduced diffusion kinetics were the factors for the AlCrTaTiZr/AlCrTaTiZr-N films to act as a very effective barrier material for inhibiting the diffusion of Cu.

In supercapacitors, high-entropy composites can be used as electrodes. Xu et al. [57] uniformly deposited quinary HEA-NPs (high-entropy alloy nano-particles) on super aligned electro-spun carbon nanofibers (ACNFs) and adopted as electrodes of supercapacitors. They found that the FeNiCoMnMg HEA-NPs/ACNFs electrodes with 5 mM precursor concentration show a high capacitance of 203 F/g and a specific energy density of 21.7 Wh/kg, thus indicating the promising potential for energy storage applications.

In biomaterials, HEMs with improved biocompatibility, wear, and corrosion resistance are explored. Li et al. [58] fabricated the high-entropy $Ca_{20}Mg_{20}Zn_{20}Sr_{20}Yb_{20}$ bulk metallic glass with enhanced mechanical properties and corrosion resistance to replace the $Ca_{65}Mg_{15}Zn_{20}$ bulk metallic glass. The *in vitro* tests showed that the high-entropy $Ca_{20}Mg_{20}Zn_{20}Sr_{20}Yb_{20}$ bulk metallic glass could stimulate the proliferation and differentiation of cultured osteoblasts. *in vivo* animal tests showed that the high-entropy $Ca_{20}Mg_{20}Zn_{20}Sr_{20}Yb_{20}$ bulk metallic glass did not show any obvious degradation after four weeks of implantation, and they can promote osteogenesis and new bone formation after two weeks of implantation.

In catalysis, the typical requirements of catalysts include high activity and selectivity, good stability, long lifetime, and low cost. HEMs provide new remedy to develop high-performance catalysts. The multi-principal elements with distinct sizes and chemical properties guarantee that the property of particle surface varies from regions to regions. In addition, the entropy effect suppresses the tendency of enthalpy minimization, thereby indicating high-energy state of surface regions. Thus, HEMs may be more active with a relatively lower reaction barrier due to the high energy surface. In addition, high-entropy effect may improve the stability of catalysts, and the cost can be controlled by element design. HEMs have been designed as catalysts for ammonia and carbon monoxide oxidation, water splitting, etc. As suggested by Sun and Dai [11], tuning composition and configuration of HEMs introduce untapped opportunities for accessing better catalysts and resolving issues that are considered challenging in conventional, simple systems. Readers who are interested in opportunities of HEMs in catalytic applications can refer to their review paper.

## 8.8 Summary and Perspectives

In this chapter, the possible applications of HEMs are briefly introduced, including structural applications, thermal protection and management, and thermoelectricity.

In any of the discussed field, HEMs show promising combination of good properties that are necessary in the corresponding field. For example, in T/EBCs, high-entropy oxides exhibit lower thermal conductivity, slower grain growth rate, better corrosion resistance, etc.

The concept of HEMs was first proposed in metallic materials (HEAs) in 2004 and extended to ceramic materials (HECs) in 2015. The research on HEMs is still in the nascent beginning, especially for HECs. In real-life applications, a material that can be used needs to meet a series of properties, e.g. cost, mechanical, physical, and chemical, thereby making the material design a multi-objective optimization problem. Thus, pushing a new material to market needs a long period. Even though many HEMs with intriguing properties have been reported, no HEM has been applied in real-life thus far. Nevertheless, the multi-element nature of HEMs gives rise to tailorable and unpresented properties, thereby making HEMs promising candidates in many industrial applications. It can be expected that more and more works on HEMs will emerge and be applied in various regions.

## References


1 Gludovatz, B., Hohenwarter, A., Catoor, D. et al. (2014). A fracture-resistant high-entropy alloy for cryogenic applications. *Science* 345: 1153–1158. https://doi.org/10.1126/science.1254581.

2 Senkov, O.N., Wilks, G.B., Scott, J.M., and Miracle, D.B. (2011). Mechanical properties of $Nb_{25}Mo_{25}Ta_{25}W_{25}$ and $V_{20}Nb_{20}Mo_{20}Ta_{20}W_{20}$ refractory high entropy alloys. *Intermetallics* 19: 698–706. https://doi.org/10.1016/j.intermet.2011.01.004.

3 Zhang, Y., Stocks, G.M., Jin, K. et al. (2015). Influence of chemical disorder on energy dissipation and defect evolution in concentrated solid solution alloys. *Nat. Commun.* 6: 8736. https://doi.org/10.1038/ncomms9736.

4 Bérardan, D., Franger, S., Dragoe, D. et al. (2016). Colossal dielectric constant in high entropy oxides. *Phys. Status Solidi RRL* 10: 328–333. https://doi.org/10.1002/pssr.201600043.

5 Bérardan, D., Franger, S., Meena, A.K., and Dragoe, N. (2016). Room temperature lithium superionic conductivity in high entropy oxides. *J. Mater. Chem. A.* 4: 9536–9541. https://doi.org/10.1039/C6TA03249D.

6 Braun, J.L., Rost, C.M., Lim, M. et al. (2018). Charge-induced disorder controls the thermal conductivity of entropy-stabilized oxides. *Adv. Mater.* 30: 1805004. https://doi.org/10.1002/adma.201805004.

7 Han, L., Rao, Z., Filho, I.R.S. et al. (2021). Ultrastrong and ductile soft magnetic high-entropy alloys via coherent ordered nanoprecipitates. *Adv. Mater.* 9: 2102139.

8 Yeh, J.-W. and Lin, S.-J. (2018). Breakthrough applications of high-entropy materials. *J. Mater. Res.* 33: 3129–3137. https://doi.org/10.1557/jmr.2018.283.

9 Xiang, H., Xing, Y., Dai, F. et al. (2021). High-entropy ceramics: present status, challenges, and a look forward. *J. Adv. Ceram.* 10: 385–441. https://doi.org/10.1007/s40145-021-0477-y.

**10** Chen, Y., Fu, H., Huang, Y. et al. (2021). Opportunities for high-entropy materials in rechargeable batteries. *ACS Mater. Lett.* 3: 160–170. https://doi.org/10.1021/acsmaterialslett.0c00484.

**11** Sun, Y. and Dai, S. (2021). High-entropy materials for catalysis: a new frontier. *Sci. Adv.* 7: eabg1600. https://doi.org/10.1126/sciadv.abg1600.

**12** Lee, J., Park, H., Kim, M. et al. (2021). Role of hydrogen and temperature in hydrogen embrittlement of equimolar CoCrFeMnNi high-entropy alloy. *Met. Mater. Int.* 27: 166–174. https://doi.org/10.1007/s12540-020-00752-3.

**13** Senkov, O.N., Jensen, J.K., Pilchak, A.L. et al. (2018). Compositional variation effects on the microstructure and properties of a refractory high-entropy superalloy $AlMo_{0.5}NbTa_{0.5}TiZr$. *Mater. Des.* 139: 498–511. https://doi.org/10.1016/j.matdes.2017.11.033.

**14** Gild, J., Samiee, M., Braun, J.L. et al. (2018). High-entropy fluorite oxides. *J. Eur. Ceram. Soc.* 38: 3578–3584. https://doi.org/10.1016/j.jeurceramsoc.2018.04.010.

**15** Zhao, Z., Xiang, H., Dai, F.-Z. et al. (2019). $(La_{0.2}Ce_{0.2}Nd_{0.2}Sm_{0.2}Eu_{0.2})_2Zr_2O_7$: a novel high-entropy ceramic with low thermal conductivity and sluggish grain growth rate. *J. Mater. Sci. Technol.* https://doi.org/10.1016/j.jmst.2019.05.054.

**16** Zhao, Z., Chen, H., Xiang, H. et al. (2020). High entropy defective fluorite structured rare-earth niobates and tantalates for thermal barrier applications. *J. Adv. Ceram.* 9: 303–311. https://doi.org/10.1007/s40145-020-0368-7.

**17** Zhao, Z., Chen, H., Xiang, H. et al. (2020). $(Y_{0.25}Yb_{0.25}Er_{0.25}Lu_{0.25})_2(Zr_{0.5}Hf_{0.5})_2O_7$: a defective fluorite structured high entropy ceramic with low thermal conductivity and close thermal expansion coefficient to $Al_2O_3$. *J. Mater. Sci. Technol.* 39: 167–172. https://doi.org/10.1016/j.jmst.2019.08.018.

**18** Zhou, L., Li, F., Liu, J.-X. et al. (2020). High-entropy thermal barrier coating of rare-earth zirconate: a case study on $(La_{0.2}Nd_{0.2}Sm_{0.2}Eu_{0.2}Gd_{0.2})_2Zr_2O_7$ prepared by atmospheric plasma spraying. *J. Eur. Ceram. Soc.* 40: 5731–5739. https://doi.org/10.1016/j.jeurceramsoc.2020.07.061.

**19** Li, F., Zhou, L., Liu, J.-X. et al. (2019). High-entropy pyrochlores with low thermal conductivity for thermal barrier coating materials. *J. Adv. Ceram.* 8: 576–582. https://doi.org/10.1007/s40145-019-0342-4.

**20** Xue, Y., Zhao, X., An, Y. et al. (2022). High-entropy $(La_{0.2}Nd_{0.2}Sm_{0.2}Eu_{0.2}Gd_{0.2})_2Ce_2O_7$: a potential thermal barrier material with improved thermo-physical properties. *J. Adv. Ceram.* https://doi.org/10.1007/s40145-021-0563-1.

**21** Mao, H.-R., Guo, R.-F., Cao, Y. et al. (2021). Ultrafast densification of high-entropy oxide $(La_{0.2}Nd_{0.2}Sm_{0.2}Eu_{0.2}Gd_{0.2})_2Zr_2O_7$ by reactive flash sintering. *J. Eur. Ceram. Soc.* 41: 2855–2860. https://doi.org/10.1016/j.jeurceramsoc.2020.11.052.

**22** Ren, K., Wang, Q., Shao, G. et al. (2020). Multicomponent high-entropy zirconates with comprehensive properties for advanced thermal barrier coating. *Scr. Mater.* 178: 382–386. https://doi.org/10.1016/j.scriptamat.2019.12.006.

**23** Liu, Y., Jia, D., Zhou, Y. et al. (2020). $Zn_{0.1}Ca_{0.1}Sr_{0.4}Ba_{0.4}ZrO_3$: a non-equimolar multicomponent perovskite ceramic with low thermal conductivity. *J. Eur. Ceram. Soc.* 40: 6272–6277. https://doi.org/10.1016/j.jeurceramsoc.2020.07.054.

24 Zhao, Z., Chen, H., Xiang, H. et al. (2020). High-entropy $(Y_{0.2}Nd_{0.2}Sm_{0.2}Eu_{0.2}Er_{0.2})AlO_3$: a promising thermal/environmental barrier material for oxide/oxide composites. *J. Mater. Sci. Technol.* 47: 45–51. https://doi.org/10.1016/j.jmst.2020.02.011.

25 Chen, H., Zhao, Z., Xiang, H. et al. (2020). High entropy $(Y_{0.2}Yb_{0.2}Lu_{0.2}Eu_{0.2}Er_{0.2})_3Al_5O_{12}$: a novel high temperature stable thermal barrier material. *J. Mater. Sci. Technol.* 48: 57–62. https://doi.org/10.1016/j.jmst.2020.01.056.

26 Zhao, Z., Xiang, H., Chen, H. et al. (2020). High-entropy $(Nd_{0.2}Sm_{0.2}Eu_{0.2}Y_{0.2}Yb_{0.2})_4Al_2O_9$ with good high temperature stability, low thermal conductivity, and anisotropic thermal expansivity. *J. Adv. Ceram.* 9: 595–605. https://doi.org/10.1007/s40145-020-0399-0.

27 Ren, X., Tian, Z., Zhang, J., and Wang, J. (2019). Equiatomic quaternary $(Y_{1/4}Ho_{1/4}Er_{1/4}Yb_{1/4})_2SiO_5$ silicate: a perspective multifunctional thermal and environmental barrier coating material. *Scr. Mater.* 168: 47–50. https://doi.org/10.1016/j.scriptamat.2019.04.018.

28 Wang, X., Cheng, M., Xiao, G. et al. (2021). Preparation and corrosion resistance of high-entropy disilicate $(Y_{0.25}Yb_{0.25}Er_{0.25}Sc_{0.25})_2Si_2O_7$ ceramics. *Corros. Sci.* 192: 109786. https://doi.org/10.1016/j.corsci.2021.109786.

29 Sun, L., Luo, Y., Tian, Z. et al. (2020). High temperature corrosion of $(Er_{0.25}Tm_{0.25}Yb_{0.25}Lu_{0.25})_2Si_2O_7$ environmental barrier coating material subjected to water vapor and molten calcium–magnesium–aluminosilicate (CMAS). *Corros. Sci.* 175: 108881. https://doi.org/10.1016/j.corsci.2020.108881.

30 Li, C., Meng, B., Fan, S. et al. (2021). Rare-earth-tantalate high-entropy ceramics with sluggish grain growth and low thermal conductivity. *Ceram. Int.* https://doi.org/10.1016/j.ceramint.2021.12.332.

31 Zhao, Z., Xiang, H., Dai, F.-Z. et al. (2019). $(TiZrHf)P_2O_7$: an equimolar multicomponent or high entropy ceramic with good thermal stability and low thermal conductivity. *J. Mater. Sci. Technol.* 35: 2227–2231. https://doi.org/10.1016/j.jmst.2019.05.030.

32 Zhao, Z., Chen, H., Xiang, H. et al. (2019). $(La_{0.2}Ce_{0.2}Nd_{0.2}Sm_{0.2}Eu_{0.2})PO_4$: a high-entropy rare-earth phosphate monazite ceramic with low thermal conductivity and good compatibility with $Al_2O_3$. *J. Mater. Sci. Technol.* 35: 2892–2896. https://doi.org/10.1016/j.jmst.2019.08.012.

33 Wang, J., Wu, F., Zou, R. et al. (2021). High-entropy ferroelastic rare-earth tantalite ceramic: $(Y_{0.2}Ce_{0.2}Sm_{0.2}Gd_{0.2}Dy_{0.2})TaO_4$. *J. Am. Ceram. Soc.* 104: 5873–5882. https://doi.org/10.1111/jace.17932.

34 Chen, H., Xiang, H., Dai, F.-Z. et al. (2019). Porous high entropy $(Zr_{0.2}Hf_{0.2}Ti_{0.2}Nb_{0.2}Ta_{0.2})B_2$: a novel strategy towards making ultrahigh temperature ceramics thermal insulating. *J. Mater. Sci. Technol.* 35: 2404–2408. https://doi.org/10.1016/j.jmst.2019.05.059.

35 Chen, H., Xiang, H., Dai, F.-Z. et al. (2019). High porosity and low thermal conductivity high entropy $(Zr_{0.2}Hf_{0.2}Ti_{0.2}Nb_{0.2}Ta_{0.2})C$. *J. Mater. Sci. Technol.* 35: 1700–1705. https://doi.org/10.1016/j.jmst.2019.04.006.

36 Chen, H., Zhao, Z., Xiang, H. et al. (2020). Effect of reaction routes on the porosity and permeability of porous high entropy $(Y_{0.2}Yb_{0.2}Sm_{0.2}Nd_{0.2}Eu_{0.2})B_6$ for

transpiration cooling. *J. Mater. Sci. Technol.* 38: 80–85. https://doi.org/10.1016/j.jmst.2019.09.006.

**37** Yang, R.W., Liang, Y.P., Xu, J. et al. (2022). Rare-earth-niobate high-entropy ceramic foams with enhanced thermal insulation performance. *J. Mater. Sci. Technol.* 116: 94–102. https://doi.org/10.1016/j.jmst.2021.10.050.

**38** Roychowdhury, S., Ghosh, T., Arora, R. et al. (2018). Stabilizing n-type cubic GeSe by entropy-driven alloying of $AgBiSe_2$: ultralow thermal conductivity and promising thermoelectric performance. *Angew. Chem. Int. Ed.* 57: 15167–15171. https://doi.org/10.1002/anie.201809841.

**39** Qiu, Y., Jin, Y., Wang, D. et al. (2019). Realizing high thermoelectric performance in GeTe through decreasing the phase transition temperature *via* entropy engineering. *J. Mater. Chem. A.* 7: 26393–26401. https://doi.org/10.1039/C9TA10963C.

**40** Zhou, Y., Zhao, B., Chen, H. et al. (2021). Electromagnetic wave absorbing properties of TMCs (TM = Ti, Zr, Hf, Nb and Ta) and high entropy ($Ti_{0.2}Zr_{0.2}Hf_{0.2}Nb_{0.2}Ta_{0.2}$)C. *J. Mater. Sci. Technol.* 74: 105–118. https://doi.org/10.1016/j.jmst.2020.10.016.

**41** Zhang, W., Zhao, B., Xiang, H. et al. (2021). One-step synthesis and electromagnetic absorption properties of high entropy rare earth hexaborides (HE $REB_6$) and high entropy rare earth hexaborides/borates (HE $REB_6$/HE $REBO_3$) composite powders. *J. Adv. Ceram.* 10: 62–77. https://doi.org/10.1007/s40145-020-0417-2.

**42** Zhang, W., Dai, F.-Z., Xiang, H. et al. (2021). Enabling highly efficient and broadband electromagnetic wave absorption by tuning impedance match in high-entropy transition metal diborides (HE $TMB_2$). *J. Adv. Ceram.* 10: 1299–1316. https://doi.org/10.1007/s40145-021-0505-y.

**43** Zhang, W., Zhao, B., Ni, N. et al. (2021). High entropy rare earth hexaborides/tetraborides (HE $REB_6$/HE $REB_4$) composite powders with enhanced electromagnetic wave absorption performance. *J. Mater. Sci. Technol.* 87: 155–166. https://doi.org/10.1016/j.jmst.2021.01.059.

**44** Chen, H., Zhao, B., Zhao, Z. et al. (2020). Achieving strong microwave absorption capability and wide absorption bandwidth through a combination of high entropy rare earth silicide carbides/rare earth oxides. *J. Mater. Sci. Technol.* 47: 216–222. https://doi.org/10.1016/j.jmst.2020.02.015.

**45** Zhang, H., Zhao, B., Xiang, H. et al. (2021). Electromagnetic wave absorbing properties of $Cr_2AlB_2$ powders and the effect of high-temperature oxidation. *J. Am. Ceram. Soc.* 104: 2213–2224. https://doi.org/10.1111/jace.17601.

**46** Zhang, W., Xiang, H., Dai, F.-Z. et al. (2022). Achieving ultra-broadband electromagnetic wave absorption in high-entropy transition metal carbides (HE TMCs). *J. Adv. Ceram.* 11: 545–555. https://doi.org/10.1007/s40145-021-0554-2.

**47** Sarkar, A. (2018). High entropy oxides for reversible energy storage. *Nat. Commun.* 9: 3400.

**48** Qiu, N., Chen, H., Yang, Z. et al. (2019). A high entropy oxide ($Mg_{0.2}Co_{0.2}Ni_{0.2}Cu_{0.2}Zn_{0.2}O$) with superior lithium storage performance. *J. Alloys Compd.* 777: 767–774. https://doi.org/10.1016/j.jallcom.2018.11.049.

**49** Wang, Q., Sarkar, A., Wang, D. et al. (2019). Multi-anionic and -cationic compounds: new high entropy materials for advanced Li-ion batteries. *Energy Environ. Sci.* 12: 2433–2442. https://doi.org/10.1039/C9EE00368A.

**50** Lun, Z., Ouyang, B., Kwon, D.-H. et al. (2020). Cation-disordered rocksalt-type high-entropy cathodes for Li-ion batteries. *Nat. Mater.* https://doi.org/10.1038/s41563-020-00816-0.

**51** Lin, L., Wang, K., Sarkar, A. et al. (2022). High-entropy sulfides as electrode materials for Li-ion batteries. *Adv. Energy Mater.* 12: 2103090. https://doi.org/10.1002/aenm.202103090.

**52** Wang, D., Jiang, S., Duan, C. et al. (2020). Spinel-structured high entropy oxide $(FeCoNiCrMn)_3O_4$ as anode towards superior lithium storage performance. *J. Alloys Compd.* 844: 156158. https://doi.org/10.1016/j.jallcom.2020.156158.

**53** Chen, H., Qiu, N., Wu, B. et al. (2020). A new spinel high-entropy oxide $(Mg_{0.2}Ti_{0.2}Zn_{0.2}Cu_{0.2}Fe_{0.2})_3O_4$ with fast reaction kinetics and excellent stability as an anode material for lithium ion batteries. *RSC Adv.* 10: 9736–9744. https://doi.org/10.1039/D0RA00255K.

**54** Chen, T.-Y., Wang, S.-Y., Kuo, C.-H. et al. (2020). *In operando* synchrotron X-ray studies of a novel spinel $(Ni_{0.2}Co_{0.2}Mn_{0.2}Fe_{0.2}Ti_{0.2})_3O_4$ high-entropy oxide for energy storage applications. *J. Mater. Chem. A* 8: 21756–21770. https://doi.org/10.1039/D0TA06455F.

**55** Yan, J., Wang, D., Zhang, X. et al. (2020). A high-entropy perovskite titanate lithium-ion battery anode. *J. Mater. Sci.* 55: 6942–6951. https://doi.org/10.1007/s10853-020-04482-0.

**56** Jiang, C., Li, R., Wang, X. et al. (2020). Diffusion barrier performance of AlCrTaTiZr/AlCrTaTiZr-N high-entropy alloy films for Cu/Si connect system. *Entropy* 22: 234. https://doi.org/10.3390/e22020234.

**57** Xu, X., Du, Y., Wang, C. et al. (2020). High-entropy alloy nanoparticles on aligned electronspun carbon nanofibers for supercapacitors. *J. Alloys Compd.* 822: 153642. https://doi.org/10.1016/j.jallcom.2020.153642.

**58** Li, H.F., Xie, X.H., Zhao, K. et al. (2013). in vitro and in vivo studies on biodegradable CaMgZnSrYb high-entropy bulk metallic glass. *Acta Biomater.* 9: 8561–8573. https://doi.org/10.1016/j.actbio.2013.01.029.

# 9

# Challenges and Future Directions of High-Entropy Materials

## 9.1 Introduction

In previous chapters, we have systematically reviewed concepts related to high-entropy materials (HEMs), including theoretical aspects, perspective applications, and various properties. In this chapter, challenges and future directions of the investigation on HEMs will be discussed. The challenges and future directions for specific HEM sets have been well discussed in some review papers [1–6]. Readers who are interested in related topics can refer to these papers.

## 9.2 Vastness of Tunable Elements, Microstructures, and Properties

The first idea that makes HEMs popular is the vastness of element combinations in the new, unexplored central region of phase diagrams, which may guarantee new materials with unprecedented properties and physics that cannot be achieved in conventional materials. The vastness of composition, on the one hand, makes HEMs the most compelling field of research due to possibility of finding new materials; on the other hand, it raises the challenges of dimensional explosion, which hinders the efficiency of exploration. As shown by Miracle and Senkov [2], 37 metal elements yielded 435 897 possible five-element combinations and a total of 2 834 496 alloy systems with three to six elements. If 72 metal elements are considered, the number of five-element systems expands to 13 991 544, and the number of systems with three to six elements explodes to 171 318 882. In the estimation, the variation in composition has not been considered, which will further substantially increase the challenge of exploration. In addition, the application of a material does not solely depend on the composition. The vastness has now grown to include three dimensions: compositional complexity, microstructural complexity, and properties complexity [2]. At the beginning of the research on HEMs, attention was usually paid to develop new materials with single-phase solid-solute state, such as high-entropy alloys (HEAs) or high-entropy ceramics (HECs). Then, motivated by tailoring of

*High-Entropy Materials: From Basics to Applications,* First Edition.
Huimin Xiang, Fu-Zhi Dai, and Yanchun Zhou.

properties, microstructurally complex HEMs were developed, e.g. multiphase microstructures with two or more phases, including solid solutions, intermetallic compounds, and/or ceramics phases. The microstructural complexity is particularly important in alloys and other structural materials and can be controlled by composition selection, heat treatment, mechanical processing, etc. Furthermore, in real applications, a material needs to meet multiple properties, e.g. strength, plasticity, creep resistance, oxidation resistance, and thermal conductivity. However, it is hard to improve different properties simultaneously. For example, it is well known that there is usually a dilemma between strength and plasticity, where improving one usually sacrifices the other. To push HEMs to industrial applications, challenges originating from the complexities in composition screening, microstructure design, and tailoring of properties should be resolved.

## 9.3 Preparation, Characterization, and Modeling

The properties of a material depend both on its composition and microstructure. The distribution of elements, the charge state of elements, and the detailed microstructure are essential to the understanding of materials' properties. For example, it has been reported that yield strength of an HEA can be significantly increased by short-range orders [7]. Grain boundary segregation and spinodal-like decomposition may lead to the loss of ductility in HEAs [8]. Ordered oxygen complexes can enhance both the strength and ductility of HfNbTiZr HEA, thus breaking the long-standing strength–ductility trade-off [9]. A novel crystal-glass high-entropy nanocomposite can be developed by doping glass-forming elements Si and B into CoCrFeNi HEA, taking advantage of the low stacking fault energy (SFE) of the HEA and the compliant nature of the metallic glass phase at the nanoscale, which results in a yield strength approaching the theoretical shear strength limit [10]. Eutectic HEAs with a dual-phase heterogeneous lamella microstructure, which combines the high strength of body-centered cubic (BCC) phase and the high ductility of face-centered cubic (FCC) phase, lead to materials with both high strength and ductility [11]. Tailoring the charge compensation elements to tuning oxygen vacancy density can significantly alter Li-ion conductivity of high-entropy oxides [12]. Therefore, we need novel preparation methods to control the microstructure and distribution of elements, advanced characterization methods to obtain the microstructure and distribution of elements, and advanced modeling methods to explore the relation among composition, processing, microstructure, and properties.

Preparation methods are usually materials specific, for example, whether the material is metallic, ceramic, bulk, coating, or particle. Now, methods adopted to prepare HEMs are similar to conventional materials, e.g. arc melting, high-energy ball milling, pulsed laser deposition, hot pressing, and solution-based methods. In the future, highly efficient, low cost, and scalable preparation methods need to be developed.

Characterization methods include X-ray diffraction (XRD); extended X-ray absorption fine structure (EXAFS); scanning electron microscopy (SEM) usually coupled with energy dispersive spectroscopy (EDS) and electron back-scattered diffraction (EBSD); transmission electron microscopy (TEM) coupled with EDS or electron energy loss spectroscopy (EELS); and state-of-the-art TEM techniques such as aberration-corrected scanning transmission electron microscopy (AC-STEM) and atom probe tomography (APT). These techniques are still in development toward high speed, high resolution, and high dimensional detection and reconstruction.

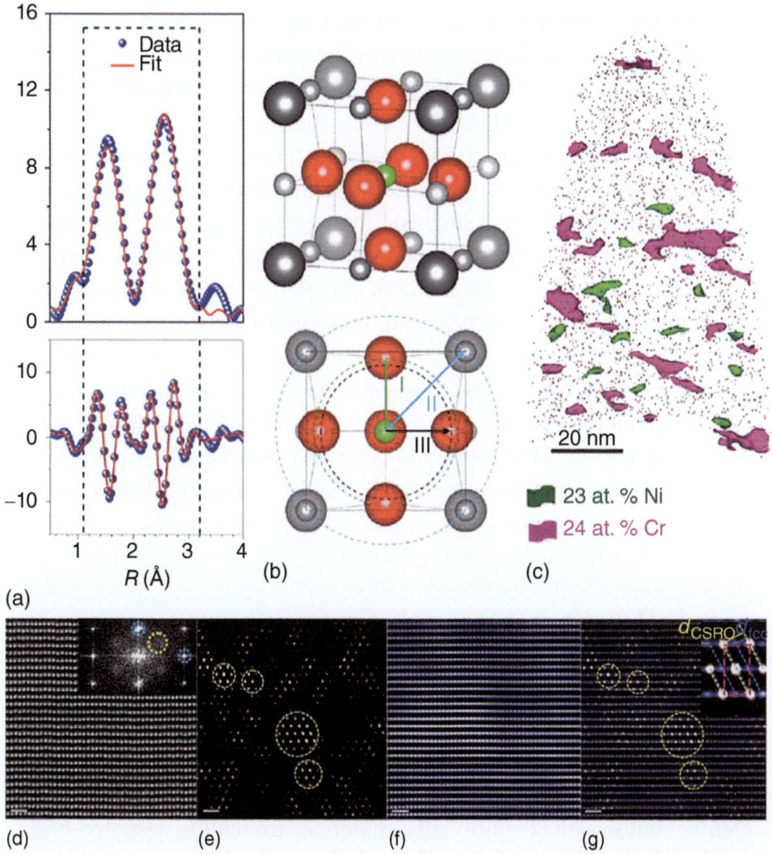


**Figure 9.1** Advanced methods for the characterization of HEMs. (a) EXAFS data and models about cobalt for $Mg_xNi_xCu_xCo_xZn_xCr_xO$ ($x = 0.1667$). Data and best-fit model are shown for the magnitude (top) and imaginary portion of the real space function $\chi(R)$ as a function of radius away from the cobalt absorber ($R$). (b) Modified local structure of $Mg_xNi_xCu_xCo_xZn_xCr_xO$ ($x = 0.1667$) from the EXAFS result in (a). (c) Atom map of a tip showing grain boundary segregation of impurities: Cr (24 at.%) and Ni (23 at.%) iso-composition surfaces viewed from the grain boundary normal. (d–g) Detection of short-range order in VCoNi by STEM. (d) Lattice image of the FCC phase with the [112] zone axis and the corresponding FFT pattern (inset). This pattern again displays the extra diffuse reflections (see, for example, inside the yellow circle) at $\{\bar{3}\,11\}/2$ positions, besides the sharper Braggs spots from the FCC phase (blue circles). (e and f) Inverse FFT image showing the chemical short-range order (CSRO) regions (several are circled) and the FCC lattice, respectively. These two images are superimposed in (g), with a close-up view of a CSRO region in the inset. Overlapping in this way produces bright sites that highlight the extra CSRO lining up on $\{\bar{3}\,11\}$ planes (yellow dashed lines). $d_{fcc}$ denotes the spacing of $\{\bar{3}\,11\}$ planes in the normal FCC lattice, whereas $d_{CSRO}$ displays the spacing corresponding to the extra chemical order. The red dashed box outlines this unit period for the local CSRO configuration. All scale bars are 0.5 nm except in (c). Source: (a–b) Reproduced with permission from [13]/John Wiley & Sons. (c) Reproduced with permission from [8]/American Association for the Advancement of Science.

Conventional methods such as XRD, SEM, and TEM are usually adopted to characterize HEMs to provide information on crystal structures, microstructures, and element distributions. Advanced methods like EXAFS, STEM, and APT can provide more detailed information. For example, the lattice distortion of HEMs is usually illustrated by analyzing EXAFS [13]. Short-range orders can be well detected by STEM by analyzing the contrast in atomic resolution high angle annular dark field (HAADF) and fast Fourier transformation (FFT) techniques [7]. Element segregations are usually analyzed by APT [8]. Figure 9.1 shows some examples of these advanced techniques applied in HEMs.

To understand the relation of "composition–microstructure–properties," modeling of HEMs need to be explored in the future, for example, modeling diffusion in HEMs to clarify whether diffusion is sluggish, modeling strength of HEMs based on the lattice distortions, modeling twinning induced plasticity (TWIP) or transformation-induced plasticity (TRIP) effects, modeling element distributions in real HEMs to understand the effect of short-range order and segregation effects, and modeling the microstructure of HEMs based on the combination of composition and processing. To develop the models, one can take the bottom-up schema, for example, simulation methods such as density functional theory (DFT)-based method, molecular dynamic (MD) simulations, or methods based on continuum theory. Alternatively, data-driven schema can be developed from machine learning (ML) methods.

## 9.4 Materials Database, Materials Screening, and Design

As stated above, though the vastness in compositions, microstructures, and properties of HEMs brings compelling opportunities, the complexities in the combination of compositions, microstructures, and properties lead to significant difficulties in material screening and design. Traditionally, materials are mainly designed based on experience and "trial and error" paradigm, which is less efficient and hard to apply in HEMs. Materials Genome Initiative (MGI) was proposed in 2011 [14], which changes the traditional "trial and error" research schema to a new paradigm that accelerates the discovery, design, development, and application of materials by synergistically combining experiment, theory, and computation in a tightly integrated, high-throughput manner. Thus, to develop HEMs in a highly efficient manner, high-throughput preparation methods, high-throughput characterization methods, high-throughput prediction models, and other possible materials screening and design methods need to be developed, and databases on HEMs need to be built.

On the experimental point of view, high-throughput preparation and characterization methods need to be explored for different HEMs. In the past, some high-throughput experimental tools have been developed. Combinational and high-throughput preparation methods offer efficient experimental strategies for rapidly surveying a wide array of material compositions and processing parameters coupled to the screening of microstructure and properties in parallel [15]. The concept of combinational methods is transferred from organic chemistry to the research of solid-state materials, where the preparation and processing methods are usually more complicated, especially for structural materials. The technologies are usually adopted to study films and particles and can be seldom applied to bulk materials. Rajan [15] in his review paper categorized the combinational methods roughly into two groups: continuous category and discrete category, depending on whether the variable is continuous or not. Figure 9.2 shows the classifications given by Rajan [15]. Figures 9.3 and 9.4 illustrate typical high-throughput methods extracted from a review paper by Ellendt et al. [16]. With the high-throughput preparation methods producing samples in a highly efficient way, high-throughput characterization methods offer highly efficient methods to obtain microstructures and properties of materials. Those characterization methods that are easy and fast to implement can be designed to be carried out in a high-throughput manner, for example, XRD, EBSD, hardness test or compression test (usually carried out by nano-indentation), and thermal conductivity measurement by fitting the temperature

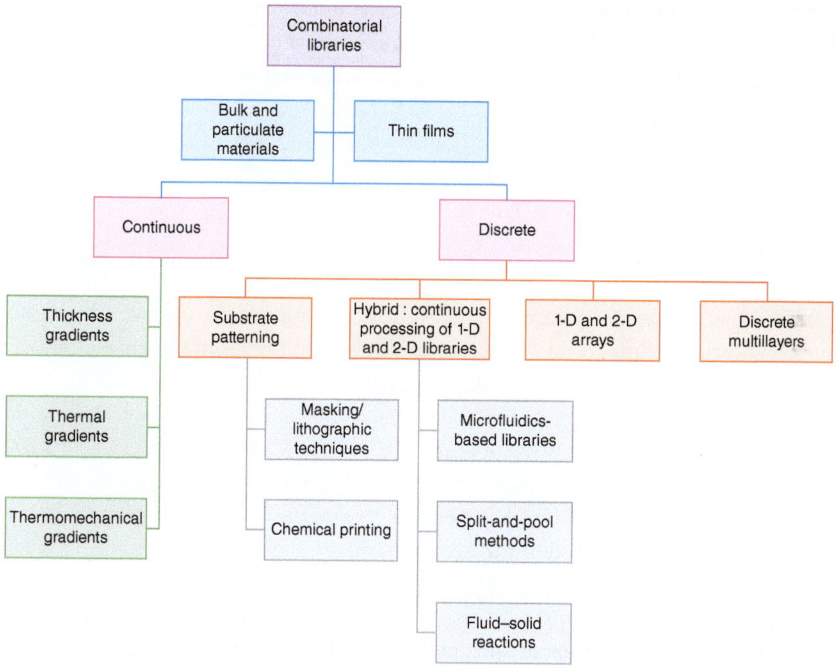


**Figure 9.2** (a) A classification chart of combinational methods for bulk, particulate, and film materials. The methods are divided into two broad categories: continuous and discrete variation in parameters. The continuous category involves experimental strategies to create controlled gradients in parameters (e.g. temperature, composition) in a material, which permits the investigation of variations in microstructures and properties with respect to the parameters. The discrete category involves more flexible experimental procedures, e.g. manipulate materials in small volumes and deposit them in discrete arrays. Moreover, hybrid methods can also be developed. For example, application of gradient fields to material arrays. Source: Reproduced with permission of Rajan [15]/Annual Reviews.

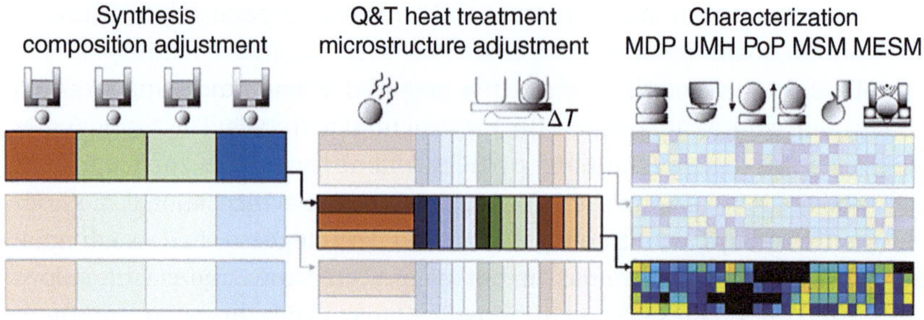


**Figure 9.3** Process steps in a high-throughput method for structural materials: synthesis, heat treatment, and characterization. Process chains can run in parallel. MDP, micro-compression test; UMH, universal micro-hardness measurement; PoP, particle-oriented peening; MSM, micro-sample machining; MESM, micro-extrusion strain measurement. Source: Ellendt et al. [16]/Springer Nature/CC BY-4.0.

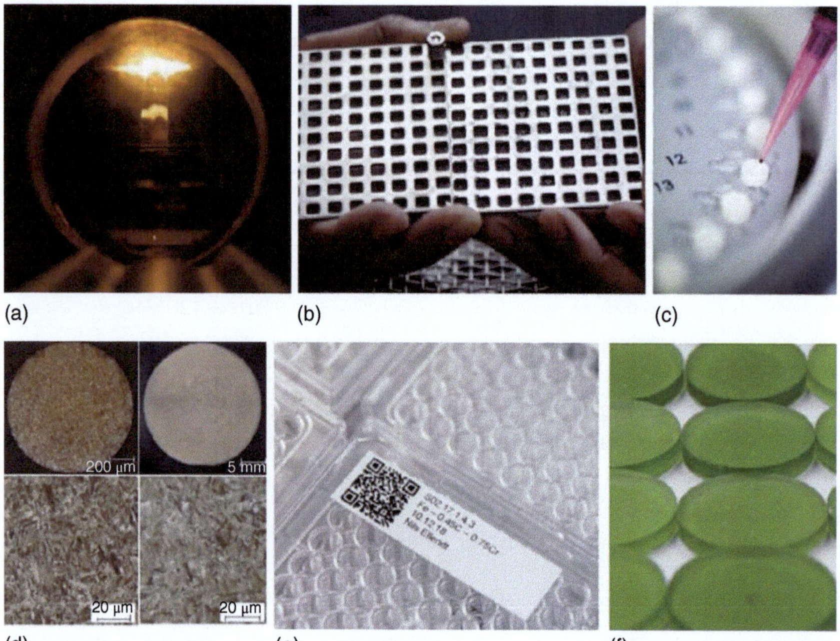


**Figure 9.4** Synthesis, heat treatment, and preparation process chain: (a) sample synthesis with a high-temperature droplet generator (sample diameter 900 µm), (b) batch-wise heat treatment in a furnace, (c) insertion of samples in differential scanning calorimetry (DSC) containers for fully automated heat treatment, (d) comparison of the microstructure of AISI 5140 heat-treated spheres and 20-mm rod, (e) samples in high-throughput container, (f) samples prepared as half-spheres. Source: Reproduced with permission from Ellendt et al. [16], Springer Nature, CC BY 4.0.

profile (e.g. the method by Braun et al. [13]). Presently, there are very few studies on high-throughput preparation and characterization on HEMs, even though parallel comparison of a set of samples has been carried out in some papers [13, 17, 18]. In the future, there is no doubt that more and more high-throughput experimental works on HEMs will be carried out. Furthermore, the development of high-throughput experimental tools is still in high demand, regardless of HEMs or conventional materials.

From the theoretical point of view, demand-driven material design has long been the objective of materials science and technology. To achieve this goal, two fundamental aspects need to be addressed. On the one hand, models on "composition–microstructure–property" relations are necessary, which then can screen materials microstructures and compositions to give suggestions on candidate materials. On the other hand, predictive models on microstructures with respect to composition and processing are necessary, which give suggestions on how to synthesize and how to manufacture a new material. Computational methods that can be used include DFT-based method, MD simulations, Monte Carlo (MC) simulations, CALculation of PHAse Diagrams (CALPHADs), phase field simulations, discrete dislocation dynamics models, finite element analysis (FEA), surrogate models from MLs, or simplified models based on physical assumptions. With the increase in computational capability and the development in artificial intelligent (AI) and ML, parallel simulations in a high-throughput manner and ML-assisted approaches are playing increasing roles in the design of materials. In the chapter "Theoretical Design Aspects in High-Entropy Materials," we have comprehensively reviewed the simulation methods and other models that have been applied in HEMs. High-throughput simulations have been applied in some research studies. For example, high-throughput DFT calculations have been applied to predict whether high-entropy carbides can be synthesized [17], to investigate phase stability and order–disorder transformations in refractory HEAs [19], to assist the design of high-performance HEAs with TWIP or TRIP effect [20], to search for HEAs [21], and to design composition of Li-ion battery materials [18] (see Figure 9.5). High-throughput CALPHAD has been usually applied to tailoring compositions of HEAs based on a target microstructure, e.g. to design high-performance lightweight high-entropy superalloys [22] (see Figure 9.6) and to accelerate the exploration of high-entropy superalloys [23]. Though some high-throughput simulations have been successfully adopted to screen compositions for HEMs, there is still a long way for target-oriented design of HEMs based on theoretical calculations.

Presently, we are in the age of AI. AI models (ML models) can extract deep interrelations between different factors from data, which can be viewed as a surrogate model of "composition–microstructure–property" relations. With the development of AI and data science, new research paradigm emerges, which is the data-driven materials design with the aid of AI models. In this paradigm, data set and AI models are the key factors, and more and more databases have been built, e.g. materials project [24], AFLOWLIB [25], and OQMD [26]. High-throughput experiments and calculations are effective methods to produce data for these databases, especially the data from high-throughput calculations. However, most of the databases include only data calculated by high-throughput DFT calculations, where the data can be generated in a unified process. In contrast, data from other methods are harder to collect due to the lack of unified processes or the presence of significant noises in the data. However, only DFT data are not sufficient; various data obtained by other methods are necessary in the future to

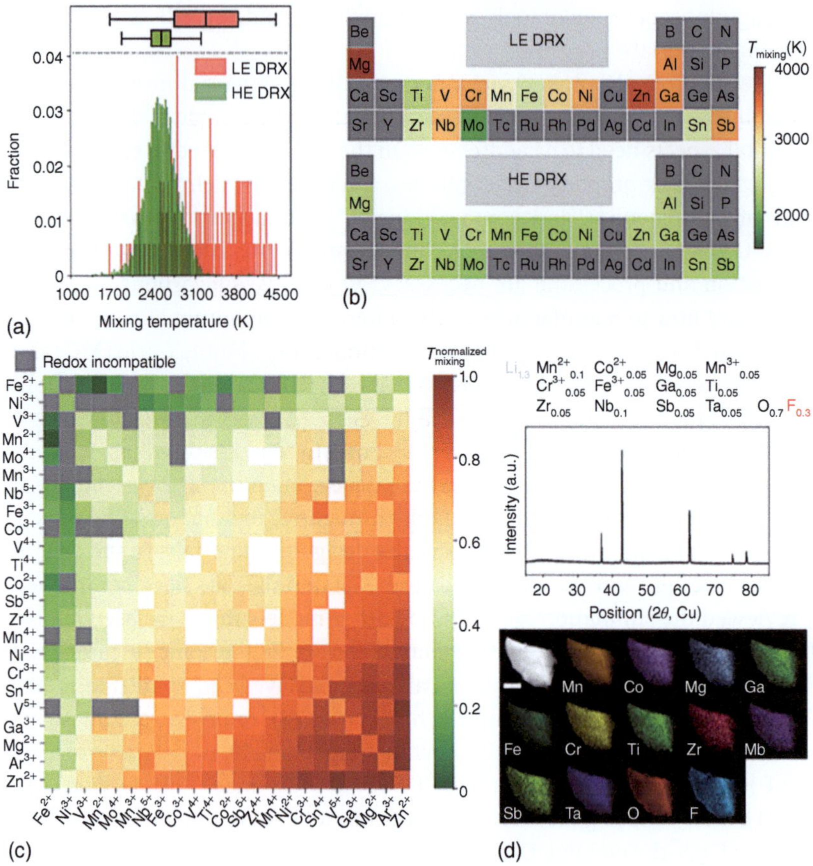


**Figure 9.5** Compatibility of metals in high-entropy (HE) DRX (disordered rock salt) cathodes. (a) Probability distribution of mixing temperature for low-entropy (LE) and HE DRX cathodes. The box plot (top) shows the median and interquartile range of mixing temperatures for LE and HE DRXs. (b) Periodic table-type heat map of mixing temperatures of different elements in LE and HE DRXs. The elements shaded gray were not included in the calculations. (c) Normalized mixing temperature of different TM (transition metal) species in DRX compounds. The color bar indicates increasing incompatibility from green to red. Gray fields are redox-incompatible couples. White squares are TM couples that were not included in their compositions. (d) Demonstration of an HE DRX containing 12 TM species (TM12). The XRD pattern (top) confirms the phase purity and the STEM/EDS mapping (bottom) shows the uniform distribution of different elements. Scale bar, 400 nm. Source: (d) Lun et al. [18], Springer Nature.

achieve target-oriented materials design. Though a high-throughput method in combination with databases and surrogate AI models can greatly improve the efficiency of materials development, the design of HEMs still faces the curse of dimensionality. When the high-throughput methods are still too expansive to be used in reality, active learning or adaptive learning methods may be helpful, which can intelligently inform what combination of parameters will improve our knowledge best.

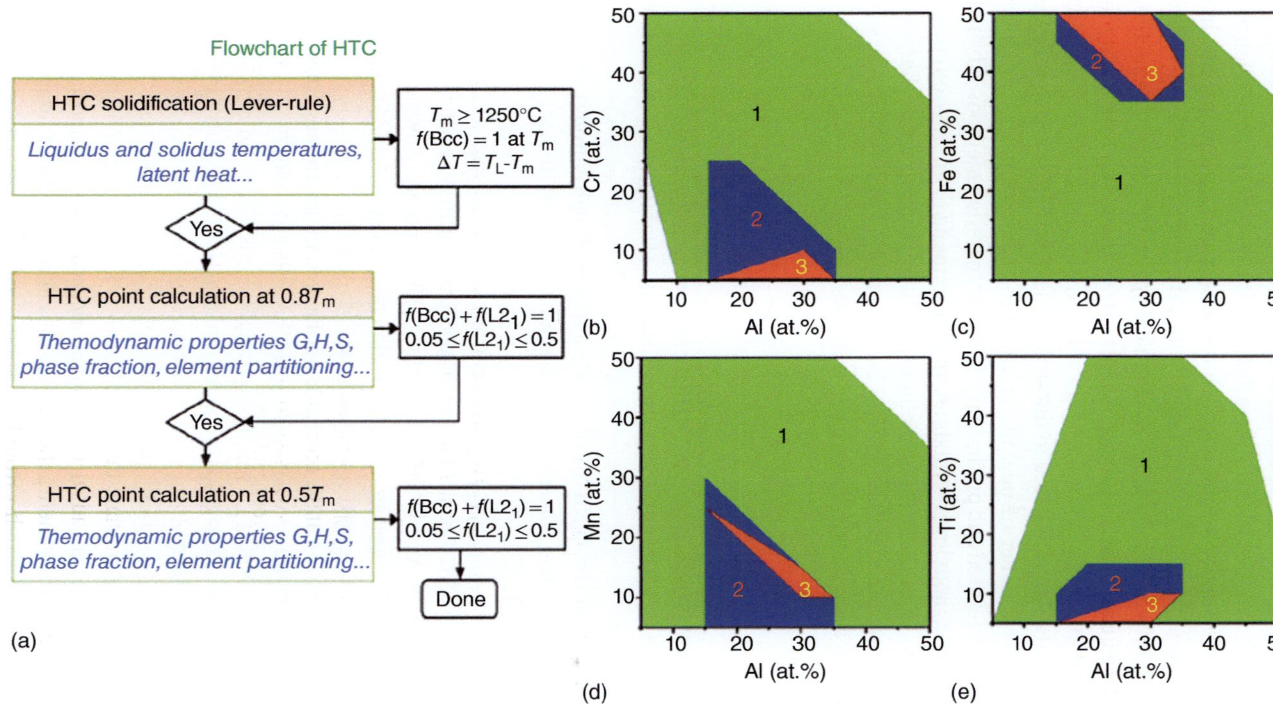


**Figure 9.6** High-throughput CALPHAD screening optimal alloy compositions in the Al–Cr–Fe–Mn–Ti system. (a) Flowchart of the high-throughput calculations (HTC in the figure). (b) Al–Cr projection. (c) Al–Fe projection. (d) Al–Mn projection. (e) Al–Ti projection. Source: Feng et al. [22]/Springer Nature/CC BY-4.0.

## 9.5 Conclusions

A coin always has two sides. The vastness in the combination of compositions and microstructures associated with HEMs not only offers opportunities to explore vast new materials with unprecedented properties, but it also brings the challenge of the curse of dimensionality. In the past two decades, significant progresses have been made, where numerous new materials with intriguing properties have been discovered. Only limited combination in chemical space and processing methods have been explored until now. In principle, most conventional solid-state materials can be redesigned under the concept of high entropy, and the associated investigation methods can also be transferred to the study of HEMs. HEMs are an immense field with countless number of new materials. Our understanding of these new materials is still very limited and preliminary. Novel preparation methods, advanced characterization, and modeling methods are in high demand to improve our understanding of HEMs. High-throughput methods, including preparation, characterization, and modeling, in combination with big data and AI techniques may serve as powerful means to challenge the curse of dimensionality.

## References

1 Xiang, H., Xing, Y., Dai, F. et al. (2021). High-entropy ceramics: present status, challenges, and a look forward. *J. Adv. Ceram.* 10: 385–441. https://doi.org/10.1007/s40145-021-0477-y.

2 Miracle, D.B. and Senkov, O.N. (2017). A critical review of high entropy alloys and related concepts. *Acta Mater.* 122: 448–511. https://doi.org/10.1016/j.actamat.2016.08.081.

3 Ye, Y.F., Wang, Q., Lu, J. et al. (2016). High-entropy alloy: challenges and prospects. *Mater. Today* 19: 349–362. https://doi.org/10.1016/j.mattod.2015.11.026.

4 George, E.P., Curtin, W.A., and Tasan, C.C. (2020). High entropy alloys: a focused review of mechanical properties and deformation mechanisms. *Acta Mater.* 188: 435–474. https://doi.org/10.1016/j.actamat.2019.12.015.

5 Li, W., Xie, D., Li, D. et al. (2021). Mechanical behavior of high-entropy alloys. *Prog. Mater Sci.* 118: 100777. https://doi.org/10.1016/j.pmatsci.2021.100777.

6 Yeh, J.-W. and Lin, S.-J. (2018). Breakthrough applications of high-entropy materials. *J. Mater. Res.* 33: 3129–3137. https://doi.org/10.1557/jmr.2018.283.

7 Chen, X., Wang, Q., Cheng, Z. et al. (2021). Direct observation of chemical short-range order in a medium-entropy alloy. *Nature* 592: 712–716. https://doi.org/10.1038/s41586-021-03428-z.

8 Ming, K., Li, L., Li, Z. et al. (2019). Grain boundary decohesion by nanoclustering Ni and Cr separately in CrMnFeCoNi high-entropy alloys. *Sci. Adv.* 5: eaay0639. https://doi.org/10.1126/sciadv.aay0639.

9 Lei, Z., Liu, X., Wu, Y. et al. (2018). Enhanced strength and ductility in a high-entropy alloy via ordered oxygen complexes. *Nature* 563: 546–550. https://doi.org/10.1038/s41586-018-0685-y.

**10** Wu, G., Balachandran, S., Gault, B. et al. (2020). Crystal-glass high-entropy nanocomposites with near theoretical compressive strength and large deformability. *Adv. Mater.* 32: 2002619. https://doi.org/10.1002/adma.202002619.

**11** Shi, P., Ren, W., Zheng, T. et al. (2019). Enhanced strength-ductility synergy in ultrafine-grained eutectic high-entropy alloys by inheriting microstructural lamellae. *Nat. Commun.* 10: 489. https://doi.org/10.1038/s41467-019-08460-2.

**12** Bérardan, D., Franger, S., Meena, A.K., and Dragoe, N. (2016). Room temperature lithium superionic conductivity in high entropy oxides. *J. Mater. Chem. A* 4: 9536–9541. https://doi.org/10.1039/C6TA03249D.

**13** Braun, J.L., Rost, C.M., Lim, M. et al. (2018). Charge-induced disorder controls the thermal conductivity of entropy-stabilized oxides. *Adv. Mater.* 30: 1805004. https://doi.org/10.1002/adma.201805004.

**14** National Science and Technology Council (2011). Materials Genome Initiative for Global Competitiveness. https://www.mgi.gov/sites/default/files/documents/materials_genome_initiative-final.pdf (accessed 27 July 2022).

**15** Rajan, K. (2008). Combinatorial materials sciences: experimental strategies for accelerated knowledge discovery. *Annu. Rev. Mater. Res.* 38: 299–322. https://doi.org/10.1146/annurev.matsci.38.060407.130217.

**16** Ellendt, N., Clausen, B., Mensching, N. et al. (2021). Experimental methods to enable high-throughput characterization of new structural materials. *JOM* 73: 3347–3355. https://doi.org/10.1007/s11837-021-04901-w.

**17** Sarker, P., Harrington, T., Toher, C. et al. (2018). High-entropy high-hardness metal carbides discovered by entropy descriptors. *Nat. Commun.* 9: 4980. https://doi.org/10.1038/s41467-018-07160-7.

**18** Lun, Z., Ouyang, B., Kwon, D.-H. et al. (2020). Cation-disordered rocksalt-type high-entropy cathodes for Li-ion batteries. *Nat. Mater.* https://doi.org/10.1038/s41563-020-00816-0.

**19** Wang, Y., Yan, M., Zhu, Q. et al. (2018). Computation of entropies and phase equilibria in refractory V–Nb–Mo–Ta–W high-entropy alloys. *Acta Mater.* 143: 88–101. https://doi.org/10.1016/j.actamat.2017.10.017.

**20** Li, Z., Körmann, F., Grabowski, B. et al. (2017). Ab initio assisted design of quinary dual-phase high-entropy alloys with transformation-induced plasticity. *Acta Mater.* 136: 262–270. https://doi.org/10.1016/j.actamat.2017.07.023.

**21** Lederer, Y., Toher, C., Vecchio, K.S., and Curtarolo, S. (2018). The search for high entropy alloys: a high-throughput ab-initio approach. *Acta Mater.* 159: 364–383. https://doi.org/10.1016/j.actamat.2018.07.042.

**22** Feng, R., Zhang, C., Gao, M.C. et al. (2021). High-throughput design of high-performance lightweight high-entropy alloys. *Nat. Commun.* 12: 4329. https://doi.org/10.1038/s41467-021-24523-9.

**23** Senkov, O.N., Miller, J.D., Miracle, D.B., and Woodward, C. (2015). Accelerated exploration of multi-principal element alloys for structural applications. *Calphad* 50: 32–48. https://doi.org/10.1016/j.calphad.2015.04.009.

**24** Materials Project https://materialsproject.org (accessed 27 July 2022).

**25** Aflow - Automatic FLOW for Materials Discovery. http://aflowlib.org (accessed 27 July 2022).

**26** OQMD. https://www.oqmd.org (accessed 27 July 2022).

# Index



*High-Entropy Materials: From Basics to Applications,* First Edition.
Huimin Xiang, Fu-Zhi Dai, and Yanchun Zhou.